TESTING OF COMMUNICATING SYSTEMS

IFIP – The International Federation for Information Processing

IFIP was founded in 1960 under the auspices of UNESCO, following the First World Computer Congress held in Paris the previous year. An umbrella organization for societies working in information processing, IFIP's aim is two-fold: to support information processing within its member countries and to encourage technology transfer to developing nations. As its mission statement clearly states,

> IFIP's mission is to be the leading, truly international, apolitical organization which encourages and assists in the development, exploitation and application of information technology for the benefit of all people.

IFIP is a non-profitmaking organization, run almost solely by 2500 volunteers. It operates through a number of technical committees, which organize events and publications. IFIP's events range from an international congress to local seminars, but the most important are:

- the IFIP World Computer Congress, held every second year;
- open conferences;
- working conferences.

The flagship event is the IFIP World Computer Congress, at which both invited and contributed papers are presented. Contributed papers are rigorously refereed and the rejection rate is high.

As with the Congress, participation in the open conferences is open to all and papers may be invited or submitted. Again, submitted papers are stringently refereed.

The working conferences are structured differently. They are usually run by a working group and attendance is small and by invitation only. Their purpose is to create an atmosphere conducive to innovation and development. Refereeing is less rigorous and papers are subjected to extensive group discussion.

Publications arising from IFIP events vary. The papers presented at the IFIP World Computer Congress and at open conferences are published as conference proceedings, while the results of the working conferences are often published as collections of selected and edited papers.

Any national society whose primary activity is in information may apply to become a full member of IFIP, although full membership is restricted to one society per country. Full members are entitled to vote at the annual General Assembly, National societies preferring a less committed involvement may apply for associate or corresponding membership. Associate members enjoy the same benefits as full members, but without voting rights. Corresponding members are not represented in IFIP bodies. Affiliated membership is open to non-national societies, and individual and honorary membership schemes are also offered.

TESTING OF COMMUNICATING SYSTEMS

Proceedings of the IFIP TC6
11th International Workshop on
Testing of Communicating Systems (IWTCS'98)
August 31- September 2, 1998, Tomsk, Russia

edited by

Alexandre Petrenko
Computer Research Institute of Montreal (CRIM)

Nina Yevtushenko
Tomsk State University

SPRINGER SCIENCE+BUSINESS MEDIA, LLC

Library of Congress Cataloging-in-Publication Data

A C.I.P. Catalogue record for this book is available
from the Library of Congress.

ISBN 978-1-4757-6703-2 ISBN 978-0-387-35381-4 (eBook)
DOI 10.1007/978-0-387-35381-4

Printed on acid-free paper.

CONTENTS

PREFACE

This volume contains the proceedings of the IFIP WG 6.1 International Workshop on Testing of Communicating Systems (IWTCS'98), held in Tomsk, Russia, from August 31 to September 2, 1998. This Workshop continues the IFIP International Workshop on Protocol Test Systems and it is the eleventh of a series of the annual meetings sponsored by the IFIP Working Group 6.1. The ten previous workshops were held in Vancouver, Canada, 1988; Berlin, Germany, 1989; Mclean, USA, 1990; Leidschendam, the Netherlands, 1991; Montreal, Canada, 1992; Pau, France, 1993; Tokyo, Japan, 1994; Evry, France, 1995; Darmstadt, Germany, 1996; Cheju Island, Korea, 1997.

As in the years before, the workshop aims at bringing together researchers and practitioners in the field of testing of communicating systems. Forty three papers were submitted to the IWTCS'98 and reviewed by the members of the Program Committee and by additional reviewers. 18 papers have been selected for the workshop. These papers and three invited papers are reproduced on this volume.

IWTCS'98 was organized under auspices of IFIP WG 6.1 by Centre de Recherche Informatique de Montréal (CRIM, Computer Research Institute of Montreal), Canada and Tomsk State University (TSU), Russia. It was financially supported by these organizations and the Commission of the European Communities.

We would like to thank everyone who has contributed to the success of the workshop. In particular, we owe our gratitude to the authors for writing and presenting their papers, the reviewers for assessing and commenting on these papers, the members of the Program Committee, M. Kim for sharing his experience in organizing IWTCS'97 with us, and to all people from CRIM and TSU, especially Alan Bernardi, Alain Gravel, Johanne Dumont, Yves Bélanger, Irina Koufareva, who provided support for this conference.

Alexandre Petrenko and Nina Yevtushenko

Tomsk, Russia, September 1998

CONFERENCE CO-CHAIRS

Alexandre Petrenko
Centre de Recherche Informatique de Montréal
(CRIM, Computer Research Institute of Montreal), Canada
petrenko@crim.ca

Nina Yevtushenko
Tomsk State University, Russia
yevtushenko@elefot.tsu.tomsk.su

PROGRAM COMMITTEE

G. Agibalov (TSU, Russia)
N. A. Anisimov (Academy of Sciences, Russia)
B. Baumgarten (GMD, Germany)
G. v. Bochmann (University of Ottawa, Canada)
E. Brinksma (University of Twente, Netherlands)
R. Castanet (University of Bordeaux, France)
A. R. Cavalli (INT, France)
S. T. Chanson (Hong Kong Univ. of Science & Technology, China)
V. Chapenko (IECS, Latvia)
A. T. Dahbura (DEC, USA)
S. Dibuz (Ericsson, Hungary)
R. Dssouli (University of Montreal, Canada)
J. P. Favreau (NIST, USA)
R. Groz (CNET, France)
I. S. Grunsky (Institute of Applied Math. and Mechanics, Ukraine)
T. Higashino (Osaka University, Japan)
D. Hogrefe (University of Luebeck, Germany)
G. J. Holzmann (Bell Laboratories, USA)
M. Kim (Information and Communications University, Korea)
H. Koenig (Technical University of Cottbus, Germany)
V. B. Kudrjavcev (Moscow State University, Russia)
E. Kwast (KPN Research, Netherlands)
V. G. Lazarev (Institute for Information Transm. Problems, Russia)
D. Lee (Bell Laboratories, USA)
M. T. Liu (Ohio State University, USA)
J. C. Maldonado (ICMSC/USP, Brazil)
J. de Meer (GMD, Germany)
K. Parker (Telstra Research Laboratories, Australia)

H. Pranevicius (Kaunas University of Technology, Lithuania)
O. Rafiq (University of Pau, France)
D. Rayner (NPL, UK)
K. Stoettinger (Deutsche Telekom, Germany)
K. Suzuki (KDD R&D Labs, Japan)
K. C. Tai (North Carolina State University, USA)
K. Tarnay (KFKI, Hungary)
J. Tretmans (University of Twente, Netherlands)
H. Ural (University of Ottawa, Canada)
S. T. Vuong (University of British Columbia, Canada)
J. Wu (Tsinghua University, China)

FURTHER REVIEWERS

E. M. Aboulhamid
C. Besse
J. Bi
L. Cacciari
J. Chiang
C. Desroches
P. Dini
A. Ennouaary
O. Festor
H. Fuchs
P. Godfroid
J. Grabowski
L. Heerink
O. Henniger
S. Heymer
M. Higuchi
P. Hunel
A. Idoue
S. Kang
B. Koch
I. Koufareva
A. Loureiro
L. Lima
M. Mori
T. Ogishi
H. S. Oh
M. Schmitt
A. Shaff
J. Shin
Q. M. Tan
R. de Vries
T. Walter
C. J. Wang
F. Zaidi
J. Zhu

PART ONE

Past and Present of Research on Automata Testing
Invited Talk

Testing of automata: from experiments to representations by means of fragments

Igor S. Grunsky

Institute of Applied Mathematics and Mechanics,
National Academy of Sciences of Ukraine,

74, R. Luxemburg str., Donetsk, 340114, Ukraine
e-mail: math @ iamm.ac.donetsk.ua

Abstract

This paper discusses some developments and results of the theory of test generation from automata. These developments are driven by the needs to better understand the nature of automata testing and thus to make the testing theory more applicable to real systems. We provide an overview of some important results in automata testing recently obtained in the Soviet Union and in the countries that have arisen from it.

Keywords
Automaton (finite state machine), testing problems, checking experiments and sequences, checking test, test complexity, experiments on automata, automata fragment

1 INTRODUCTION

Automata or finite state machines have been widely used as mathematical models of discrete systems in diverse areas such as computer hardware and software and more recently, communication protocols [2,3,4,7,18,19,29,31]. In this paper, we study the fault-detection problems on automata. We are given the state diagram of an automaton A, and we have a «black box» automaton *B* which is supposed to implement *A*. We can test *B* by applying input sequences (tests) and observing the produced output sequences. We want to design the tests to determine whether or not *B* is an implementation of *A*. This problem has been referred to as the «fault-detection» or «checking problem» in automata theory. In the recent literature, this problem is also called conformance testing (of communication protocols).

There is an extensive literature on problems of automata testing. One may state that the theory of automata testing has been built over the last 40 years. There exist several excellent books [15,18,29] and surveys [2,20,31] in this theory, but many interesting results were omitted in these publications. The aim of this paper is to represent some important results recently obtained in the USSR and in the countries arisen from the Soviet Union (mainly, in Russia and Ukraine).

2 BACKGROUND

2.1 Basic notations

A deterministic finite state machine *(FSM)* or an automaton *A* is quintuple $A=(S,X,Y,\delta,\lambda)$, where S,X,Y are finite and nonempty sets of states, input symbols, and output symbols, respectively; δ: $Dom_A \rightarrow S$ is the state transition function and λ : $Dom \rightarrow Y$ is the output function; Dom_A is a specification domain of *A*, i.e. a subset of $S \times X$. If $Dom_A = S \times X$ then A is complete, otherwise it is partial. Let n,m,r be the cardinality of S,X,Y, respectively. If $p=x_1...x_k$ is an input word, then $\delta(s,p)$ is the state reached by *A* from state *s* when *p* is applied to *A*, and $\lambda(s,p)=y_1...y_k$ is the corresponding output word. The pair $(p, \lambda(s,p))$ is called an input-output word produced by the state *s*. We use $w=(x_1,y_1)...(x_k,y_k)$ to denote this word. The word *w* determines the unique partial «string» automaton $R(w)$ with the state set $\{0,1,...,k\}$, transition function Δ, and output function Λ: $\Delta(i,x_{i+1})=(i+1)$, $\Lambda(i,x_{i+1})=y_{i+1}$ for $0 \le i \le k-1$ and these functions are undefined in all other cases.

We denote λ_s and λ_s^k the sets of all input-output words produced by *s* which are of finite length and length equal to or less than *k*, respectively. The automaton *A* is reduced if $\lambda_s \neq \lambda_t$ for every pair of distinct states *s* and *t*. The reduced automaton is uniquely characterized by the set $L_A=\{\lambda_s\}$, $s \in S$. An important characterization of *A* is the set $\Phi_A = \cup_{s \in S} \lambda_s$.

Let s_0 be a designated initial state of *A*. The automaton *A* is called connected if for all states $s \in S$ exists an input word *p* such that $\delta(s_0,p)=s$. The automaton is called strongly connected if it is connected for all $s_0 \in S$.

Two automata *A* and *B* are said to be equivalent if $L_A=L_B$. If s_0 and t_0 are the initial states of *A* and *B*, then *A* and *B* are equivalent if $\lambda_{s0}=\lambda_{t0}$.

An input word *p* is a distinguishing word for *A*, if $s \neq t$, $s,t \in S$, implies $\lambda(s,p) \neq \lambda(t,p)$. The automaton *A* is called definitely diagnosable of order *k* if every *p* of length *k* is a distinguishing word (DDA-k, for short). It is known that *k* in this case is equal to or less than $n(n-1)/2$. A word *p* is a homing word for *A* if $\lambda(s,p)=\lambda$

(t,p) implies $\delta(s,p)\neq\delta(t,p)$. If every p of length k is a homing sequence for A, then A is the finite memory of order k automaton. The order k of finite memory is also equal to or less than $n(n-1)/2$.

2.2 Testing problems

In testing problems, we have a machine about which we lack some information, and we would like to deduce this information from input-output words obtained by experimenting with the machine. An experiment is a process when we apply input words to the machine, observe the produced output words and deduce the missing information about the machine. The applied input words are called a test suite and obtained input-output words are called an experiment.

We discuss the machine verification problem also known as the fault-detection or conformance testing problem. We are given a *specification* machine A, i.e. we have its state transition and output function. We are also given an *implementation* machine B that is a «black box» and we can only observe its input-output behaviour. We want to design a test (an experiment) to determine whether B conforms (is equivalent) to A. Obviously, without any assumption the problem is undecidable. There are a number of «classical» assumptions that are usually made in the literature:

a) the specification machine A is complete, reduced and strongly connected;
b) an implementation machine B has the same input alphabet as A;
c) B belongs to some known class F of «faulty machines».

The class F is often assumed to coincide with the class F_n of all machines with the number of states not more than n. The specification machines considered in the machine verification problem usually have a designated initial state (initialized machines).

Let $W=\{(p_i,q_i)\}$, $1\leq i\leq k$, be a set of input-output words. The set W is called a checking experiment, if $W\subseteq\lambda_{s0}$ and, for all $B\in F_n$ and any state t of B, $W\subseteq\lambda_t$ implies that the machine B is equivalent to A. The parameter k is called the multiplicity of the experiment W, the length of the longest word from W is called the length of W, and the total length of all words in W is the size of W. The experiment is called simple when $k=1$ and multiple otherwise. A simple checking experiment (p,q) is called a checking sequence and p is called a checking test.

There is an extensive literature on testing automata, the fault-detection problem, in particular. It is convenient for us to distinguish three periods in this research: primary, classical, and modern periods. The primary period was opened by the famous Moore's paper [23] in which he studied a related, but harder problem of machine identification: given a machine with a known number of states, determine its state diagram. This period is characterised by the combinatorial nature of offered solutions, i.e. the obtained results are implied by counting the number of automata in a certain class.

Moore proved that there exists a multiple checking test for A with n states and F_n containing all input words of length $2n-1$. He also provided an exponential algorithm for constructing a checking sequence of an exponential length and an exponential lower bound for this problem on the basis of the class F cardinality. Books [7,18,29] give a good exposition of the major results of the primary period.

The classical period was started by the influential Hennie's paper [14]. He shown that if the automaton A has a distinguishing input sequence of length l, then one can construct a checking sequence of a length polynomial in l and nm. Unfortunately, not every machine has a distinguishing sequence. Furthermore, only exponential algorithms are known for determining the existence such sequences. In [27] it is shown that the problem is PS-complete [1]. Rystsov [26] proved that for the length l of shortest distinguishing sequences the inequality

$3^{n(1-\varepsilon)/6} < l < 3^{n(1+\varepsilon)/6}$ holds, where ε is any positive real number and $n \to \infty$. Nevertheless, the main Hennie's idea is most fruitful. It lies in embedding a distinguishing sequence in a test sequence in a special way to:

a) obtain the response of each state of A to the distinguishing sequence and,

b) check each transition of A by applying a proper input at the start state, observing the produced output, and verifying the tail state of the transition by using the distinguishing sequence.

Based on this idea, many papers were published in which various subsequences were used to verify the start and tail states of transitions, distinguishing sequences, adaptive distinguishing sequences, locating sequences, identifying sequences, homing sequences and others. To the classical period an important Vasilevskii's paper [30] belongs. He provided a polynomial algorithm for constructing a multiple checking experiment, and proved polynomial upper and lower bounds on the length of checking sequences. Books [3,4,15,18] and papers [20,31] give a good exposition of the major results of the classical period. During this period, many algorithms were constructed for special classes of specification automata, diagnosable, definitely diagnosable of order k [18] etc. In [25,28], the exact upper bound of adaptive distinguishing sequences was obtained. In [8], so-called checking sequences for state of A were investigated. The period has continued till recent days.

3 MODERN RESEARCHES

The results obtained during the primary and classical periods give a basis and a possibility to build a general theory of automata checking. A variant of the theory is stated below. Consider an automaton A and a (possibly partial) automaton $R=(T,X,Y,\Delta,\Lambda,D_R)$. A mapping $\varphi\colon T \to S$ is a homomorphism of R to A, if $\varphi(\Delta(t,x))=\delta(\varphi(t),x)$, $\Lambda(t,x)=\lambda(\varphi(t),x)$ for all $(t,x)\in D_R$. The automaton R is called a fragment of A if such a mapping exists. This fact is denoted $R \leq A$, and in the case, when φ is one-to-one mapping, it is denoted $R \subseteq A$. It is obvious, that if $w \in \Phi_A$, then $R(w)$ is a fragment of the automaton A.

Let F be a class of reduced complete automata over the alphabets X, Y and τ be a similarity relation on $F \cup \{A\}$. $\tau(A)$ is the class of automata similar to A. An automaton R is said to be a representation of A with respect to F and τ (a representation of (A,F,τ), for short), if $R \leq A$ and if $R \leq B$, $B \in F$, implies $B \in \tau(A)$. It is obvious, that if w is a checking sequence for A and F, then the automaton $R(w)$ is the representation of (A,F,τ), where τ is the automata equivalence relation, that is $(A,B) \in \tau$ iff $L_A = L_B$. If W is a checking experiment, then the tree-like automaton $R(W)$ defined analogously to $R(W)$ is a representation of A, as well. The notion of a representation is a nontrivial and useful generalisation of checking experiments and it enables us to construct a unified profound theory of automata checking. The representation theory is widely stated in [11]. In the same book, the application of the representation theory to problems of technical diagnostics is discussed.

3.1 Existence of representations

Theorem 1 [11]

The following statements are equivalent:
1. a representation of (A,F,τ) exists,
2. the automaton A is a representation of (A,F,τ),
3. if $A \subseteq B$, $B \in F$, then $B \in \tau(A)$.

Given a checking experiment W, the tree-like automaton $R(W)$ is the representation. The existence condition for this important class of tree representations is given by:

Theorem 2 [11]

A tree representation of (A,F,τ) exists iff there exists a natural k such that for each $B \in F - \tau(A)$ there exists a state t of B such that $\lambda_t^k \neq \lambda_s^k$ for all $s \in S$.

The statement 3 of the Theorem 1 for a finite F may be checked effectively. The class F in Theorem 2 may be either finite or infinite. Any finite class F has this property but the converse is not always true. From this it follows that multiple checking experiments exist for any finite F.

In [11] several existence conditions are found for several types of representations of various (A,F,τ).

3.2 Representation structure

For the design of checking experiments, as it has been stated above, special sequences (distinguishing, locating etc.) play a significant role. Let us introduce a notion of state identifiers to generalize such sequences. A fragment R with a fixed state t is an identifier of state s of A if $R \leq A$ and if each homomorphism of R to A maps t to s. The fragment R is a state identifier of A, if it is an identifier of some state s. Let $R \leq A$ and I be a state identifier of A. The identifier I is said to be verified in R, if $R \leq B$, $B \in F$ implies that I is a state identifier of B.

Let J be a set of the state identifiers verified in R. Consider an equivalence relation on J: $(I_1,I_2) \in \sigma$ if I_1,I_2 are identifiers of the same state for all B

$\in F$, $R \leq B$. The pair (J,σ) generates on the state set of R a reflexive and symmetric relation: $(t_1,t_2) \in \beta$ iff there exist $I_1,I_2 \in J$ for some states t_1,t_2, respectively, $I_1,I_2 \leq R$, and $(I_1,I_2) \in \sigma$. The smallest congruence relation $\rho \supseteq \beta$ is called a closure of β. The closure ρ generates the fragment $[R]=R/\rho$ which is called a closure of R by (J,σ).

Theorem 3 [11]

If $[R]=A$ for (J,σ), then R is a representation of (A,F,τ).

An input-output word $w=(x_1,y_1)...(x_k,y_k)$, $w \in \Phi_A$ is called an initial (final) identifier of A if the fragment $R(w)$ with the state o (k, respectively) is a state identifier of A.

It follows from this definition that if p is a distinguishing (homing) word for A, then $R(p,\lambda(s,p))$ is an initial identifier of s (a final identifier of $\delta(s,p)$, respectively) of the automaton A. Taking into account this definition we may say that the results of Hennie and his successors are the corollaries to *Theorem 3.*

Consider the cases when the condition of Theorem 3 is both, sufficient and necessary. Let $F=F_n$, τ be an isomorphism relation, and A be a *DDA-k.*

Theorem 4 [11]

A word $w \in \Phi_A$ is a checking experiment for DDA-k, where $k \leq 11$, iff $[R(w)]=A$, where J is the set all initial state identifiers verified in $R(w)$.

Consider the case when A is *DDA-k*, $k=1$. Let Φ_A^1 be the set of all input-output words $w \in \Phi_A$ of length 1, and J be the set of all initial state identifiers of length 1 verified in R. Define a non oriented graph $G(R)=(V,E)$, where V is equal to the state set of $[R]$, and $(V_1,V_2) \in E$ if $V_1 \neq V_2$, $\Lambda(V_1,x) \neq \Lambda(V_2,x)$ for some $x \in X$. A mapping φ: V onto $\{1,2,...,n\}$ is called a colouring of $G(R)$, if $(V_1,V_2) \in E$ implies $\varphi(V_1) \neq \varphi(V_2)$. The graph $G(R)$ is said to be uniquely colourable iff all colourings of $G(R)$ are isomorphic.

V.A. Kozlovsky has proved the following important results.

Theorem 5 [11]

A fragment R is a representation of $(A,F_n,=)$ iff the following conditions hold:

1. *$R \leq A$,*
2. *$\Phi_A^1=\Phi_R^1$,*
3. *$G(R)$ is uniquely colourable.*

On the basis of the Theorem 5 he has proved

Theorem 6 [11,16]

The problem «is a word w a checking experiment of $(A,F_n,=)$?» is NP-complete.

We note that the checking of conditions *1,2* of Theorem 5 may be performed by a polynomial algorithm. In [17] V.A. Kozlovsky has proved that this problem is *NP*-complete for a special class of machines, so-called group *DDA-1*.

3.3 Important subclasses of F_n.

The results discussed above indicate the fundamental difficulties in constructing tests for $(A,F_n,=)$. The difficulties stimulate investigation of subclasses of F_n for which the test derivation can be done more efficiently. A number of such classes are known [2, 11, 13, 16, 24]. We consider here the two classes: a locally generated class [11, 16] and a class generated by a fault-function [5, 6, 10, 11, 13, 31].

Given an automaton A, we define a class $F(A)$ generated by local transformations of A. The neighbourhood of state $s\in S$ in A is the set $O_A(s)$ of states t such that $\delta(s,x)=t$ or $\delta(t,x)=s$ for some $x\in X$. If $\delta(s,x)=z$ for some $s,z\in S$ and $x\in X$, then replacing z by some $t\in O_A(v)$ we obtain an automaton B which we call as the one directly generated by A. An automaton $B\in F(A)$ iff there exists a sequence of automata $B_0=A$, $B_1,\dots,B_k=B$ such that B_i is directly generated by B_{i-1}, $i=1,\dots,k-1$. Clearly, $F(A)\subseteq F_n$.

Theorem 7 [16]

A word $w=(x_1,y_1)\dots(x_n,y_n)$, $w\in\Phi_A$ *is a checking experiment of* $(A,F_n,=)$, *where A is a DDA-1, iff* $\Phi_A^{\,1}=\{(x_1,y_1),\dots,(x_k,y_k)\}$, *and* (x_k,y_k) *occurs in w at least twice.*

Corollary 8 [16]

1. There is a polynomial algorithm solving the decision problem «is a word w a checking experiment of $(A,F(A),=)$*»;*

2. The length d(w) of the shortest checking experiment w of $(A,F(A),=)$ *satisfies the inequality* $mn+1\le d(w)\le mn+(m-1)n(n-1)/2$, *where both the lower and upper bounds are reachable.*

In the book [11], a special case of the local transformations, so-called input faults on A, is considered. For these faults and a so called inversible A, an algorithm for constructing checking sequences of length at most $(2n-1)m$ is proposed.

Consider now the class generated by a fault-function [10]. Let S, X, Y be some alphabets of states, input, and output, respectively. Consider the pair of functions (f,g), where $f(p)\subseteq S$ and $g(p)\subseteq Y$ for each input word p. For an empty word e we define $f(e)=s_0\subseteq S$, $g(e)=e$. Let F be a class of automata in these alphabets. The pair (f,g) is said to be an evaluating function for F, if $\delta(s_0,p)\in f(p)$ and $\lambda(\delta(s_0,p),x)\in g(px)$ for each $x\in X$, input word p and each automaton $A\in F$ with the initial state s_0. Each pair (f,g) generates a maximal class F for which the pair is an evaluating function. Assuming that an automaton A in F serves as a specification (it is a correct automaton), all the others are considered as faulty automata, and the pair (f,g) is called a fault function for A and F. In [13] a fault function with $g(p)=\lambda(s_0,p)$ was considered. It is obvious, that such a fault function determines the class $F(A)$, if $f(e)=s_0$ and $f(px)=O_A(\delta(s_0,p))$.

Fault functions are a powerful tool for defining automata classes. In [10,13,32] the methods for constructing multiple checking tests were proposed. These methods improve the Vasilevskii results [30] and may yield simpler tests. S. Yu. Boroday [5,6] has considered a subclass of the class F of automata with single transition faults, (automata whose transition and output functions differ from those of A only for one pair (s,x) or for one state s). As shown in [5,6], in these classes test derivation is simplified.

3.4 Finite-definable classes of automata

Consider now the testing problems when a class F can be infinite but it is defined by a finite means. In this section, we study two ways of defining an infinite class, by a set $M \subseteq X \times Y$ [9,12] and by nondeterministic automata [2,21,22,24].

Let $F(X,Y)$ be the class of all initialized reduced automata over inputs X and outputs Y. Given a subset $M \subseteq X \times Y$, consider a class $F(M)$ of all automata from $F(X,Y)$ in which every $(x,y) \in M$ can be produced by at most one state. Such M exists in practice (for example, protocol machines with a status message [19,31]). It is easy to see that the class $F(M)$ may be either finite or infinite. Let M_x be the set of all $(x,y) \in M$, $y \in Y$, and m_x be the cardinality of M_x.

Theorem 9 [12]

The following statements are equivalent:

1. The class $F(M)$ is finite,

2. $m_x=r$ for some $x \in X$,

3. there exists some $x \in X$ such that for each $A \in F(M)$, x is a distinguishing word.

The statement 2 can be checked by a polynomial algorithm. Let $G(M) \subseteq F(M)$ be a subclass of all automata in which every cycle of the transition graph has a state producing some $(x,y) \in M$.

Theorem 10 [12]

A multiple checking experiment of $(A,F(M),=)$ exists iff $A \in G(M)$.

The condition of Theorem 10 can be verified by a polynomial algorithm. The polynomial condition of checking sequence existence can be found in [12] as well.

Let R be a fragment of $A \in F(M)$. A closure $[R]_M$ of R by M is a (possible partial) automaton constructed from R by identifying all states of R producing the same $(x,y) \in M$ and its next states according to x.

Theorem 11 [9]

1. Let $n<r$ or $m_x<r$ for all $x \in X$. A fragment R is a representation of $(A,F(M),=)$ iff the closure $[R]_M$ is equivalent to A.

2. Let $n=r$ and $m_x=r$ for some $x \in X$. The fragment R is a representation of $(A,F(M),=)$ iff the following conditions hold:

a) $R \leq A$,

b) $\Phi_A^{\,1}=\Phi_R^{\,1}$,

c) $[R]_M$ is uniquely colourable.

Another way of defining automata classes uses a nondeterministic automaton. Let $N=(V,X,Y,h,v_0)$ be an initialized nondeterministic observable automaton with h: $V \times (X \times Y) \rightarrow V$. Given state v and input-output word w, $v'=h(v,w)$ denotes a state reached when the automaton produces w. In the general case, the function h is partial, i.e. $h(v,w)$ may be undefined for some w. Let L_N be a set of all words w for which $h(v_0,w)$ is defined. An automaton N defines a class $F(N)$ of automata A from $F(X,Y)$ such that $\lambda s_0 \subseteq L_N$. The automaton $A \in F(N)$ is called an implementation of N. A state v of N is deterministic if for each p a unique output word q exists such that $h(v,(p,q))$ is defined. All deterministic states form a kernel of A.

Theorem 12 [22]

The class is finite iff the automaton N has a cycle outside of its kernel.

In [21] B.Lukyanov has found a sufficient condition for checking the test existence for $(A,F(N),=)$, where $A \in F(X,Y)$, and proposed a method for deriving a test. S. Yu. Boroday [6] has found a necessary and sufficient condition for checking the test existence and has given an algorithm. Moreover, he has given a method for checking whether the automaton is contained in $F(N_1)$ or in $F(N_2)$, where N_1, N_2 are nondeterministic automata.

4 CONCLUSION

The key issues in automata testing are the structure of experiments and the analysis of the experiments. The first problem consists in determining what kind of information about the specification machine must be present in checking experiments. The second problem consists in finding all automata generating a given experiment. Both problems are closely related and are very hard. This paper deals with the first problem. A framework has been introduced for this problem exploration on the base of the state identifiers. Note that in [11] some more types of identifiers are studied (input and output identifiers, for example). The results presented in the paper (and in [11]) show us that the representations and the identifiers of non-observable components of automata are powerful means for this problem solution.

5 ACKNOWLEDGEMENTS

This research was partly supported by the Russian Found for Fundamental Research (grant 98-01-00113). The author wishes to thank Alex Petrenko (CRIM, Canada) for his stimulating help at all the stages of the preparation of this paper.

6 REFERENCES

[1] A. V. Aho, J.E. Hopcroft, J. D. Ullman, The Design and Analysis of Computer Algorithms. Reading, MA: Addison-Wesley, 1974.

[2] G. v. Bochmann and A. Petrenko, Protocol Testing: Review of Methods and Relevance for Software Testing, Proc. of the 1994 Int. Symp. on Software Testing and Analysis (ISSTA), Seattle, Wash., T. Ostrand Ed., pp. 109-124, 1994.

[3] A. M. Bogomolov, A. S. Barashko, I. S. Grunsky, Experiments on Automata. Kiev: Naukova Dumka, 1973 (in Russian).

[4] A. M. Bogomolov, I. S. Grunsky, D. V. Speransky, Checking and Transformation of Discrete Automata. Kiev: Naukova Dumka, 1976 (in Russian).

[5] S. Yu. Boroday, Checking of Single Faults of Finite State Machine Defined by a Fault Function, Kibernetica, no.6, pp 65-74, 1995.

[6] S. Yu. Boroday, Experiments on Effectively Defined Classes of Automata, Ph.D. Thesis. Saratov University, 1997.

[7] A. Gill, Introduction to the Theory of Finite-State Machines. New York: Mc Graw-Hill, 1962.

[8] S. M. Goberstein, Check Words for the States of Finite Automaton, Kibernetika, no 1, pp. 46-49, 1974.

[9] I. S. Grunsky, Structure of Fragments Uniquely Representing Automata, Intellectual Systems (Moscow), v.2, pp. 249-258, 1997 (in Russian)

[10] I. S. Grunsky, Checking of Automata Faults Defined with an Explicitly Given Fault Function, Proc. of Institute of Applied Math. and Mechanics, Ukrainian Acad. Of Sci. (Donetsk), vol.1, pp. 38-43, 1997.

[11] I. S. Grunsky, V. A. Kozlovsky, G. G. Ponomarenko, Finite Automata Representation by Means of Behaviour Fragments. Kiev: Naukova Dumka, 1990 (in Russian).

[12] I. S. Grunsky, I. I. Maximenko, On Experiments on Automata without an Upper Bound on the Number of their States, Dopovidi of Nat. Acad. Of Sci. (Kiev, Ukraine), no.6, pp. 31-35, 1996.

[13] I. S. Grunsky, A. Petrenko, Design of Checking Experiments on Automata Describing Protocols, Automatika I vychislitelnaya tehnika (Riga), no.4, pp. 7-14, 1988 (in Russian), Automatic Control and Computer Sciences, Allerton Press, Inc., USA, (in English).

[14] F. C. Hennie, Fault Detecting Experiments for Sequential Circuits, Proc. IEEE 5-th Ann. Symp. On Switch. Circuits Theory and Logical Design, 1964, pp. 95-110.

[15] Z. Kohavi, Switching and Finite Automata Theory. Mc Graw-Hill, 1970.

[16] V. A. Kozlovsky, On the Recognition of an Automaton Relative to a Locally Generated Class, Soviet Math. Dokl, vol. 23, no.3, pp. 625-628, 1981.

[17] V. A. Kozlovsky, On Representation of the Group Automata, Kibernetika i Sistemny Analiz, no.2, pp. 21-28, 1996.

[18] V. B. Kudryvtsev, S. V. Aleshin, A. S. Podkolzin, Introduction to Automata Theory, Moscow: Nauka, 1985 (in Russian).

[19] D. Lee, and D. Su, Modelling and Testing of Protocol Systems, Proc. IFIP TC6 10th Int. Workshop on Testing of Commun. Syst., Chapman & Hall, Myungchui Kim and Sungwon Kang Eds., pp. 339-364, 1997.

[20] D. Lee, M. Yannakakis, Principles and Methods of Testing Finite State Machines - A Survey, Proc. of the IEEE, vol. 84, no.8, pp. 1090-1123, Aug. 1996.

[21] B. D. Lukyanov, On Distinguishing and Checking Experiments on Nondeterministic Automata, Kibernetica, no.5, pp.69-76, 1995.

[22] B. D. Lukyanov, Deterministic Implementations of Nondeterministic Automata, Kibernetika, no.4, pp.34-50, 1996.

[23] E. F. Moore, Gedanken-Experiments on Sequential Machines, Automata Studies, Prinston, NY: Prinston Univ. Press, 1956.

[24] A. Petrenko, G. v. Bochmann, M. Yao, On Fault Coverage of Tests for Finite State Specifications, Computer Networks and ISDN Systems, vol. 29, pp. 81-106, 1996.

[25] I. K. Rystsov, A Proof of Accessible Bound of Adaptive Diagnostic Experiment Bound for Finite Automata, Kibernetika, no.3, pp. 20-22, 1976.

[26] I. K. Rystsov, On Asymptotic Bound of Diagnostic Word Length for Finite Automata, Kibernetika, no.2, pp. 31-35, 1980.

[27] I. K. Rystsov, An Investigation of Solutions Complexity for Finite Automata Theory Problems. Kandidat Degree dissertation, Kiev: Institute of Cybernetics Ukr. Acad. of Sci., 1980, 121p.

[28] M. N. Sokolovskii, Diagnostic Experiment with Automata, Kibernetika, no.6, pp. 44-49, 1971.

[29] B. A. Trakhtenbrot, Y. M. Barzdin, Finite Automata, Behaviour and Synthesis. Amsterdam: North-Holland, 1973.

[30] M. P. Vasilevskii, Failure Diagnosis of Automata, Kibernetika, no.4, pp. 98-108, 1973.

[31] M. Yannakakis, D. Lee, Testing Finite State Machines: Fault Detection, J. of Comput. and Syst. Sci., vol. 50, pp. 209-227, 1995.

[32] A. Petrenko, and N. Yevtushenko, Test Suite Generation for a FSM with a Given Type of Implementation Errors, IFIP Transactions Protocol Specification, Testing, and Verification XII (the Proceedings of IFIP TC6 12th International Symposium on Protocol Specification, Testing, and Verification) 1992, pp. 229-243.

7 BIOGRAPHY

Igor S. Grunsky received the Diploma degree in Radiophisics and Electronics from the University of Kiev, the USSR, in 1963 and Ph.D. (Candidate) degree in discrete mathematics and computer science from the Computing Centre of the Soviet Academy of Sciences, Moscow, in 1976. Since 1965, he has been with the Institute of Applied Mathematics and Mechanics of the Ukrainian Academy of Sciences. He is currently the Head of Applied Discrete Mathematics Laboratory of this institute. His research interests include discrete mathematics, automata theory, testing problems, checking and diagnostics of hardware, software and protocols.

PART TWO

New Areas for Testing

TMN and Telecommunication Networks Testing

Mahamat Guiagoussou[1], Michel Kadoch[2]

1. Centre de Recherche Informatique de Montréal
1801, avenue Mcgill College, Montréal, Québec, H3A 2N4
Canada
mguiagou@crim.ca

2. École de Technologie Supérieure
1100, rue Notre-Dame Ouest, Montréal, Québec H3C 1K3
Canada
kadoch@ele.etsmtl.ca

Abstract

The objectives of the paper is to review the conflicting requirements and needs of existing telecommunication network testing approaches, services and equipment, in order to provide a simple but powerful solution that allows automated and standardized testing activities. We present in the paper a progressive transition from the current proprietary practices, to a distributed and fully TMN compliant management of telecommunication networks testing.

Keywords

TMN, OSI, Network testing, CMIS/CMIP, Network Performance, RTU

1. INTRODUCTION

Before the deregulation and the privatization of telecommunication industries, telecommunication network management was simpler. With the telecommunication areas rapid changes, multiple vendors and technologies interoperability, new services

and technologies integration became the biggest challenge of telecommunication networks and services management [11]. A completely automated management is almost impossible and unrealistic. The provisioning of an automated system removes the crucial need for human operator intervention.
Telecommunication Network Management (TMN) addresses the need for automation by providing standardized machine-to-machine interfaces that replace current manual functions and allows a management of heterogeneous equipment [1, 2]. By providing general and flexible interfaces, it also addresses the need to support rapid evolution of technology and integration of new services [3, 6]. For the interest of testing, two recommendations have been proposed by OSI/TMN standards : the *X.745* (Test Management Function) [7] and the *X.737* (Categories of Diagnostic and Confidence Tests) [8].
The automation of management aspects such as testing and faults diagnosis is a key issue in Telecommunication Network and Service Maintenance. Current telecommunication maintenance and fault management activities consist in continually monitoring the network elements and services. When troubles are experienced by users, they are notified to the network operation centers or to a help desk. Field personnel are sent to make manual measurements and tests in order to troubleshoot the problems and restore the affected services.
The objectives of the paper is to review the conflicting requirements and needs of existing telecommunication network testing approaches, services and equipment, in order to provide a simple but powerful solution that allows automated and standardized testing activities. We present in the paper a progressive transition from the current proprietary practices to a distributed and fully TMN compliant management of telecommunication networks testing.

2. REVIEW OF TMN FROM TESTING PERSPECTIVES

This section presents an overview of TMN principles, standards, and architectures. It also presents standards recommendations defined for testing management. Then driven forces for TMN deployment are discussed.

2.1 Overview of TMN

2.1.1 Evolution and Objectives

TMN have been introduced by ITU-T, and defined in recommendations M.3000 [1], and M.3010 [2]. Standardization studies started in 1985 with the definition of interfaces and the specification of interface protocols between Operation Systems (OSs) and transmission terminals. In 1988 the first recommendation M.30 was included as part of the blue book. Extension was then provided to include the management of all telecommunication networks and services. In 1992 M.3010 [2], the revised version was published. TMN also provides a structured architecture for the interconnection of OS (Operation system) and/or telecommunication equipment in order to allow management information exchanges. These interconnections use standardized interfaces including protocols and messages. The basic objectives of TMN focus on: the use of generic telecommunication network models for the management of heterogeneous network services and equipment; the operation across multiple vendors and different technologies; the inter-working among the multiple management and operation systems; and the management of inter-working among separately managed networks or domains.

2.1.2 Functional Architecture

Functional architecture describes the appropriate distribution of functionality within TMN [2, 4]. These functionality were defined as functional blocks, namely : Operation System Function (OSF) processes information related to the telecommunications management; Network Element Function (NEF) which communicates with the TMN for the purpose of being monitored and/or controlled; Q-Adapter Function (QAF) is used to connect as part of the TMN those non-TMN entities which are NEF-like and OSF-like; Mediation Function (MF) acts on information passing between an OSF and NEF (or QAF) to ensure that the information conforms to the expectations of the function blocks attached to the MF; Work Station Function (WSF) provides the means to interpret TMN information for the management information user; and Data Communication Function (DCF) for the transfer of Telecommunication Network Management Information.
At the service boundaries of TMN function blocks, reference points are defined as access to services. They represent conceptual points of information exchanges between non overlapping functional blocks. Three main interfaces are defined within TMN : the q-class reference points (between functional blocks within TMN that can contain a management application) ; the f-class reference points (between a Work Station and TMN); and the x-class reference points (between OSF blocks of two TMNs or OSF block of a TMN and the equivalent function of another network). Figure 1 presents examples of reference points between functional blocks. The upper part illustrates reference points between a TMN block and a non TMN OSF (i.e., reference point m). The lower part gives more specific cases of reference points : the left one between users, a WSF, an OSF and a NEF, and the right one between an OSF and a QAF. In each case, the OSF use a q3 reference point to communicate via a MF which communicates in turn with the final blocks (NEF or QAF) using a qx reference point (a q interface with reduced functionality).

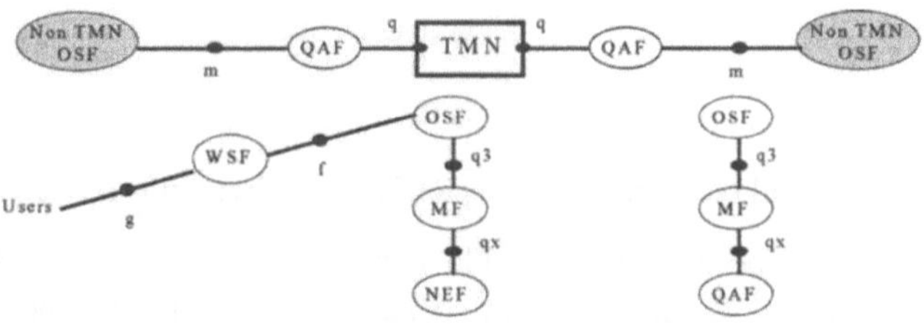

Figure 1 - Example of function blocks and reference Points

TMN identified a need for a hierarchy of management responsibilities. Such hierarchy can be described in terms of management layers. The scope of each layer is wider than the layer below it. In general, it is expected that upper layers will be more generic in functionality while lower layers are more specific. The Logical Layer Architecture

(LLA) [2] implies the clustering of management functionality into layers. It uses a recursive approach to decompose a particular management activity into a series of nested functional domains. Each functional domain forms a management domain under the control of an operation system function (OSF) and thus each domain is called an OSF domain. A domain may contain other OSF domains to allow further layering and/or it may represent resources (logical or physical) as managed objects (MOs) within that domain. All interactions within a domain take place at generic q reference points. However, interactions between peer domains, i.e. crossing an OSF domain boundary, can take place at a q or x reference points depending upon the business strategy applicable for that interaction. When providing network services it is common for management to cross the boundaries of an Administration, hence arrangements are made for inter-TMN interactions.

2.1.3 Information architecture

In order to allow effective definition of managed resources, the TMN methodology makes use of the OSI system management principles and is based on an object-oriented paradigm. Management systems exchange information modeled in terms of managed objects [9]. As illustrated in Figure 2, managed objects collected within a Management Information Base (MIB) are conceptual views of the resources that are being managed or may exist to support certain management functions (e.g. event forwarding or event logging). Thus, a managed object is the abstraction of such a resource that represents its properties as seen by (and for the purposes of) management processes. A managed object may also represent a relationship between resources or a combination of resources (e.g. a Network). It must be noted that object oriented principles apply to the information modeling, i.e. to the interfaces over which communicating management systems interact and should not constrain the internal implementation of the telecommunications management system.

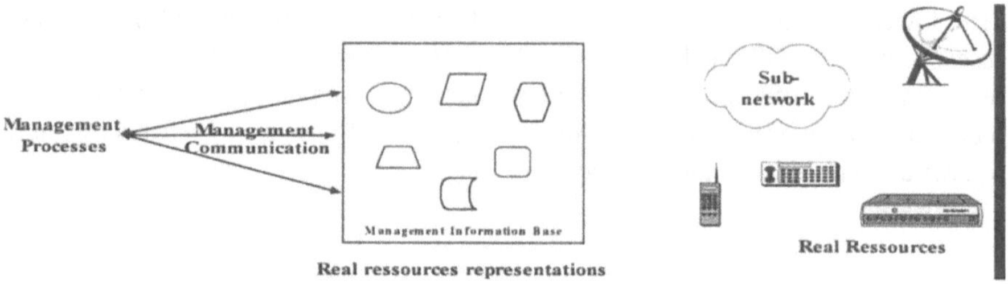

Figure 2 : Real resources modeling and management information base

An object in the perspective of TMN/OSI management is defined by: the attributes visible at its boundary; the management operations which may be applied to it; the behavior exhibited by it in response to management operations or in reaction to other types of stimuli and notifications that are emitted when some internal (e.g. threshold crossing) or external (e.g. interaction with other objects) occurrence affecting the object is detected. The value of these attributes and the interactions the object can have with its management environment define its behavior. For example, such interaction could occur when an operation requests is received from a manager via an agent or when

results are sent back in response to a given request after an action have been performed. The GNIM (Generic Network Information Model) [6] provides management of the inter-operability of services, networks, NE, OS, etc. It also provides a uniform management information model. It is a technology independent model that can be applied to different types of equipment that use common and standardized interfaces.
Management of a telecommunications environment is an information processing application. Because the environment being managed is distributed, network management is also a distributed application. This involves the exchange of management information between management processes for the purpose of monitoring and controlling the various physical and logical networking resources. The Manager/Agent concepts, developed for OSI systems management, are introduced in TMN framework. The concepts necessary for the organization and interworking of complex managed systems (e.g. networks) are also introduced under the headings of management domains and shared management knowledge. For a specific management association, the management processes will take on one of two possible roles defined in X.701 [4]. The Manager role describes the part of the distributed application that issues management operation directives and receives notifications, while the agent role describes the part of the application process that manages the associated managed objects. The role of the Agent will be to respond to directives and requests issued by a Manager. It will also reflect to the Manager a view of these objects and emit notifications reflecting the behavior of these objects. A Manager is the part of the distributed application for which a particular exchange of information has taken the manager role. Similarly, an Agent is the part that has taken the agent role. The managing entity uses the Common Management Information Protocol (CMIP) to access managed information provided by an agent residing either in a stand alone open system or in the managed resource.

2.1.4 Physical Architecture

Physical architecture describes implementation of reference points (i.e., interfaces) and examples of physical components within TMN. For each functional block a physical blocks can be implemented thus leading to a physical architecture. These physical blocks are : Operation System (OS); Network Element (NE); Q-Adapter (QA); Mediation Device (MD); Work Station (WS); and the Data Communication Network (DCN). Reference points are realized within the physical architecture by physical interfaces within system or equipment. They are represented by capital letters (Q, F, X) and form the common boundary between associated building TMN blocks. The F interfaces can be found between WS and the TMN, the Q (Q3) interface between TMN devices and the X interface between TMN devices of one TMN with devices of another TMN via DCN.

2.2 Standards for Telecommunication Networks Testing

Two standards for the purpose of telecommunication networks tests are presented : X.745 (Test management function) and X.737 (Categories of Diagnostic and Confidence Tests).

The X.745 [8] specification describes the needs for the remote control of testing activities and defines a framework for the specification of tests to be applied to resources including open systems. A test is considered as the operation and monitoring of open systems or their parts, in a specified environment in order to collect information on the functionality and/or the performance of the considered system(s). A test is defined with the creation of the environment for the test to be performed, the control and the monitoring of the considered systems (i.e. operations of the test), the modification of a normal operational environment. The test control includes for example the need to suspend, resume and terminates the test. Actually, tests equipment available in the market does not allow to automatically suspend, nor to resume the tests they perform.

Each test is defined with a unique identifier in such a way that data generated by the same test can be traced and correlated. Parts or aspects of the system environment that can require alteration for the purpose of the test are : the connections to other open systems; the configuration of the tested system; the traffic load required by the tested system; and the configuration of the testing systems and instruments. In some cases, test scheduling mechanisms are required. A test can also be specified in such a way that it is activated when some predefined conditions are satisfied (i.e. a threshold is reached) or when a specific event is detected (i.e., an Alarm Indication Signal).

Specification of a test includes the description of its *objectives*, its *environment*, its *controllability* (synchronous, asynchronous), and the *testing procedures* and *tests states.* The execution of a test involves two or several application processes. The manager-agent paradigm is used to define these processes: the managing process is concerned with the test initiation (it is called *test conductor*), and the agent process is allocated the task of executing the tests (it is called *test performer*). The test performer is requested by the test conductor to realize the test. A simplified architecture of the test management is depicted in Figure 3. A test request is sent to a managed object in the test performer. This later has the ability to receive and respond to the test request. Such a functionality is called TARR (Test Action Request Receiver). The managed objects that provide management view of the object under test (MOT) such as the telecommunication circuits to be tested, are identified within the test request. Each test must use one or several MOT(s). Additional information representing management or managed resources that are used for the need of testing (insertion and test signal reception points) such as Remote Testing Units (RTU) defined as AO (Associated Object). Information collected during the execution of the tests are specified and stored in support objects called TO (Test Object).

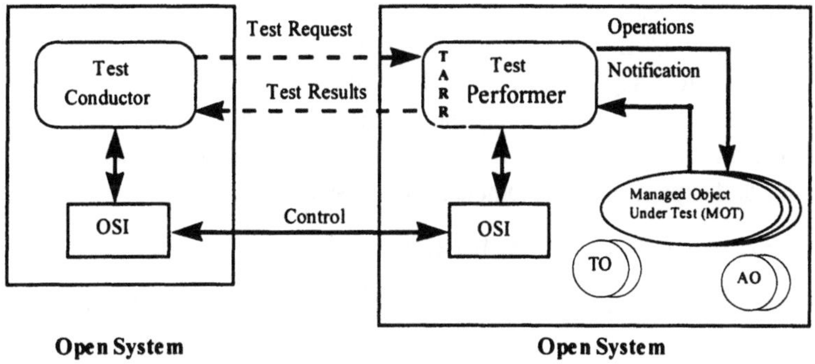

Figure 3 : Simplified Test Management Architecture

The X.737 specification [7] defines a set of basic tests for the categories of tests for the service introduction and faults diagnostic. These test are used to verify the ability of a specific equipment, a system or a network to realize correctly its allocated function (i.e., the tested entity continues to behave properly as defined in its specification). The specifications also define the way these tests are achieved and notifications are used to detect faults in order to isolate the cause of problems. These tests can be classified as follow: resources tests category (*resource boundary test, resource self test),* communication path tests category (*loop-back test, far-end connection test, connectivity test, connection test, delay, data integrity test)*; protocol integrity test; and test infrastructure test.

In order to use these generic tests in current telecommunication environment, additional specialization and refinement are required. Explicit tests selection must be specified. The advantage of an appropriate tests classification is that is ease the matching of test classes with well Known fault types, the reduction of faults diagnosis and troubleshooting complexity.

2.3 Motivations and driven forces for TMN deployment

2.3.1 General motivations

Several key drivers are forcing the deployment of TMN solutions in telecommunication networks management [10]. TMN standards and other related recommendations are becoming more and more stable. Telecommunication industries, R&D community, and software developer efforts reached so far an acceptable level of practical applicability. Distributed computing standards are now emerging. Computing technology such as CORBA have actually real time application oriented implementations (e.g., Orbix, Obeline, etc.) that are used in several domains. Software technologies and concepts such as: manager-agent mechanism, object oriented paradigm, intelligent multi-agent technologies, mobile agent theory, software reengineering and databases are widely used and deployed. Rapid prototyping, reduced development cost, and ease of use are some of the attracting opportunities provided by Web-Based Management for TMN. Intranet and Internet remote access facility are becoming the a promising tool for remote and standardized management activities. Even if SNMP based management are more popular than CMIP, proxy products are available to allow the usability of TMN based protocols. Actually, number of TMN compliant OSs and NEs are made available by vendors, service provider, and equipment manufacturers. Artificial Intelligence and Expert System theory are now being actively used to help implementing human knowledge and expertise, and automate routine tasks.

2.3.2 Specific testing motivations

As telecommunication networks are growing in complexity, types, and dimension, their maintenance and specifically their testing poses a big challenge. From the later perspectives, several issues drive the advance of distributed and opened maintenance in telecommunication networks [14]. Wide area digital transmission facilities are used in corporate networks. Thus the requirements for appropriate network testing, drive the way these facilities are managed. Network quality is one of the national and international success and market competitiveness criteria for enterprises and organization. Network missions are critical; any trouble that disrupts a service can seriously affect productivity. The shift from voice to data traffic introduces a new requirement for better performance and high quality of transmission services. Delay, downtime, or performance degradation are not tolerated by real-time and interactive data traffic. Users are more and more dependent on higher speed and multimedia applications. In addition, with the growing number of high speed facilities carrying more traffics and linking large number of users, any network problem will impact an increasing number of users and severely degrade their confidence on the network. Such complex situation requires highly qualified network management experts to appropriately handle the testing management problem. Another motivation of TMN testing is the need for maintenance costs reduction. The need to suppress the risk of network quality degradation is an economic need that focus on the reduction of networks operation costs. There is a considerable pressure and stress on network management personnel to drastically reduce operation center costs.

From the testing perspective, an appropriate solution for the specification, design and implementation of tests management must be worked out with these considerations in mind.

3. TESTS MANAGEMENT IN TELECOMMUNICATION NETWORKS

This section focuses on software testing tools, gives a brief overview of existing testing equipment and draw a picture of current telecommunication network testing.

3.1 What is to be managed ?

The environment to be considered while addressing the telecommunication network management testing is composed of: (i) the customers/users of services provided by these networks, (ii) the operators/owners of the networks, (iii) the managed telecommunication networks.

From a management perspective, the customers and users of the networks services are viewed as human monitors of managed system they use. The information they provide when notifying any abnormal conditions that affect the services is useful for management purposes. Another important issue to be considered is the organizational policies defined by the owners of the managed telecommunication networks. They intervene at a higher level of the management by ensuring the application of policies they defined to preserve a high quality of the network and the services provided to their customers. Finally the managed telecommunication network is the main component of the management environment to be controlled from several perspectives.

The telecommunication network example presented in Figure 4 depicts a generic network. A typical telecommunication network is composed of terminating points connected by physical paths (link), multiplexor equipment or digital cross connect system (DCS). Several circuits (voices and data) are multiplexed onto smaller number of higher speed circuits carrying data between multiplexers. These circuits are composed of sequences of physical connection links (e.g., cables, fiber, wireless links) and number of equipment (e.g., CSU, DACS, NUI, Repeaters).

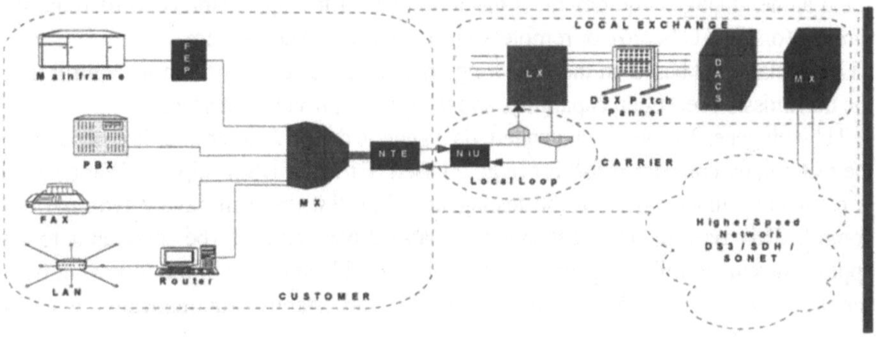

Figure 4 : Example of Managed Telecommunication Network

The description criteria of the presented network is based on ownership (i.e., customer network part local loop, and carrier network). The customer part of the network contains multiplexing system used to feed into the carrier's network customer applications. All type of traffic like voice, data, video, and fax are multiplexed and carrier by higher speed networks. The customer network incorporates: mainframe, LAN, PBX, Fax, stand alone PCs and terminals. It also incorporates interconnection components such as Gateways, Routers, and Network Terminating Equipment (NTE). Some of these managed elements have some management capabilities. For example, added to its basic function (e.g., electric isolation between the network and the customer equipment), NTE is capable of monitoring telecommunication access point (e.g., E1 access point), to provide testing facility (e.g., loop back for bit error rate testing), and to generate alarms (e.g., AIS: Alarm Indication Signal). At the local loop, Network Interface Units (NIU) limit the border between the customer network and the carrier network and provide several management capability (e.g., full test access, loop back facility, etc.). Line repeaters are used to regenerate the signal on the loop and also provide loop back and network troubles detection facilities. Finally the carrier domain is composed of local exchange, cross-connect and multiplexing system. The local exchange is composed of office equipment that regenerates multiplexed signal before their routing. It is also composed of DSX patch panel, or Digital Access Cross-connect (DACS). Data traveling on the networks circuits are monitored, amplified by the intermediate equipment. Telecommunication networks that use Synchronous Digital Hierarchy (SDH) have capability to detect faults on different levels via embedded overhead information such as check sum (CRC-4) and trial labels.

3.2 Telecommunication Networks Tests Management

3.2.1 Tests Units

Number of instruments, devices and equipment explicitly dedicated for the monitoring and testing of network components are widely used in telecommunication network maintenance [12]. We call an entity with testing capabilities a *Testing Unit* (TU). It is used in local management site or remotely. While used remotely it is called *Remote Testing Units* (RTU). A TU is a software tool that remotely control testing equipment or stand alone managed equipment. It has a number of testing capability to be requested directly (on the TU board), or remotely via a control software system.

TUs are classified based on the following criteria : the category of remote control used, the transmission/reception capability, and the type of provided interface.

A TU with generic driver is controlled via a generic driver. Its user interface depends on the technology considered for test (e.g., T1 Bert, E1 Jitter, DS-3 Slip and Wander), but is completely independent of the testing unit. The test unit can have a terminal type control. This type of TU does not have any driver, and can be integrated to test applications in the following manner: associate an address of one of the following type : serial port, modem, X.25 or TCP/IP to the TU and establish automatically an ASCII terminal session (e.g., Telnet service in TCP/IP) between the client and the TU. A TU can have a Windows/DOS, or an X Windows/UNIX type of control. This type of TU is typically a Personal Computer equipped with interface cards that allow to perform network testing. The software that run within the TU which controls its functionality is often a Windows 16 bits software or a UNIX/Xwindows software. Finally an undergoing class of TU is that one with opened type of control. Current development will lead in a near future to a new generation of TUs that are controlled in an opened way. The standard used are CORBA based and ITU-T CMIS/CMIP protocols.

The transmission/reception characteristic of TUs performing tests at physical layer is described in terms of number of transmission/reception ports. A half duplex TU has one transmission port and one reception port, a dual monitor TU has one transmission port and two reception ports, and the full duplex TU has two transmission ports and two reception ports.

The last TU classification criteria is based on type of Interfaces that the unit is able to connect to and perform testing on. TUs have often more than one test interface. Almost all the TUs can use only one of these interfaces during a test session. For example a FIREBERD 6000 which is a TU from Telecommunication Techniques and Corporation (TTC) can in theory be equipped with eight different interfaces such as T1, E1, DS-3, V.35, X.21, 232 TTL. Some TUs such as those from WANDEL & GOLTERMANN (WG) are manufactured with a certain number of interfaces that can never be modified by the user. However other TUs such as FIREBERD can be modified by the user, but only in a cold manner (i.e. the device must be powered off before the modification). Thus when a software connects on such a device, what is detected will be available for all the duration of the session.

3.2.2 Testing Realities

Telecommunication networks testing activities include : tests configuration (connections, setups), tests scheduling/execution, tests results collection/analysis, and fault isolation. In general manual practices are more frequent in testing. They often lead to human errors and mismatches. Such a testing lack clear documented cases and appropriate guidelines for tests results interpretations/usage. There is no clear test objectives neither good tests classifications. The only available interpretation is that relying on highly experienced and skilled operators. The control of tests is also done manually. Known tests technology are proprietary and depend on vendor, manufacturer, organization policy, or test equipment. The classic manual practices of troubles causes isolation are based on the collection and usage of tests results such as: bit errors, bite error rate (BER), Far end Alarm Signal (FAS) errors, pattern slips, error-free seconds (EFS), percent error-free seconds (%EFS), etc.

Two types of testing are identified : *out-of-service* (OOS) *testing* and *In-Service* (IS) *testing*. The OOS testing is used when installing network components (e.g., T1 Circuits) and verifying end-to-end continuity. It is also used for fault isolation by inserting pseudorandom patterns that simulate a live data exchange and allows to collects tests results to be analyzed for faults isolation. Some acceptance and conformance testing including time and stress tests are also performed while the tested system is out of service. IS testing are performed while the system is on line and is provided the service for which it's tested. The later type of test can be disruptive thus can affect services provided by the managed network.

Basically there is two generic testing configuration approaches: *loopback testing, and end-to-end testing*. Loopback testing is performed with only one TU connected to a element of the network segment to be tested. It is characterized by a limited faults detection capability. For example, while testing a circuit in a loop back configuration, only one direction can be checked. Due to their management capability, most NEs remove code errors they receive (fault tolerance) before transmitting the data. This affect the analysis of results and by the way the faults diagnosis. End to End testing overcome most of the enumerated drawback of loopback testing. It is performed with two TUs located at both sides of the network segment or circuits to be tested. This testing configuration allows analysis in both directions of the segment under test. This approach is better than loopback because the direction of errors may easily and quickly be found.

3.2.3 Specific Test Categories

As X.737 test categories are generic, we refine and specify telecommunication network tests classes [12]. These specific tests classes are :

- ***Continuity Test :*** This test is aimed to show that continuity of bit information transmission will not be corrupted or abnormally delayed on a path between the TU and a looped-back NE.

- ***Signal Delay Test :*** two types of delay measurements are identified. (1) Equipment Delay Test allows to verify if the delay time between the receipt of an information bit and the retransmission of the same bit on any channel of the NE shall not exceed a given tresholded value (0.5ms). (2) Path Delay Test allows to verify if the delay time between the receipt of an information bit and the retransmission of the same bit on any channel of the transmission path shall not exceed Xms. It shows that the signal delay on a looped-back transmission path is less than Xms (X=60ms for FO, and X=600ms for Satellite).
- ***Bit Error performance Test :*** this type of test is used to verify that a NE does not introduce bit errors into any channel at a rate greater than 1 in 10^{10} for any equipment configuration when measured for 24 hours. A bit error is defined as any output bit which does not have the same logic state as the corresponding input bit. The objective of this test is to show that the tested circuit will introduce no more than 1 bit error on any channel over a 24 hours interval.
- ***Jitter Measurement Test :*** three type of jitters measurements are described. (1) Input Jitters tolerance Test consists in identifying the maximum allowable number of jitters generated by a NE without any incrementation of bit error. (2) Allowed Output Jitter Test shows a NE Jitter value is not greater than that specified by CCITT Recommendations (G.743 for T1 1.544 kb/s interfaces and G.703 for E1 2.048 kb/s interfaces). With no jitters at the input to the multiplexer and demultiplexer, the Jitter at the demultiplexer output should not exceed 1/3 unit interval peak-to-peak. The objective of this test is to shows that output from an NE is less than a given limit (1/3 UI for 2.048 kb/s disgroup) when measured at a number of frequencies in a given interval (58 frequencies from 10Mh to 100 MHz). (3) Jitters transfer test consist of the determining at what extent a NE passes Jitter to its output. The objective of this test is to shows how much NE amplifies (or attenuates) jitters present at its input.
- ***VF Tone Generation Test :*** it is an analog continuity like test. It shows that a voice frequency tone can be audibly generated. VF Tone Frequency Response Test is an analog loopback test. It shows that a VF frequency response can be obtained from a loop back.

With this classification and specialization of tests on hand, the remaining question is : what is the distribution and mapping of test classes versus faults classes? The issue here is to be able to appropriately monitor the telecommunication networks in order to detect the impairments and the faults symptoms, then select the appropriate tests types, category/class, or suites that can accurately locate the source of the fault. Another issue is the management of these testing activities that could include more complex problems such as the analysis of alarms and faults symptoms, the collection of test, the storage of tests results and also the automatic interpretation of these results. TMN Testing solution discussed in the next section addresses some of these issues.

4. TOWARD A FULL TMN COMPLIANT TESTS MANAGEMENT

4.1 TMN Current State

Despite the emerging research works and available TMN products [13], several barriers and obstacles are encountered while applying TMN principles. There is a big gap between what TMN standards say and what is currently implemented. TMN tells where to start but never indicates where to stop. This leads to several conflicting interpretations and implementation of standards. As stated before, the first constrains faced by TMN application developers is the huge number of standard documents and the framework complexity and instability. TMN architects also have to handle the full TMN/OSI Protocol stack implementation complexity and high cost. Even thus, existing TMN implementations have several performance problems. Simple Network Management Protocol (SNMP) [10] is the widely used and predominant solution in the market because of its simplicity and it availability. Due to the OSI/TMN protocols complexity and a large availability of SNMP based NEs, TMN visibility is limited. TMN Information models also fail to meet existing and future Operation Administration Maintenance & Provisioning (OAM&P) needs. It is unclear how to apply concepts such as service and business layer management because the Logical Layered Architecture (LLA) is difficult to implement: it is hard to identify the border between layers. Furthermore R&D costs are higher and developers are resisting to shift to TMN solutions. There is a limited time, restricted initiative in product development and delivery putting stress and lot of pressure to researchers and developers. These limitations motivate progressive evolution toward the full TMN compliant management solution.

4.2 TMN requirements for Testing

There is a need for providing infrastructure to globally manage telecommunication networks testing activities from one end to another. This implies several considerations ranging from user/provider, time schedule, to development platforms considerations. Another requirement is related to the need for a high level and generic testing capability and instrumentation features. Practical and easy to implement solutions, flexible and compromising approach that integrate current practices and TMN concepts are needed. If a TMN test environment has to be deployed, it has to provide a global management infrastructure of the testing activities. This would allow to test the complete behavior of the telecommunication networks and systems. The approach must have a simple but powerful tests and test management description and information model detailed enough to be easily implemented in existing platforms. The automatic collection and analysis capabilities of test results is also highly required. Graphical presentation of tested network and also the ongoing testing activities are required. In addition, browsing and visual presentation of testing MIB (Management Information Base) are also required. Support for test data manipulations (export/import) are needed. Facility to monitor and control the test are also required. From simple tests, a test management system environment should be able to allow an easy and/or dynamic extension to complex test without adding too much effort. A hypertext help facility can be useful for training and online documentation.

4.3 Directions for Intermediate Test Management

The approach taken consist simply to apply progressively TMN approach and principles to the management of tests activities. The solution proposed start while designing the tests management application by a limited compliance with TMN then progressively upgrade the level of compliance to TMN. The following three tables sketch a guideline design approach for the transition from the completely proprietary solution to a full TMN test management system based to the TMN architectures.

Informational and Functional Architecture

Design issues	From	To
Network Element	NE Entity	Managed Object
Reference points (test and tested)	Relation	Exchanged Object
Test Units (software or devices)	TU Entity	Associated Objects
Tests categories	Tables and procedures	Test Object
Test session	Set of Test Entities	Session Object
Management protocols	SNMP, CMOT/L	CMIP
Communication protocol	TCP/IP	OSI Stack
Work Station	Proprietary	Generic GUI

Functional Architecture

Design issues	From	To
Test Management OSF (Conductor)	Function Entity	Manager (Object oriented)
Performance Management OSF	Function Entity	Manager (Object oriented)
Mediation function (Tests Performers)	Function Entity	Agent (Object oriented)
Interfaces between OSFs	Relation	Exchanged Manage Object
Global Management	Centralized	Hierarchical (LLA)

Physical Architecture

Implementation issues	From	To
Test Management Information Base (MIB)	SQL based	Object Oriented
Blocks (WS, OS, NE, MD, QA, DCN)	Proprietary	Standards
Interfaces (F, Q, X)	Proprietary	Standards

4.4 A Hybrid Test Management System

The application of the directions presented in the previous section can lead to an hybrid performance and test management system for telecommunication networks (Figure 5).

The depicted system is developed within the VIVALDI project[1] . It is aimed to automate and integrate current testing techniques using some TMN guidelines. The system is initially developed as a centralized management application (management server) witch implements two Operation Systems Functions: *tests management OSF* and *performances management OSF*. Client application has been developed in the early stage of the project as a demonstration prototype. It provides Work Station facilities (Graphical User Interface) that allow a Human Machine Interaction to access, in a user friendly way, services provided by the tests and performances manager. This management station is designed to run on Windows 95/NT. It is locally or remotely used by human operators to initiate, execute, control and monitor tests operations and access the telecommunication network performance data.

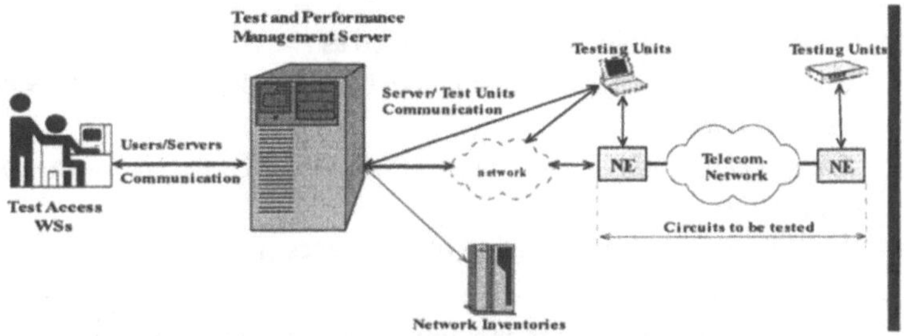

Figure 5 - Simplified Test management Architecture

The management entities have the capacity to access the configuration and inventory database as well as the topology of the managed telecommunication network. The later feature allows to identify the network components to be tested (e.g., circuits relational database) and to provide routing information associated with them. It also provides the available testing and performance measurement resources. The test management system is based on X.745 and X.737 recommendations. Categories of tests defined by X.737 are considered together with the well known tests from telecommunication network people. Specialized Testing Units are connected to the managed network. These TUs are remotely controlled by the tests management server via some «Mediation Devices » not in the sense of TMN. Several tests drivers are defined to allow the use of specialized test equipment provided by different vendors and manufacturers (e.g., FIREBERD 6000 from TTC, PA-41 from WANDEL & GOLTERMNAN, etc.). The performance management performs continual networks polling of telecommunication networks elements using performances monitoring agents that collect performance data. When a control reached by the user threshold is attained, the management system generates alarms and sends them to the central tests and performance management. This module also handles a relational data base (MIB) containing long term (one year) statistical data and information on the performances of network managed elements.

4.5 To a Full TMN Compliance

As we have seen in the previous section, some management components implement

---

[1] VIVALDI is a project that have been undertaken by ETS (École de Technologie Supérieure) with a subsidy from MINACOM International R&D center.

well defined policies (testing, performance measurements, configurations and planning). To be fully compliant with TMN management solutions must satisfy several constrains. For example, the management information model must be based on standard OSI/TMN recommendations. Communication protocol between TMN entities must ideally be CMIP (Common Management Information Protocol) over which CMIS (Common Management Information Service). Different interactions points defined as TMN reference points between the functional blocks and implemented by various interfaces (e.g., F interface between the management server and the client WS, Q3 interface between the server and the NEs), should also follow TMN recommendations. That is the interfaces should be modeled as objects and communication based on the OSI Stack and CMIP.

Full TMN compliance conditions can be resumed as :

- Specification of an Information model that is independent from technology or vendors (GNIM)
- Clear specification of TMN blocks (WS, OS, MF/QA, NE) with separation of boundaries
- Identification and specification of interfaces (F, Q3, Qx, X, M) using TMN methodology
- Replacement of the ad-hoc communication protocol by CMIP
- Replacement of TCP/IP or Proprietary DCN by OSI Stack
- Development of Q-Adaptors to integrate legacy components
- Migration of the relational or proprietary MIB to an Object Oriented database

5. CONCLUSION

In this paper we have analyzed the conflicting requirements and needs of current telecommunication network testing. The objectif is to provide an appropriate solution to remote testing activities and to automate the procedures. In order to situate the design context, a review of existing TMN principles and framework have been presented. The current telecommunication testing activities with respect to the TMN driven forces were also reviewed. The main test management requirements were analysed and then the problem addressed by this paper: «how to fly toward a Full TMN Distributed Testing Management ?» was discussed. An intermediate solution for the development of a partially centralized test management based on TMN principles was adopted. The requirements for a transition to a full TMN implementation was then discussed. It was then shown that a smooth transition toward a full TMN can efficiently bring the realization of the telecommunication network testing. In order to validate the proposed ideas, a telecommunication network is presented as a case study.
Future work will be directed toward the enhancement of the TMN testing approach, taking into consideration existing testing work such as protocol validation and conformance testing. Furthermore a new implementation solution based on the mobile agents paradigm is under development.

6. REFERENCES

[1] ITU-T Recommendation M.3000, *Overview of TMN Recommendations*, Geneva 1994.

[2] ITU-T Recommendation M.3010, *Principles for a Telecommunications Management Network*, Geneva 1992.

[3] ITU-T Recommendation M.3200, *TMN Management Services : Overview*, Geneva 1992.

[4] ITU-T Recommendation M.3400, *TMN Management Functions*, Geneva 1992.

[6] ITU-T Recommendation M.3100, *Generic Network Information Model*, Geneva 1995.

[7] ITU-T Recommendation X.737, *Test Management Function Management function.*

[8] ITU-T Recommendation X.745, *Diagnostic and Confidence Tests Categories.*

[9] ITU-T Recommendation X.720, *Information Technology - OSI - Structure of Management Information : Management Information Model,* Geneva 1992.

[10] William Stallings, *Networking Standards - A Guide to OSI, ISDN, LAN and MAN Standards*, Addison Wesley Publishers Ltd., 1992.

[11] Roch H. Glitho and Stephen Hayes, *Telecommunications Management Network : Vision vs Reality*, IEEE Communications Magazine, March 1995.

[12] Bellcore, GR-818-CORE, *Network Maintenance : Access and Testing - Generic Test Architecture,* Bell Communication Research, 1995.

[13] Bernard S. Ku, *Use of TMN for SONET/SDH Network Management,* NOMS'96 - IEEE/IFIP 1996 Network Operations and Maintenance, Conference Proceeding, Volume 2, April 1996.

[14] TTC, *Implementing a Distributed Test Solution, Telecommunication Techniques* Corporation, 1996.

Conformance testing methodology of Internet protocols Internet application-layer protocol testing – the Hypertext Transfer Protocol

Roland Gecse
Conformance Center, Ericsson Ltd.
H-1037 Budapest, Laborc u. 1., Hungary,
Tel: +36 1 437 7618, Fax: +36 1 437 7219,
e-mail: roland.gecse@lt.eth.ericsson.se

Abstract

This paper examines the applicability of OSI conformance test methodology to Internet protocols. It summarizes the differences between them and introduces the Internet Reference Model along with a new abstract test method, which was designed for the practical purposes of conformance testing of TCP/IP protocols. Some interesting test cases, that were chosen from HTTP, demonstrate the facilities of the model and give impression of testing Internet protocols.

Keywords

conformance test, Internet, HTTP

1 INTRODUCTION

Up to now, in the Internet community, conformance testing was an unknown concept. However, the need for recommendation conforming TCP/IP implementations grows as the application of Internet protocols in business telecommunication systems is becoming reality. It is probable that more and more vendors are going to provide Internet products, whose reliability and interoperability with other products have to be assured.

Although conformance testing methodology (X.290–X.296, 1994-95) was originally intended for OSI based systems, there are ongoing discussions about its applicability to the TCP/IP protocol stack. Numerous articles and conference contributions justify that these questions present a current topic.

(Bi1, 1997) founds theoretical base of relay system testing, which is then used, among others, for the testing of Simple Mail Transfer Protocol (Bi2, 1997) and IP router. (Kato1, 1997) and (Kato2, 1997) focus on detailed analysis of Transmission Control Protocol's flow control algorithms that are expected to be used on measuring and fixing the majority of implementation problems listed in (Paxton, 1998). On the other hand, (Malek, 1998) deals with interoperability test suite derivation that may be used for the purpose of Internet testing.

The following issues, beside others, will be argued in this paper. Sections 2-5 give an overview of the Internet protocol structure, introduce the Internet Reference Model and suggest a new abstract test method. Also, similarities and differences in layering, data flow and configuration are fetched in comparison to the OSI Basic Reference Model (BRM) (X.200, 1994). After the presentation of a possible test realization (section 6) and a short overview of the Hypertext Transfer Protocol (section 7), section 8 gives some practical testing examples from the field of client, server and proxy testing.

2 COMPARING INTERNET AND OSI ARCHITECTURE

The OSI BRM has 7 layers, each of which with a well-defined task. OSI protocol stacks are designed to fit to this model. The protocol entities (PEs) of a particular protocol suite are associated to the appropriate layers. Peer-to-peer communication between two PEs of the same layer takes place in abstract protocol data units (PDUs) while physical communication with upper and lower layers' PEs is only possible via service primitives (SPs).

Unfortunately, Internet was not planned to have such a detailed abstract model. The structure of TCP/IP, which represents the actual state of Internet has evolved gradually from the beginnings (Carpenter, 1996). Internet has only four layers: link, network, transport and application. Although the general functions of these layers are not as well-defined as OSI's, they provide almost the same functionality. Disregarding that reliable service appears first only in the transport layer, network and transport layers map to their OSI counterparts. Internet link layer maps, in general, to OSI physical and data-link layers. Since the application layer holds all remaining functionality (OSI layers 5-7), applications may gain enormous complexity. Internet protocols do not have standardized SPs, thus in contrast to open systems, the communication between neighboring layers is implementation specific. This, besides the loosely specified layer characteristics, results that layer boundaries are flexible. Another feature that must be kept in mind when talking about Internet is the whole TCP/IP protocol stack should be considered as a single unit together with a set of alternative protocols. The transport layer, for example consists of two protocols: the transmission control protocol (TCP), which is a connection-mode service and the user datagram protocol (UDP) that provides a connectionless service. In a particular communication process, at most one of these services is used.

From the configuration point of view a real open system can act as end system, relay system or both simultaneously. Internet systems have also this kind of configurations with noting that relay systems are called also intermediaries. Intermediaries are further subdivided according to working aspects to proxy, gateway and tunnel systems which will be discussed later.

3 CONFORMANCE TESTING OF INTERNET PROTOCOLS

From the conformance testing perspective it is worth to distinguish between hardware and software implementations. Hardware implementations (eg. IP router, Web-TV equipments) neither implement the whole TCP/IP protocol stack, nor provide interface to protocol layers. Accordingly, they could be examined only by an external test system. Software implementations (eg. FTP client, httpd programs) on the other hand have numerous advantages over hardware systems. Besides the existing test methods (X.290–X.296, 1994-95; Bil, 1997), they imply the possibility of designing more effective new test methods. For the understanding of this methods, a particular TCP/IP implementation should be examined.

4.4BSD-Lite's Net/3 networking code (Wright, 1995) can be considered as a reference implementation of the Internet protocol suite*.

The structure of the Net/3 networking code is presented in figure 1. Application level protocols (FTP, Telnet, RIP) are distinguished from the underlying TCP/IP stack. They are running as processes in the device's user space while underlying layers protocols used to be implemented as a single unit in the operating system space.

The internal structure of this unit consists of three layers: application programming interface (API) or socket layer, protocol layer and interface layer. The public functions of this unit can be reached at the kernel entry points using system calls (SCs) which represent the operating systems' service primitives.

API, in addition to separating the application layer, provides a protocol independent interface to the entities of the underlying protocol layer. It offers a set of different networking features of the kernel that can be reached uniformly via SCs.

The protocol layer holds the Internet transport (UDP, TCP) and network (IP, ICMP, IGMP) layer protocols (Stevens, 1994). The protocol layer does not provide SCs to application layer entities.

The interface layer consists of various device drivers implementing link layer

*Besides TCP/IP, it also supports Xerox Network Systems (XNS), OSI communication protocol families and the Unix domain protocols that are provided for interprocess communication (IPC).

protocols (eg. Ethernet) and procedures that are used for address conversion between the protocol layer and itself. The code for different pseudo devices (loopback interface, BSD packet filter (BPF)) can also be found there. Interface layer functions are accessible through SCs. The packet filtering functions are further applicable for control and observation.

Now, having a global picture of the overall structure of TCP/IP, the Internet Reference Model will be introduced.

4 THE INTERNET REFERENCE MODEL

It can be stated that all of today's software TCP/IP implementations are based upon the architecture of Net/3. By considering this a model will be introduced that is suitable for conformance testing and incorporates the listed features of software implementations.

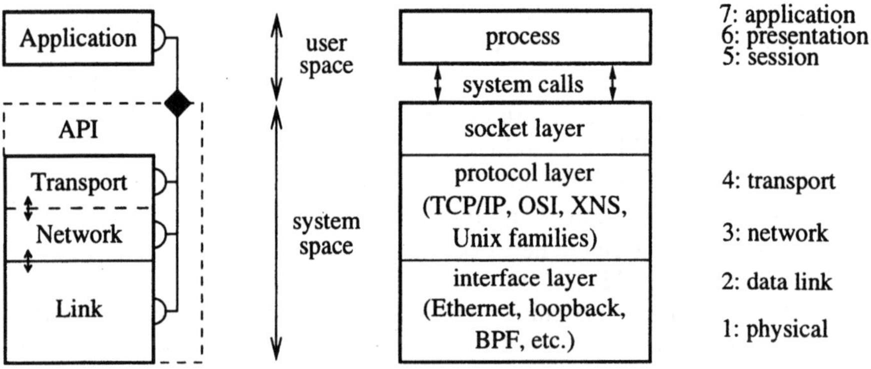

Figure 1 The Internet Reference Model (left), general organization of Net/3 networking code (right)

In the Internet Reference Model (IRM), the functions of SPs are replaced by SCs of API*. These SCs allow applications to send PDUs directly to each layers protocol entity. The API itself should be considered as a switch that connects applications to the selected underlying service via SCs. The functions of the API are provided at kernel entry points (rhombus). The semicircles present the possible destination protocol layers to which SCs provide access. The dashed line expresses that API itself is not a protocol.

Although the IRM has some minor differences from OSI BRM, which are coming from design aspects, the applicability of existing conformance test methodology is straightforward.

*In this context, API is used as a general term, which in a particular implementation (eg. Net/3) stands for both socket API and BPF. That is, because the socket API does not provide access to the interface layer.

5 ABSTRACT TEST METHODS

Considering the open structure of software implementations, the new Joint Test method (JT) will be defined, which can be uniformly applied to testing of all protocols of IRM.

JT can be applied both in Single Party Testing (SPyT) and in Multi Party Testing (MPyT) context. When used in SPyT, it resembles to the local (X.290–X.296, 1994-95) test method. Whereas the MPyT variant has similarities to the local transverse test method in (Bil, 1997).

JT is shown in figure 2, and uses the graphical notation of (Baumgarten, 1994).

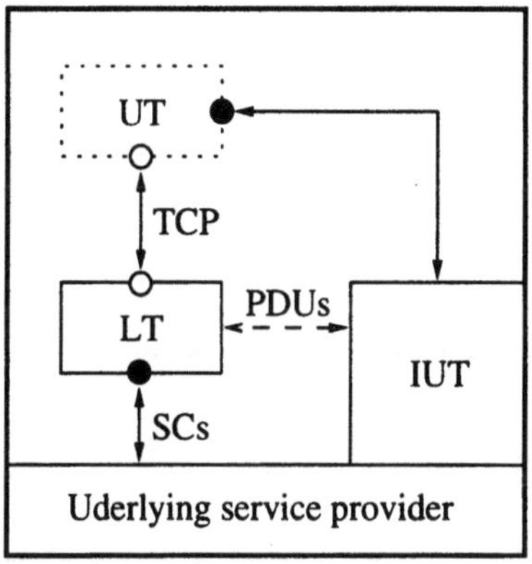

Figure 2 The joint test method.

JT has the following characteristics:

- Test system and system under test (SUT) are on the same system.
- There is an optional Upper Tester (UT), and one Lower Tester (LT) in SPyT; no UT, an arbitrary number (usually 2) of LTs and a Lower Tester Control Function (LTCF) in MPyT. UT, LT(s) and LTCF are application layer processes.
- The Points of Control and Observation (PCOs) are at the LT and UT.
- Test coordination is done using Unix IPC.
- Test events are exchanged in PDUs using SCs of API. The control and observation is provided by means of API.

The most significant difference to the ancestor test methods, which is very advantageous in practical testing of software TCP/IP implementations, is that LT(s), UT and coordination procedures are placed in the application layer regardless of the layer which is occupied by IUT. Another good feature is that JT can be applied to both end systems (SPyT) and relay systems (MPyT), thus intermediaries can be tested out of their environment.

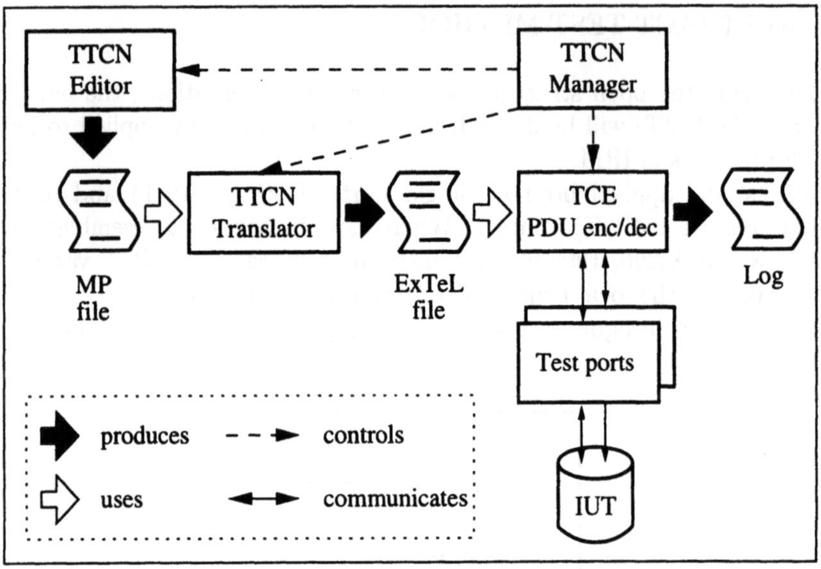

Figure 3 SCS structure.

6 TEST REALIZATION

Having an implementation to be tested and an abstract test suite (ATS), the means of testing should be provided. It consists of the implementation of tester functionalities, the derivation of ATS into executable test suite (ETS) and the production of test documents.

System Certification System (SCS) is a set of tools provided by Ericsson that can be used in a wide variety of testing: functional testing (white-box technique), conformance and interoperability testing (black-box) and performance testing (white/black-box). SCS is based on the following principles:

- Protocol independence. This means that different protocols can be tested on the same manner.
- Multiple simultaneous protocols. Not only one but many protocols can be accessed from the same test.
- Distribution. One test may be distributed (over the Internet), making it possible for each part of the test most closely related to one interface to reside in the box containing that physical interface.
- Platform independence. SCS is independent of the platform in which the SUT executes in. It can execute the same tests both against the physically real SUT and the SUT only simulated in a workstation (bypassing the lowest protocol layers).

One of the main ideas in SCS is that it is an interpreting execution platform.

This means that a TTCN test suite (an MP file) given as input to the Translator is first converted into an intermediate language, ExTeL (Executable Test Language), which then can be directly executed (interpreted) by the ExTeL Test Component Executor, TCE (see also figure 3 above). With this method there is only one phase from a TTCN test suite to the final executable format which makes it different compared to the compiling methods, where an extra compilation and linking phase has to be performed.

Another important feature in SCS is the Test Port concept. With this solution it is possible to develop the core functionality separately without affecting the existing test ports and vice versa.

There exist also two built-in PDU encoder/decoders: BER (Basic Encoding Rules) and a raw binary encoder/decoder.

TTCN Manager is the front end in SCS. It has the control over execution and monitoring. The log files for different test components can be observed in real time.

7 THE HYPERTEXT TRANSFER PROTOCOL

The Hypertext Transfer Protocol (HTTP) (Fielding, 1997) is an application-level protocol for distributed hypermedia information retrieval systems. It is used by World-Wide Web global information initiative since 1990.

HTTP is a generic, stateless, client-server protocol that can be used in wide variety of services by extending through extension of its request methods.

The first version of HTTP - HTTP/0.9 was a simple raw data transfer protocol. HTTP/1.0, as defined by RFC 1945 improved the protocol with many features (MIME-like messages and headers etc.). However, it does not provide enough facilities for handling the effects of hierarchical proxies and virtual hosts. The actual version, HTTP/1.1 offers sophisticated methods for content negotiation, cache control etc.

HTTP has three kinds of communicating parties: client or user agent, origin server and intermediary. There are three common forms of intermediary: proxy, gateway and tunnel.

A proxy is special communication party which, unlike the others, has no OSI equivalent. It can act as both client and server. It may service client requests internally or by passing them on, with possible translation, to other proxies or servers. A gateway receives requests as it were the origin server, and forwards them with possible translation. The client may not be aware that it is communicating with a gateway. A tunnel acts as a blind relay between connections, and is not considered a party of the HTTP communication.

Each party of the communication which is not acting as a tunnel may employ an internal cache for handling requests. The effect of a cache is that the request-response chain is shortened.

In the simplest case communication takes place via a single connection between user agent and origin server. However, more than one connection may be required when intermediaries are present in the request-response chain.

A significant difference between HTTP/1.1 and earlier versions is that persistent connections are the default behavior. Persistent connection means that the connection is not closed after the initial request-response pair. In this case the client can issue further requests. The advantages of persistent connections are: less communication overhead (fewer connections must be set up and released), and thus increased speed. The drawbacks contain: longer duration of connections (origin servers wait for clients to send further requests) and, possibly, conflicts of asynchronous close events (either party of the communication may choose to close the connection any time).

8 TESTING HTTP

8.1 Test documents

The main point of HTTP is information retrieval. World-Wide Web information consists of resources that have to be placed on origin servers and are addressed with Uniform Resource Identifiers (URIs). In order to test HTTP communicating parties, these resources should be provided to the IUT. The resources supplied to them are the following:

- An origin server should have a test Data Base (DB) containing resources that are available for presentation and a set of Configuration files (C) determining its internal operation *.
- A proxy is very similar to the server. It has no local DB, but it usually has and internal cache.
- A user agent has only C, but it also may contain an internal cache.

In addition to putting IUT into a test context (DB and C), the test components also should be familiar with that context. Because of the fact that DB and C play some role of UT, they will be considered so. Thus, their contents must be set up before the test campaign is launched (from test purposes, Protocol Implementation Conformance Statement (PICS), Protocol Implementation Extra Information for Testing (PIXIT)).

If OSI conformance testing methodology should be applied to any protocol of the TCP/IP stack, the Request For Comments should be accompanied by these test documents.

Selected PICS questions and test purposes will be demonstrated below.

*eg.: how many instances of the server should be created; where is the DB located inside the file system; possible mappings that should be applied to various URLs; which documents have to be considered secure and require authentication

8.2 Test configurations

As server, proxy and client have different functions, different test configurations should be defined for testing each of them. Since all common kinds of implementations (except Web-TV-like equipments) are software programs, the best choice is the application of JT. Let's examine the demonstrated configurations deeper!

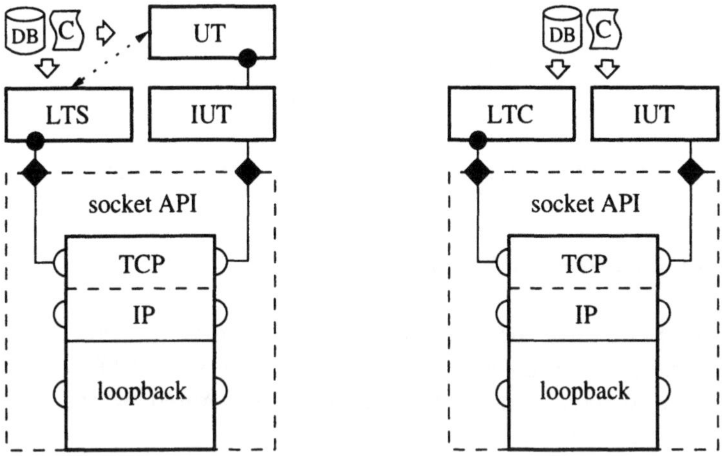

Figure 4 Test configuration for client (left) and server (right) testing.

Figure 4 shows the configuration for testing a user agent. This arrangement consists of an UT and a LT. Their PCOs are denoted with filled circles. The role of UT is played by a user who makes the IUT to issue requests for a certain document. The LT acts as the origin server, it examines and answers the received requests from its DB, according to its C. Test results are determined by the UT and LT together. LT examines whether the client has retrieved the right resource while UT investigates if IUT has presented the resource as well. Test coordination is also done internally.

The configuration for origin server tests can be seen on figure 4. The difference from the client's test configuration is that UT is absent. It has only one PCO at LT. The communication is initiated and the results are examined by LT. Physically, DB belongs to the IUT, however LT also makes use of it.

The configuration for proxy testing (figure 5) is based on the MPyT variant of JT. It has two LTs; LTC plays the client role, while LTS simulates the server.

The test realization happens using SCS. The LTs are implemented as test ports of SCS. In MPyT case, concurrent TTCN is applied. Test coordination between the test components is made using Unix IPC.

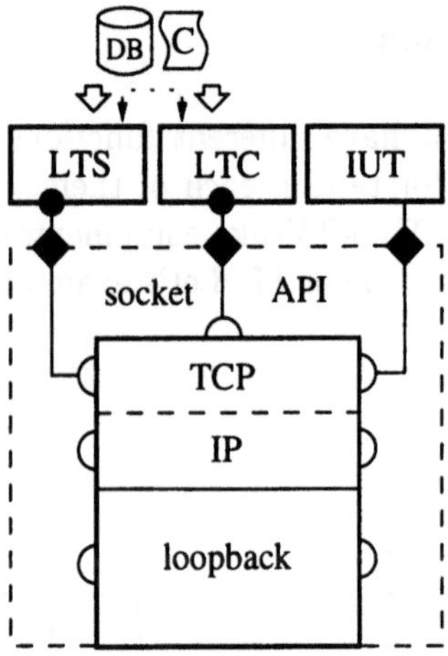

Figure 5 Test configurations for proxy testing.

8.3 Testing client behavior

The behavior of a client or user agent (UA) is, in general, controlled by human using intuitive on-line graphical interface. However HTTP requests that are necessary to retrieve desired resources are generated by the user agent independently of its user.

Let's take a look at a common example: a user browsing the Web, selects an anchor of an HTML page. Doing that, he/she makes the UA issue an HTTP request to the target URL's origin server acquiring an HTML document. After successful downloading, the client parses the document and finds numerous references to inline images and a hyperlink to a style-sheet resource containing essential definitions affecting the layout of the document. The UA should retrieve the required data without any kind of user interaction. Nowadays, HTML pages show many pictures so clients often issue a couple of requests for getting all the contents of a page.

According to the (Fielding, 1997) a client may pipeline its requests. This means it may issue multiple requests without waiting for each of the server's responses. The client, furthermore, should be prepared for expected failure of its attempt eg. when it communicates with an HTTP/1.0 server that neither supports persistent connections nor pipelining, it should cope with such variations. According to this clause, a suggested test purpose follows along with its PICS selection reference.

PICS1.1

Is the IUT (HTTP/1.1 conforming client) able to use pipelining? o YES

TP1.1

To check if the IUT (client) will not pipeline immediately after connecting to the origin server if its last pipeline attempt has failed.

s_tcp_socket ? receive	GET (ROOT, HTTP_1_1, server)	
s_tcp_socket ! send	s200_Ok (ROOT_DOC)	
s_tcp_socket ? receive (res0 := request.request_line.uri)	GET (*, HTTP_1_1, server)	
s_tcp_socket ? receive	GET (*, HTTP_1_1, server)	
s_tcp_socket ? receive	GET (*, HTTP_1_1, server)	
s_tcp_socket ! send	s200_Ok (IUT)	
s_tcp_socket ! close		
START T_SERVER		
s_tcp_socket ? receive	GET (*, HTTP_1_1, server)	
RESET T_SERVER		
? TIMEOUT T_SERVER		(P)

Figure 6 Test case for testing client behavior.

The UT makes the IUT request for a given resource (PIXIT) then the IUT issues the request. The LTS receives the request and sends a 200 'OK' response accompanied with the requested document and does not close the connection, since it waits for the IUT to send further requests. After receiving the response, the IUT parses the document and finds three hyperlinks to inline images. Now, the IUT has exactly three additional requests to issue in order to get the page contents. The LTS waits for the IUT to pipeline these three consecutive requests. After that, the LTS sends the response to the first request along with the connection close message and closes the connection. The IUT should keep track of the status of its requests, and according to the specification it should request for the two unreceived resources automatically. However, its pipeline attempt has failed, so it has to get these resources after one another. If that is the case, the IUT passed the test purpose, otherwise it does not conform to the recommendation (Fielding, 1997). Figure 6 shows the TTCN test case corresponding to the this test purpose (unnecessary timer operations, preambles, postambles, default trees and OTHERWISE statements are removed for better readability).

8.4 Testing Origin Server

There are many common features in an origin server that have to be tested for conformance; the test case that was selected for this demonstration deals with access authentication.

PICS2.1:

Does the IUT (server) provide Basic Access Authentication? o YES

TP2.1:

To verify whether the IUT (server) responds to client requests, which affect protected documents, with 401 'Unauthorized' status, accompanied with the 'WWW-Authenticate' header.

LTC issues a request for a protected resource. The IUT receives the request and finds that the retrieval of the requested document needs authentication. LTC expects IUT to respond with status 401 'Unauthorized'. Moreover, this response should contain the www_authenticate header. In this case the verdict is pass, otherwise the IUT has failed. Figure 7 shows the test case for TP2.1.

c_tcp_socket ! connect
 c_tcp_socket ! send
 START T_SERVER
  c_tcp_socket ? receive [
  response.header_set.response_header.www_authenticate=
  c_Basic_Authentication]
   c_tcp_socket ! close
   CANCEL T_SERVER
  +timeout
	
	GET (PROTECTED, HTTP_1_1, IUT)
	
	s401_Unauthorized (IUT)
		
		(P)

Figure 7 Test case for testing server behavior.

8.5 Testing proxy

A proxy is the most complicated party of HTTP communication. It receives the clients requests and tries to fulfill them from its cache. If the requested resource cannot be found in the cache or the cached copy is not fresh enough, the proxy retrieves it from another proxy of the cache hierarchy or directly from the origin server.

PICS3.1:
Does the IUT (proxy) employ internal cache for handling requests? o YES
TP3.1:
To verify whether the proxy under test is able to store retrieved resources in its cache.

The test case consists of two parallel test components. The HTTP_C_PTC acts as a client, while the HTTP_S_PTC plays the origin server's role. As a pre-test condition, the proxy's cache should be cleared. Then the two parallel test components are instantiated. The client (LTC) issues a request to the proxy for the ROOT resource located on the server. The proxy gets the request and issues an additional request to the server, since the requested resource cannot be found in its storage. The server sends the response along with the target document to the proxy. If the proxy behaves well, it should store the received document and forward it to the client. After getting the response, the client issues the same request to the proxy. The proxy should fulfill this second request from its cache, it should not turn to the server again. If the server gets another proxy request, then the IUT fails. Otherwise, if the timer expired, the verdict is a conditional pass. If the client gets the new copy of the retrieved document, the verdict is a conditional pass, too. The final verdict is calculated in the postamble. Figure 8 shows the concurrent TTCN (ISO/IEC 9646-3, 1998) test case for test purpose 3.1.

9 CONCLUSIONS

In this paper, differences between OSI and Internet systems were summarized. Then the Internet Reference Model was introduced together with the Joint Test method for conformance testing. Afterwards, a practical application of

CREATE (HTTP_C_PTC : client, HTTP_S_PTC : server		
? DONE (HTTP_C_PTC, HTTP_S_PTC)		
client		
c_tcp_socket ! send	GET (ROOT, HTTP_1_1, server)	
c_tcp_socket ? receive	s200_Ok (IUT)	
c_tcp_socket ! send	GET (ROOT, HTTP_1_1, server)	
c_tcp_socket ? receive	s200_Ok (IUT)	(P)
server		
s_tcp_socket ? receive	GET (ROOT, HTTP_1_1, server)	
s_tcp_socket ! send	RESPONSE (ROOT_DOC)	
START T_SERVER		
s_tcp_socket ? receive	GET (*, *, *)	F
? TIMEOUT T_SERVER		(P)

Figure 8 Test case for testing proxy behavior.

the new concept was demonstrated on the testing of HTTP communication parties.

We have shown a framework for testing Internet protocols, which was worked out on the basis of the conformance testing framework of (X.290–X.296, 1994-95). Experiences with testing HTTP showed that some extensions are necessary to TTCN for making it more suitable to describe test cases for testing Internet protocols. This is true especially for testing performance related features of the product.

Future work can be, for example the interoperability testing based on this concept of Internet protocols, and introduction of formal extensions that are more suitable for Internet testing.

10 ACKNOWLEDGMENTS

The author would like to thank Dr. Sarolta Dibuz and Mazen Malek for their helpful comments, and is also grateful to the anonymous reviewers providing valuable comments and suggestions that have improved the quality of this paper.

REFERENCES

Baumgarten, B. and Giessler, A.: OSI conformance testing methodology and TTCN, North. Holland, 1994.

Bi, J. and Wu, J.: Towards abstract test methods for relay system testing. Testing of Communicating Systems, Volume 10 pp 381-397, IFIP, 1997.

Bi, J. and Wu, J.: Application of a TTCN based conformance test environment on the Internet email protocol. Testing of Communicating Systems, Volume 10 pp 324-330, IFIP, 1997.

Carpenter, B. (editor): Architectural Principles of the Internet, RFC 1958 Informational, IETF Network Working Group, 1996.

Fielding, R., Irvine, UC, Gettys, J., Mogul, J., Frystyk, H. and Berners-Lee, T.: Hypertext Transfer Protocol – HTTP/1.1., RFC 2068 Standards Track, IETF Network Working Group, 1997.

ISO/IEC 9646-3, The Tree and Tabular Combined Notation (TTCN), 1998.

ITU-T X.200, Information Technology – Open Systems Interconnection – Basic Reference Model: The Basic Model, 1994.

ITU-T X.290–X.296, OSI conformance testing methodology and framework for protocol recommendations for ITU-T applications, 1994-1995.

Kato, T., Ogishi, T., Idoue, A. and Suzuki, K.: Design of Protocol Monitor Emulating Behaviors of TCP/IP Protocols. Testing of Communicating Systems, Volume 10 pp 416-431, IFIP, 1997.

Kato, T., Ogishi, T., Idoue, A. and Suzuki, K.: Intelligent Protocol Analyser with TCP Behavior Emulation for Interoperability Testing of TCP/IP Protocols. Formal Description Techniques and Protocol Specification, Testing and Verification, FORTE X/PSTV XVII '97 pp 449-464, IFIP, 1997.

Malek, M. and Dibuz S.: A Pragmatic Method for Interoperability Test Suite Derivation. EUROMICRO'98, Proceedings of the 24RD Euromicro Conference, Stockholm,Sweden, 1998. Aug 24-26.

Paxton, V. (editor), Allman, M., Dawson, S., Heavens, I. and Volz, B.: Known TCP Implementation Problems, ¡draft-ietf-tcpimpl-prob-02.txt¿ Internet Draft, IETF Network Working Group, May 1998.

Stevens, W. R.: TCP/IP Illustrated Volume 1, The Protocols. Addison-Wesley, 1994.

Wright, G. R. and Stevens, W. R.: TCP/IP Illustrated, Volume 2, The Implementation. Addison-Wesley, 1995.

11 BIOGRAPHY

Roland Gecse is a Ph.D. student of the Department of Telecommunications and Telematics at the Technical University of Budapest. He received his M.Sc. in electrical engineering at Technical University of Budapest in 1996. Since 1997, he is a member of the High Speed Networks laboratory, and joined Conformance Center at Ericsson Hungary. His research interests include conformance, performance and interoperability testing of Internet protocols, formal description techniques and software engineering.

Intelligent Protocol Analyzer for WWW Server Accesses with Exception Handling Function

Tomohiko Ogishi, Akira Idoue, Toshihiko Kato and Kenji Suzuki
KDD R&D Laboratories, Inc.
2-1-15, Ohara, Kamifukuoka-shi, Saitama 356, Japan
E-mail : {ogishi, idoue, kato, suzuki}@hsc.lab.kdd.co.jp

Abstract

Recently, the TCP/IP protocols are widely used and it is mentioned that there are problems such that the throughput is limited due to protocol procedures such as retransmission and congestion control of TCP. In order to analyze these problems, we have developed an "intelligent" protocol analyzer which can estimate what communication has taken place by emulating the behaviors of the IP, TCP and HTTP protocol entities, which are used for WWW server access communications. This analyzer also supports the handling of exceptional situations such as monitoring errors and network misbehaviors. This paper describes the details of the emulation functions of HTTP and the exception handing functions, and shows some results of applying the analyzer to actual WWW server accesses.

Keywords

Protocol Analyzer, HTTP, WWW, HTML, TCP

1 INTRODUCTION

Recently, the TCP/IP protocols [1, 2] are widely used in various computer communications. Here, most users of computers use communication functions installed in operating systems or commercial software products as they are, and do not pay any attentions to their details. However, some problems occur for TCP/IP communications, in the cases that there are some packet losses due to network congestion and transmission errors, and that the protocol parameters of communicating computers are not matched.

In order to analyze those problems, we developed an "intelligent" protocol analyzer for TCP (Transmission Control Protocol), which can estimate what communication has taken place among communicating computers [3, 4]. This analyzer maintains the specification of the state transition based behaviors of TCP and emulates the behaviors of the TCP protocol entities in communicating computers. Since modern TCP includes some internal procedures for the flow control, such as the slow start algorithm, this analyzer can emulate these procedures as well. This analyzer provides information such as mappings between data segments and ack segments and updates of congestion window used for the congestion control, and this releases the burden of analyzing TCP behaviors much better than the commercially available protocol monitors [5].

However, our intelligent protocol analyzer has two points to be improved. The first one is that, since it focuses only on protocols up to TCP, it is required to implement the analyzing functions of application protocol for the examination of actual communications. Recently, WWW server access is considered to be the most common application, and here data of Web pages are transferred according to HTTP (Hyper Text Transfer Protocol) [6] located on top of TCP. Since one Web page consists of several elements to be retrieved through different TCP connections, the related data transfers to one Web page need to be combined in the examination of the Web page access. For this purpose, the application protocols, HTTP and HTML (Hyper Text Markup Language) in this case, need to be supported by our intelligent analyzer.

The second one is to add the functions to cope with exceptional situations such that the analyzer drops segments and the sequence of observed segments is not the same as that of sent out or received segments. In order to add these functions, the analyzer needs to estimate the possible situations if it detects some errors in observed segment sequence.

We have implemented new version of intelligent protocol analyzer which supports the analysis of IP, TCP and HTTP, which is the protocol stack for WWW server access application, and can handle exceptional situations such as monitoring errors and network misbehaviors. This paper describes the design of our new analyzer, which we call the *WWW intelligent protocol analyzer*, and some results applying it to actual communications. The next section describes the design overview of our WWW intelligent protocol analyzer. Sections 3 and 4 describes the detailed design of the emulation function of HTTP and the exception handling function, respectively. Section 5 gives some results of applying our analyzer to actual WWW server access communications.

2 DESIGN OVERVIEW

(1) As depicted in Fig. 1, the WWW intelligent protocol analyzer is attached to a LAN, and captures and stores all PDUs (Protocol Data Units) transmitted over the LAN. The analyzer selects PDUs transmitted or received by a specific computer focused on from the stored PDUs, and analyzes their formats and the behaviors of TCP and HTTP protocol entities in the specific computers and the computers communicating with it, e.g. computers A, B and C in the figure.

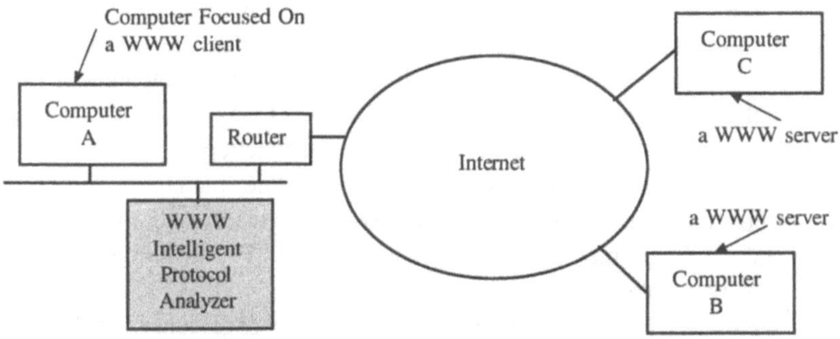

Figure 1 Configuration using WWW Intelligent Protocol Analyzer.

(2) Figure 2 depicts the software structure of the analyzer. The software is developed based on the previous intelligent protocol analyzer, and the modified or newly developed part is indicated in the figure.

- The *PDU capture module* captures PDUs transmitted over the LAN, analyzes their format and parameter values according to the IP, TCP and HTTP protocols, and saves those results in the *monitor log*.
- The *event sequence estimation module* selects PDUs which the computer focused on sent or received, and constructs the *event sequence logs*, which consist of a sequence of sent and received TCP segments and their timing, for individual computers involved in the communication by taking account of the transmission delay between the analyzer and the computers [3].
- The *TCP emulation module* emulates the TCP behaviors of these computers according to the event sequence logs. This module maintains the state transition specification of TCP for sent and received events, and estimates the state and variables in individual computers according the following procedure [3].
    - For a received TCP segment, it looks up the specification for received events, and performs a corresponding state transition. If it sends out a segment, the module checks a sent TCP segment in the event sequence and emulates the received and sent segments.
    - For a sent TCP segment, the TCP emulation module looks up the specification for sent events and checks whether the TCP protocol entity can send out the segment.

- The *HTTP emulation module*, which is newly added, maintains the state transition specification of HTTP and emulates the HTTP behaviors in individual computers being examined.

(3) In order to realize the analysis of WWW server accesses, the following functions are developed.

- The PDU capture module stores the HTTP header information and the HTML text as well as the IP and TCP headers.
- The TCP emulation module reports to the HTTP emulation module the events such that a TCP connection is established or released, and HTTP data is sent or received. We call these events *TCP primitives*. It needs to be mentioned that the TCP primitives are reported as the results of TCP processing and therefore the sequence of these primitives in a sending and receiving computers are the same even if some segments are lost and retransmitted by TCP.
- The HTTP emulation module provides the following functions;
    - *Emulation of HTTP behavior*: This module maintains the state transition specification which we have introduced, and manages the state and variables of HTTP for individual TCP connections used for WWW server accesses.
    - *Web page link*: As described above, one Web page consists of several elements such as HTML text and graphical image, and different TCP connections are used to retrieve these elements. One Web page also has several links to other Web pages, which may be accessed to using different TCP connections. Therefore, in order to analyze the communication for accessing one Web page, the HTTP emulation module links those Web page data accesses.

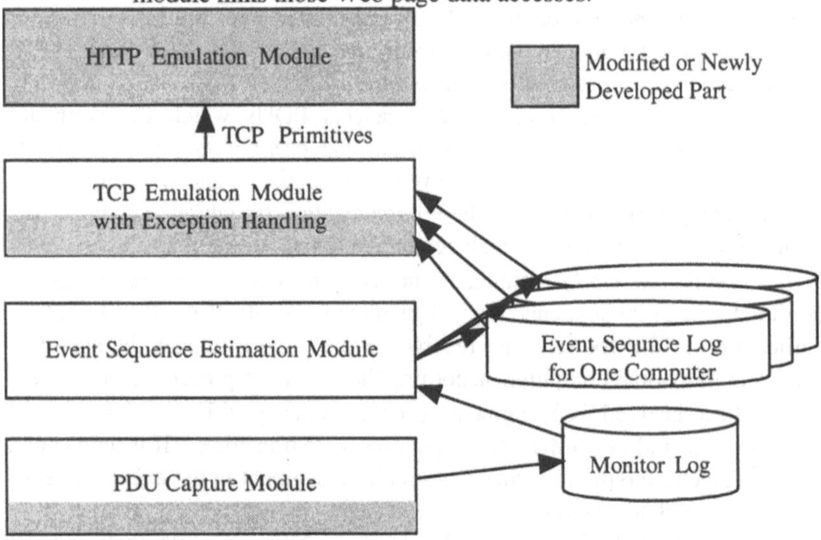

Figure 2 Software Structure of WWW Intelligent Protocol Analyzer

(4) In order to realize the exception handling, we add the following functions in the TCP emulation modules.

- We take account of the following cases in which the sequence the analyzer captures are different from the real sequence handled in the computers being examined.
    - *Drop*: The analyzer fails to capture PDUs.
    - *Timing error*: As describe above, the analyzer estimates the processing timing of PDUs in individual computers according to the transmission delay, and therefore, it is possible to make wrong estimation.
    - *PDU losses at sender side*: When a PDU is lost between the sending computer and the analyzer, the analyzer cannot observe it.
    - *PDU losses at receiver side*: When a PDU is lost between the analyzer and the receiving computer, the receiving computer will not handle it although it is observed by the analyzer.
    - *Misordering in network*: When the order of PDUs are changed in the network, the sequence observed by the analyzer is different from the real sequence in the computers.
- In those cases, it is possible that the TCP emulation module decides that the TCP entities in computers have some protocol errors. Here, the module checks whether the detected errors can be solved by assuming one of the above cases occurs. If it can be solved, the module handles that there are no protocol errors but there are exceptional situations.

3 EMULATION FUNCTION FOR HTTP

3.1. Emulation of HTTP Behavior

The HTTP procedure uses an *HTTP request message* in order to specify the identifier of a Web page and uses an *HTTP response message* in order to respond to the request [6]. Each HTTP request / response message consists of the header part including URI (Universal Resource Identifier), content type and content length, and the body part including the content like a HTML text or a graphical image. If its content length is longer than MSS (Maximum Segment Size), the content would be divided into several TCP segments.

In order to emulate these behaviors of HTTP entities, we introduce the HTTP state transition, a part of which is depicted in Fig. 2. The inputs of the state transition are TCP primitives which we defined. They includes the followings.

CNREQ : issue of TCP connection establishment request
CNIND : receipt of TCP connection establishment request
CNCNF : completion of TCP connection establishment
DTSND : issue of data segment
DTRCV : receipt of data segment
CLOSE : release of TCP connection

The states of the HTTP state transition include the followings.

CLOSED	: no TCP connection
TCON	: requested TCP connection establishment
TWFACK	: responded to connection establishment request
OPEN	: possible to transfer data in both direction, and no HTTP request / response exchanged
REQUEST SENDING	: sending HTTP request message
REQUEST SENT	: sent HTTP request message
RESPONSE RECEIVING	: receiving HTTP response message
REQUEST RECEIVING	: receiving HTTP request message
REQUEST RECEIVED	: received HTTP request message
RESPONSE SENDING	: sending HTTP response message

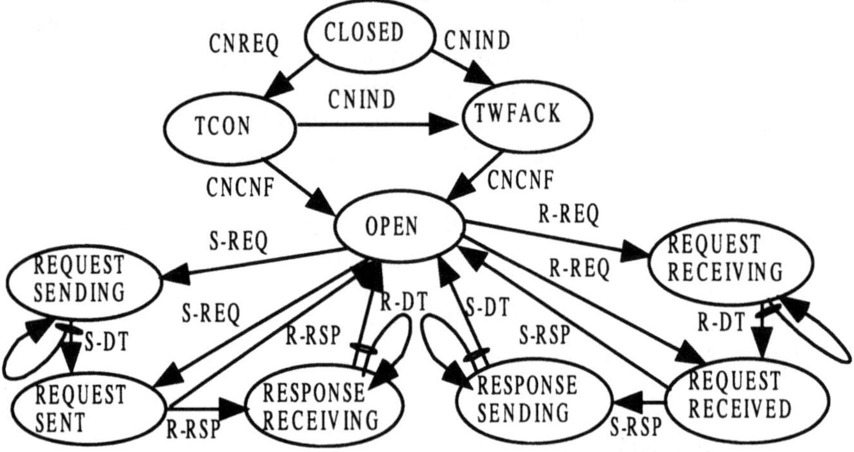

S-REQ: DTSND with REQ, S-RSP: DTSND with RSP, S-DT : DTSND without REQ/RSP
R-REQ: DTRCV with REQ, R-RSP: DTRCV with RSP, R-DT : DTRCV without REQ/RSP
(*)On receiving CLOSE event, state will change to CLOSED in all HTTP states

Figure 3 HTTP state transition

3.2. Estimation of Web Page Link

As described above, multiple pairs of requests and responses have link relationship in one Web page access. There are two types of these links. One is the case that a Web page contains more than one elements, and they are retrieved continuously. An example is that an HTML text contains keywords "<IMG SRC=" or "<FRAME SRC=" and we call this case an *active link*. The other is the case a Web page contains a hyper link to another page, e.g. an HTML text contains a keyword "<A HREF=", which we call a *passive link.*

In order to estimate the Web page link, we introduce the session ID and parent session ID, which are the identification of HTTP request and that of parent HTTP request, respectively. The estimation is performed in the following way.

(1) If an HTTP request message is new message for a Web page access, the HTTP emulation module assigns a session ID to the request message and

remember a URL (Universal Resource Locator) included in the message header when it deals with a S-REQ (DTSND primitive with REQ).

(2) When the module deals with the response message corresponding to the request message and the content type is an HTML text, it searches keywords representing active link and passive link.

(3) When it finds an active link, it considers the retrieval of the Web page element as a child session of the current one. In this case, the HTTP emulation module assigns the child session ID and registers in the Web page link table a set of URL of the Web page element to be accessed, the session ID and the parent session ID. The child session ID is 1-1, 1-2 and so on if the parent session ID is 1.

(4) When the module find a passive link, it does not assign session ID but registers URL and the parent session ID.

(5) When it deals with a new S-REQ, it checks if the URL of the request message exists in the Web page link table. When the URL is found as an active link, it gives the session ID registered in the table. When the URL is found as a passive link, it assigns a new session ID. If there are no URL in the table, the module considers that the request message is for a new Web page access.

4 EXCEPTION HANDLING FUNCTION

As described in section 2, the TCP emulation module will check the possibility of monitoring errors and network misbehaviors, i.e. drop, timing error, PDU losses at sender or receiver sides, and misordering, if it detects any protocol errors. If the protocol errors are solved by assuming that any of them have occurred, then the TCP emulating module considered that there are no protocol errors. The design of the exception handling function is performed for individual protocol errors. The following gives the details of some cases supported currently.

(1) Sending SYN+ACK segment in state CLOSED
In this case, there are two possibilities as depicted in Fig. 4 (1). One is the case that the analyzer estimated "SYN+ACK sent" has occurred prior to "SYN received." The other is the case that the analyzer dropped the SYN segment. In both cases, the actual sequence is that the SYN segment is received and the SYN+ACK segment is sent in response to it. Therefore, when the TCP emulation module detected this protocol error, it will search "SYN received" event. If it is found, the module considers that there are timing error, and otherwise, the module considers that it dropped a SYN segment.

(2) Receiving SYN segment in state CLOSED and sending no response
In this case, there are two possibilities as depicted in Fig. 4 (2). One is the case that the analyzer dropped the response SYN+ACK segment. In this case, the SYN+ACK and ACK segments are exchanged in an actual sequence, and the

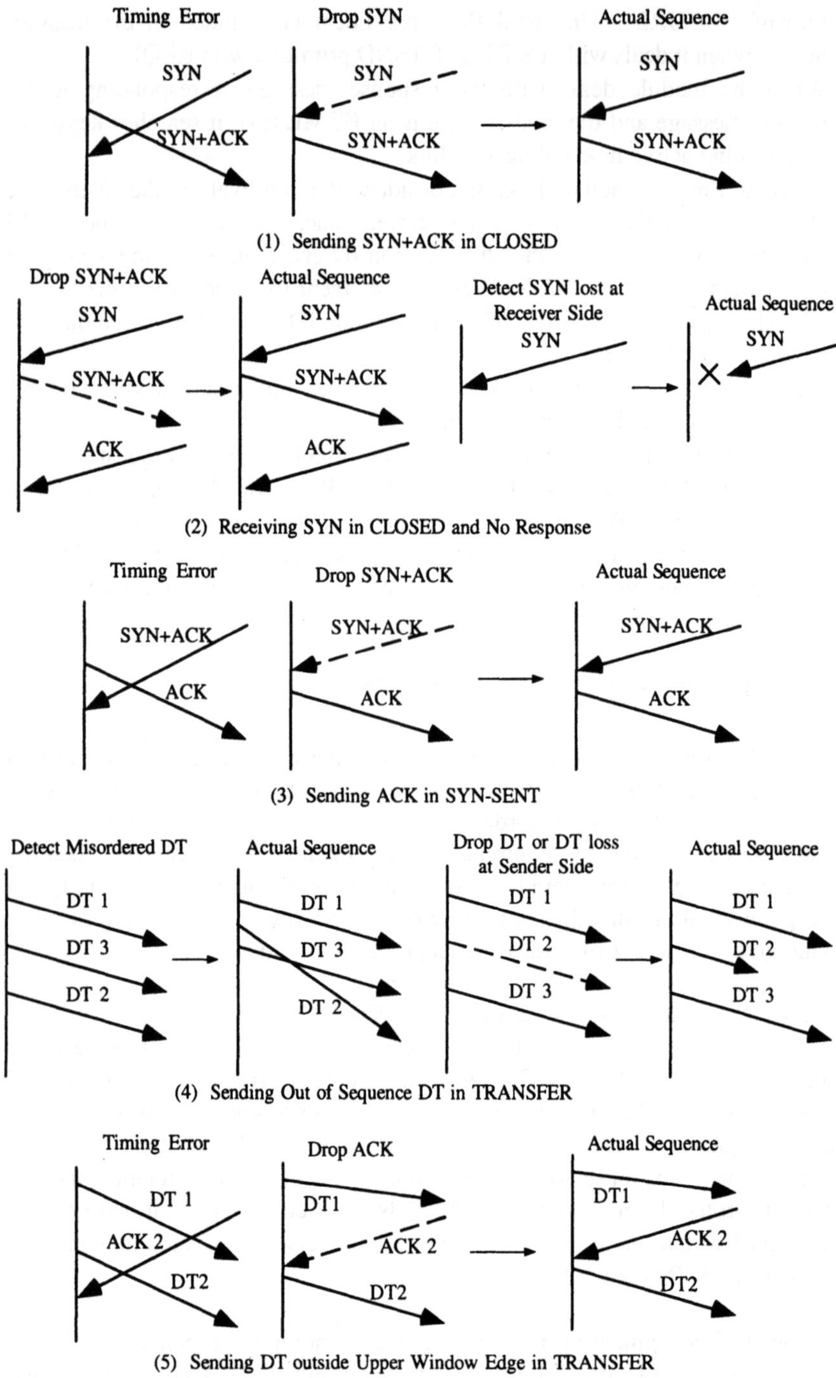

Figure 4 Exception Handling

analyzer will detect the ACK segment. The other case is that the observed SYN segment was lost between the analyzer and the receiving computer. In this case, the SYN segment is not processed and the response to it is not returned. When the TCP emulation module detected this protocol error, it will search "ACK received" event. If it is found, the module considers that it will drop the SYN+ACK segment, and otherwise, the module considers that the SYN segment is lost in the network.

(3) Sending ACK segment in state SYN-SENT
This is the similar situation with (1) and there are two possibilities, timing error and dropping SYN+ACK segment. The TCP emulation module searches "SYN+ACK received" event. If it is found, the module considers that there are timing error, and otherwise, the module considers that it dropped a SYN+ACK segment.

(4) Sending DT segment whose sequence number is out of order in state TRANSFER
This is the case that the analyzer observed a DT segment (ACK segment which contains data) whose sequence number does not match snd_nxt which means the sequence number expected to be sent next. This case has two possibilities as depicted in Fig. 4 (4). One is that the analyzer has observed the misordered sequence of DT segments. The other is that the analyzer dropped the DT segment or it is lost between the analyzer and the sending computer. When the TCP emulation module detected this protocol error, it will search "DT received" event which contains the correct sequence number. If it is found, the module considers that there was misordering in the network and resequences the DT segments for the sending computer, and otherwise, the module considers that the DT segment is dropped or lost. In the latter case, snd_nxt is updated to match the subsequent DT segments.

(5) Sending DT segment whose sequence number is larger than upper window edge
This is the case that the analyzer observed a DT segment whose sequence number plus data length exceeds the upper window edge. This is also similar with (1) and there are two possibilities. One is the case that there are timing errors for "ACK received" which opens the window and "DT sent" which violates the flow control mechanism. The other is the case that the analyzer has dropped the ACK segment which opens the window. In this case, the TCP emulation module searches "ACK received" which opens the window. If it is found, the module considers that there are timing error, and otherwise, the module considers that it dropped the ACK segment.

5 RESULTS OF APPLYING TO ACTUAL WWW SERVER ACCESS

In this section, we show the functions of our analyzer using results of applying it to an actual WWW server access. We have captured the communications between our local computer for a WWW client and the IFIP WWW server [7] in Austria. Figure 5 depicts the main window showing the emulation results. It shows the emulation result of individual computers, the left side for the client and the right side for the servers. For each event, it shows parameter values of PDUs of TCP and HTTP, the estimated values of state and variables of TCP and HTTP, and some comments concerning the protocol situation, together with the estimated time of each event.

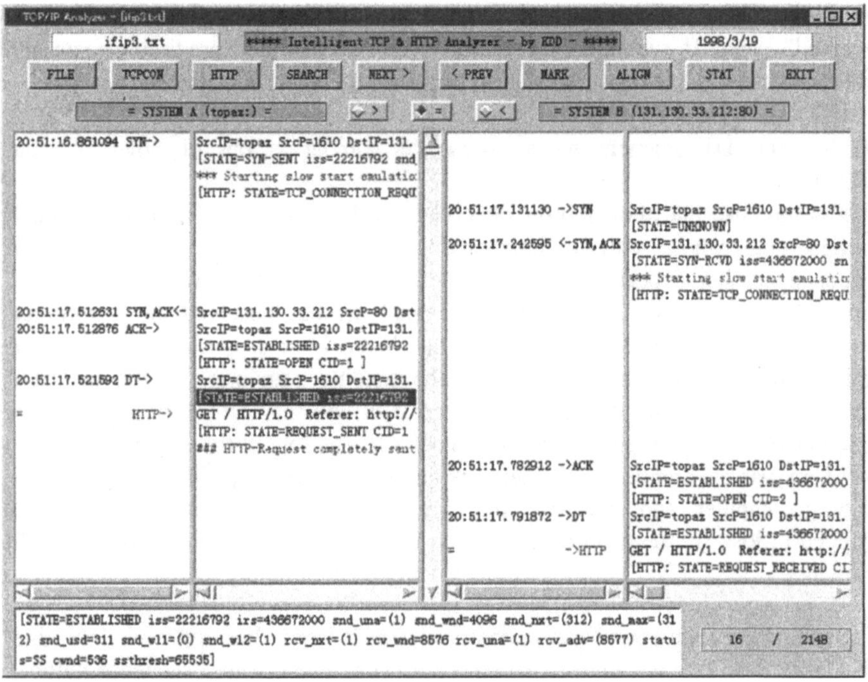

Figure 5 A Main window for Emulation Result

Although all of the analyzed information is included in this window, it would be difficult to understand because it shows the events corresponding to more than one TCP connections in a mixed way. Therefore, our analyzer prepares a window which shows the accessed Web page elements, the links among them, and the mapping between the Web page elements and TCP connections. Figure 6 depicts this window for the IFIP home page. The top home page "//www.ifip.or.at/" includes three image file as active links and has passive link to a Web page "//www.ifip.or.at/secr.htm." This window also gives the information of TCP connection, including IP address/host name, port number and the number of retransmitted TCP segments. Among them, the number of retransmitted TCP segments helps a network operator understand which TCP connection has problems. In the example of Fig. 6, there are some retransmitted segments in the access to elements "//www.ifip.or.at/image/ new.gif" and "//www.ifip.or.at/image/secr.gif."

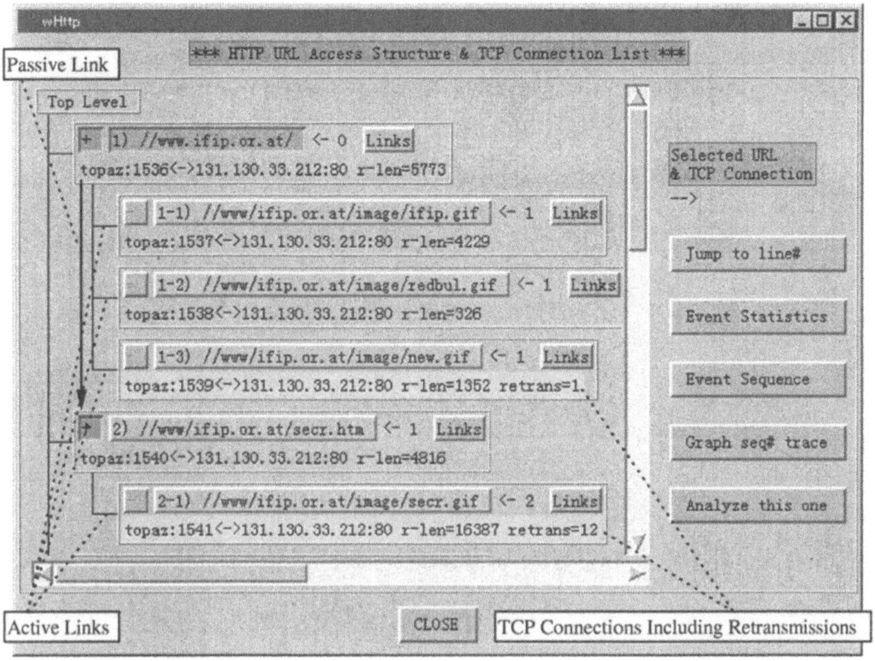

Figure 6 An Example of Web Page Links

By selecting one element, our analyzer shows the event sequence for the corresponding TCP connection. Figure 7 depicts a window showing an event sequence of the TCP connection transferring new.gif file. This window shows sending out and receiving of TCP segments and HTTP PDUs contained in TCP DT segments, in the style of sequence chart with the estimated time of sending out and receiving. As a TCP level information, this window also shows the relations between retransmitted TCP segments and its original segments, and the relations between DT segments and ACK segments.

In the case of Fig. 7, a SYN segment is retransmitted approximately 3 seconds after the first SYN segment is sent. It would be considered that the server had been too busy to respond the first request.

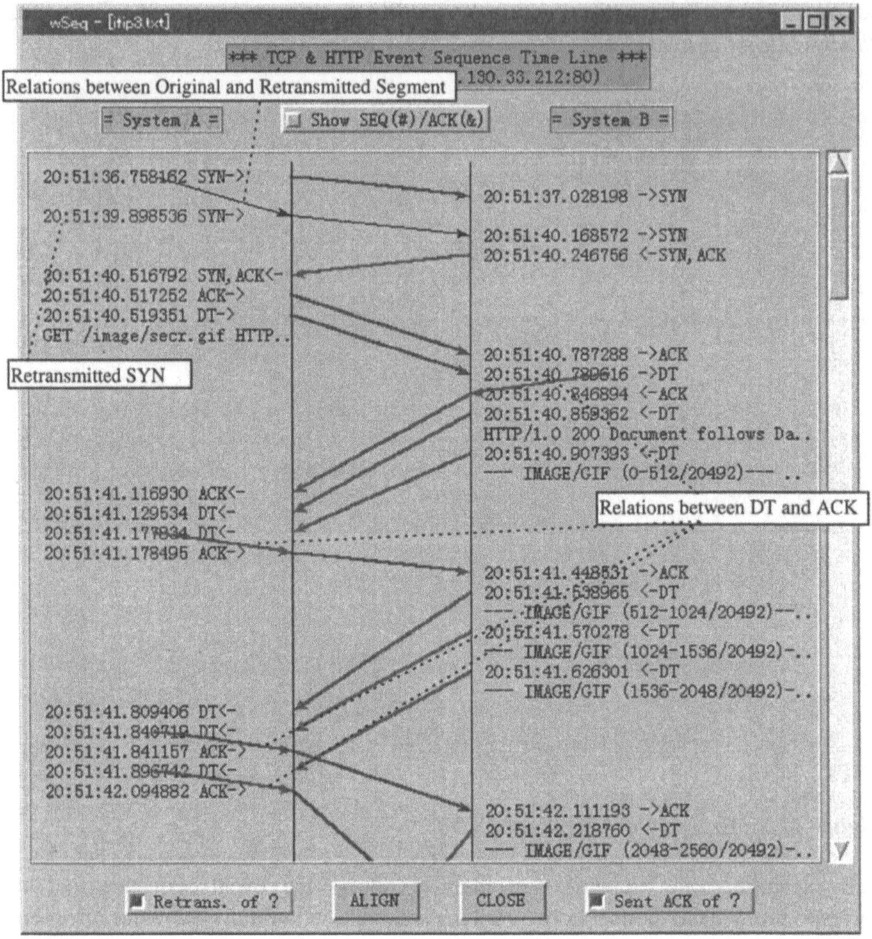

Figure 7 Example of Event Sequence Window

Figure 8 depicts an example of event sequence window, which includes retransmission of DT segments. In this case, it is considered that some data were retransmitted by fast retransmit mechanism because they were dropped in the network retransmit mechanism because they were dropped in the network. The functions showing the relations between DT and ACK segment and those between retransmitted segments and its original segments are effective in analyzing this case. Since the three ACK segments sent out by System A (the client) at 20:51:03.745422, :03.882042 and :03.893139 are duplicated ACK, System B (the server) retransmit the required DT segment according to the fast retransmit algorithm. The event sequence shows these explicitly.

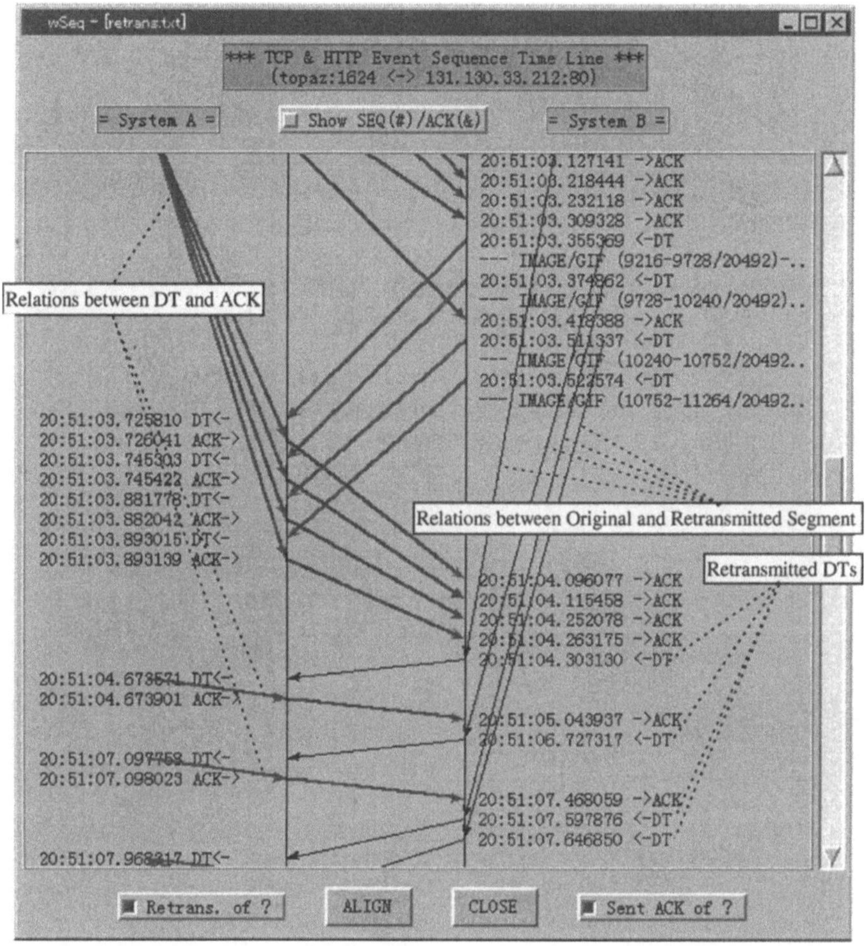

Figure 8 Example of Event Sequence Window with Retransmitted DT Segments

Our analyzer also has a function to show the transitions of the sequence number of transmitted segments, estimated congestion window (cwnd) and slow start threshold (ssthresh). Figure 9 shows those transitions for the TCP communication depicted in Fig. 8. In this case, there are some retransmission and therefore the cwnd is kept as a low value resulting in low throughput. This window is also helpful to analyzing the details of TCP level data transfer.

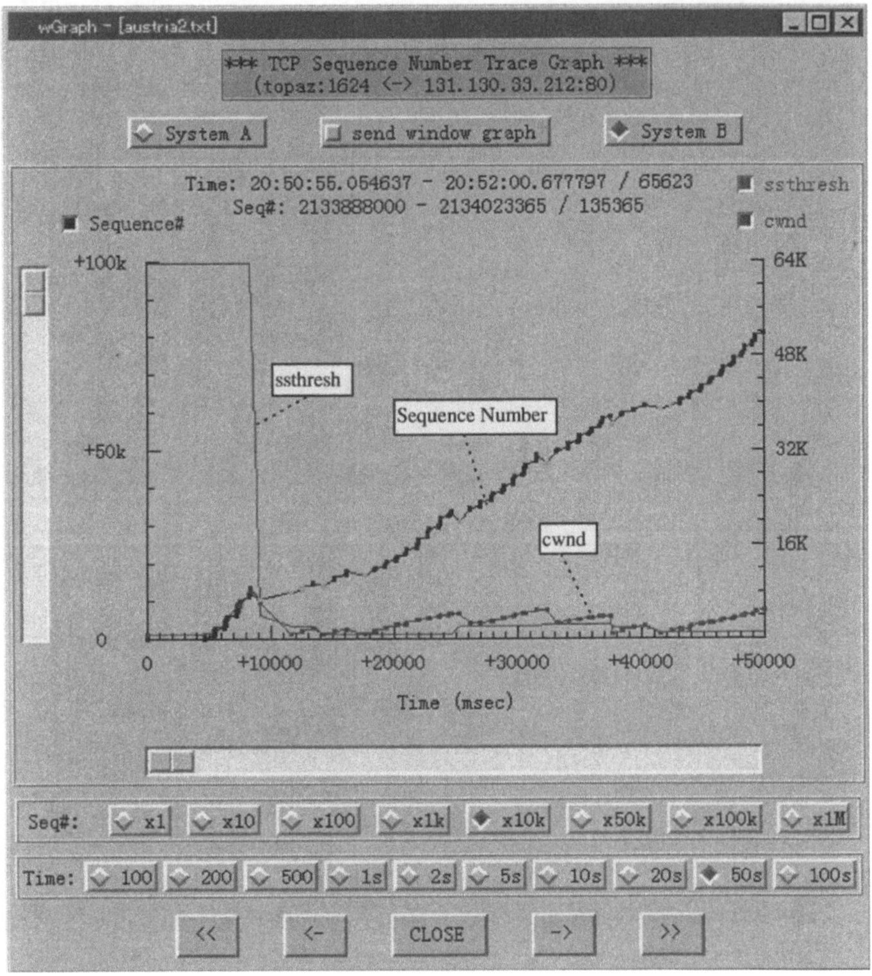

Figure 9 Transition of Sequence Number, Cwnd and Ssthresh of TCP Level

6 CONCLUSION

In this paper, we have described the design and examination results of our WWW intelligent protocol analyzer which can estimate what communication has taken place between WWW clients and servers focused on by emulating the behaviors of the IP, TCP and HTTP protocol entities in the clients and servers. The HTTP emulation function supports emulation of the HTTP state transition and estimation of Web page links by analyzing HTML text. This allows network operators to detect the structure of one Web page and the mapping between elements in the

Web page and TCP connections. The TCP emulation function supports emulation of the TCP state transition and some internal procedures for the flow control such as the slow start algorithm. It can also handle exceptional situations such as monitoring errors and network misbehaviors.

This paper has also presented some results applying the analyzer to actual WWW server accesses. Our analyzer provides a window which show the links among Web page elements and the mapping among these elements and TCP connections. By selecting one element in this window, our analyzer can show an event sequence for TCP segments and HTTP PDUs and it can also show a transition of sequence number in TCP level. These graphical interfaces help network operators analyze the WWW server access in details.

7 ACKNOWLEDGMENT

The authors wish to thank Dr. T. Muratani, President of KDD R&D Laboratories Inc., for his continuous encouragement to this study.

8 REFERENCES

[1] W. Richard Stevens, "TCP/IP Illustrated, Vol. 1 : The Protocols," Addison Wesley, 1994.

[2] DARPA Internet Program Protocol Specification, "Transmission Control Protocol," RFC 793, September 1981.

[3] T. Kato, T. Ogishi, A. Idoue and K. Suzuki, "Design of Protocol Monitor Emulating Behaviors of TCP/IP Protocols," Proc. of the International Workshop on Testing of Communication Systems, pp.416-431, September 1997.

[4] T. Kato, T. Ogishi, A. Idoue and K. Suzuki, "Intelligent Protocol Analyzer with TCP Behavior Emulation for Interoperability Testing of TCP/IP Protocols," Proc. of FORTE/PSTV, pp.449-464, November 1997.

[5] Tekelec, "Chameleon User's Manual," 1992.

[6] R. Fielding, et. al., "Hypertext Transfer Protocol -- HTTP/1.1," RFC 2068, January 1997.

[7] "http://www.ifip.or.at/," the URL of the IFIP WWW server

9 BIOGRAPHY

Tomohiko Ogishi is a research engineer of High Speed Communications Lab. in KDD R&D Laboratories, Inc. Since joining KDD in 1992, he worked in the field of computer communication. His current research interests include the protocol testing on TCP/IP communication. He received the B.S. Degree of electrical engineering from the University of Tokyo in 1992.

Akira Idoue is a senior research engineer of High Speed Communications Lab. in KDD R&D Laboratories, Inc. Since joining KDD in 1986, he worked in the field of computer communication. His current research interests include implementation of high performance communication protocols and communication systems. He received the B.S. and M.E. Degrees of electrical engineering from Kobe University, Kobe, Japan, in 1984 and 1986 respectively.

Toshihiko Kato is the senior manager of High Speed Communications Lab. in KDD R&D Laboratories, Inc. Since joining KDD in1983, he has been working in the field of OSI, formal specification and conformance testing, distributed processing, ATM and high speed protocols. He received the B.S., M.E. and Dr. Eng. Degrees of electrical engineering from the University of Tokyo, in 1978, 1980 and 1983 respectively. From 1987 to 1988, he was a visiting scientist at Carnegie Melon University. Since 1993, he has been a Guest Associate Professor of Graduate School of Information Systems, in the University of Electro-Communications.

Kenji Suzuki is the executive vice president of KDD R&D Laboratories, Inc. Since joining KDD in 1976, he worked in the field of computer communication. He received the B.S., M.E. and Dr. Eng. Degrees of electrical engineering from Waseda University, Tokyo, Japan, in 1969, 1972 and 1976 respectively. He received Maejima Award from Communications Association of Japan in 1988, Achievement Award from the Institute of Electronics, Information and Communication Engineers in 1993, and Commendation by the Minister of State for Science and Technology (Persons of scientific and technological research merit) in 1995. Since 1993, he has been a Guest Professor of Graduate School of Information Systems, in the University of Electro-Communications.

PART THREE

Test Generation

Factorized test generation for multi-input/output transition systems

Ed Brinksma, Lex Heerink and Jan Tretmans
University of Twente, NL-7500 AE Enschede
{brinksma,heerink,tretmans}@cs.utwente.nl

Abstract

In this paper we present factorized test generation techniques that can be used to generate test cases from a specification that is modelled as a labelled transition system. The test generation techniques are able to construct a sound (and complete) test suite for correctness criterion $\mathbf{mioco}_{\mathcal{F}}$ [5] by splitting up this correctness criterion into many simpler correctness criteria, and by generating tests for these simpler correctness criteria. By isolating the relevant part of the specification that is needed to generate tests for each of these simpler correctness criteria and using this part to generate tests from, test generation can be done more efficiently. These techniques are intended to keep the generation of tests from a specification feasible and manageable.

1 INTRODUCTION

Testing To assess the correctness of systems testing is a frequently applied technique. The aim of testing is to check whether an implementation is correct with respect to its specification. This is done by conducting experiments on the implementation, observing the responses of the implementation to these experiments, and comparing these responses with the ones that could be expected on the basis of the specification. An implementation is considered incorrect, or erroneous, in case the responses to an experiment are different from the ones that could be expected. Testing can only show the presence of errors in implementations, never their absence. However, it is commonly agreed that confidence in the correct operation of an implementation increases if more and more tests are conducted and no errors are found. To reason about testing in a formal framework a clear definition of the universe of experiments ($\mathcal{U}$), observations that occur when experiment $u \in \mathcal{U}$ is carried out on system p ($obs(u, p)$), and a comparison criterion ($\sqsubseteq$) for these observations must be defined. This results in an extensional definition of a correctness criterion **conforms-to** (implementation relation) between implementation i and specification s as follows

$$i \ \textbf{conforms-to} \ s \quad =_{def} \quad \forall u \in \mathcal{U} : obs(u, i) \sqsubseteq obs(u, s) \tag{1}$$

In this paper we assume that specifications and implementations can be modelled by (subclasses of) labelled transition systems. In [3, 12, 14] and others different instantiations of the relation **conforms-to** have been defined by varying $\mathcal{U}$, obs, and the comparison criterion $\sqsubseteq$.

Test generation Instead of defining implementation relations by varying $\mathcal{U}$, obs, and $\sqsubseteq$ (1) the problem in test generation is to obtain the set of experiments that are needed to distinguish between correct and incorrect implementations for a given implementation relation and given specification. Preferably, such experiments are calculated automatically from the specification and the implementation relation. Unfortunately, calculating such experiments is often too complex in space and time to be feasible.

Contribution of this paper In this paper we discuss a factorized test generation technique that can be used to generate tests from a transition system specification with respect to the correctness criterion *multi input/output conformance* $\textbf{mioco}_{\mathcal{F}}$, which was introduced in [5]. The technique is factorized with respect to the implementation relation $\textbf{mioco}_{\mathcal{F}}$ in the sense that the "complicated" correctness criterion $\textbf{mioco}_{\mathcal{F}}$ can be split up in several independent "easier-to-check" correctness criteria. Moreover, for the generation of tests for each of these simpler criteria we use a specification that is obtained by "projecting" the original specification. Such a projected specification is usually smaller in size than the original specification, and thus easier to handle by tools. In this way test generation from large-sized specifications for complicated correctness criteria becomes feasible. Some existing test tools, such as TGV [4] and AUTOLINK [13], have implemented test generation techniques that are similar to the ones of which the underlying mathematical principles are explained in this paper.

Overview Section 2 recalls [5], defining the subclass of multi-input/output transition systems. Next, section 3 presents the correctness criterion $\textbf{mioco}_{\mathcal{F}}$, and describes an algorithm that is able to generate tests from a specification with respect to $\textbf{mioco}_{\mathcal{F}}$. Section 4 investigates under which conditions specifications can be safely reduced in size without loosing the ability to generate valid tests from it. Section 5 describes two factorized test generation techniques: the technique described in section 5.1 is able to produce a sound test suite, and the one in section 5.2 produces a complete test suite. In section 5.3 implementation techniques are discussed which make the factorized way of generating tests more efficient. Finally, section 6 contains conclusions and further research.

2 MULTI-INPUT/OUTPUT TRANSITION SYSTEMS

Many behaviour description languages use the formalism of labelled transition systems as their underlying semantic model (e.g., CCS [10], LOTOS [6]). In this paper we use rigid transition systems, i.e., without internal actions, to specify and model the behaviour of systems.

DEFINITION 1 *A (labelled) transition system (LTS) over L is a quadruple* $\langle S, L, \rightarrow, s_0 \rangle$ *where S is a (countable) set of states, L is a (countable) set of observable actions,* $\rightarrow \subseteq S \times L \times S$ *is a set of transitions, and* $s_0 \in S$ *is the initial state.*

The universe of LTSs over L is denoted by $\mathcal{LTS}(L)$. Instead of $(s, a, s') \in \rightarrow$ we write $s \xrightarrow{a} s'$. We extend the relation $\rightarrow$ with labels that are sets of actions: $s \xrightarrow{A} s$ means that s cannot perform actions $a \in A$, i.e., $s \xrightarrow{A} s =_{def} \forall \mu \in A : s \xrightarrow{\mu}\!\!\!\!/\ $. Such self-loop transitions are called refusal transitions [12]. A failure trace is a finite sequence $\mu_1 \cdot \ldots \cdot \mu_n$ of actions and refusals. We write $p \xrightarrow{\mu_1 \cdot \ldots \cdot \mu_n} p_n$ for $\exists p_1, \ldots p_{n-1} : p \xrightarrow{\mu_1} p_1 \xrightarrow{\mu_2} \ldots \xrightarrow{\mu_n} p_n$, and $p \xrightarrow{\mu_1 \cdot \ldots \cdot \mu_n}$ for $\exists p_n : p \xrightarrow{\mu_1 \cdot \ldots \cdot \mu_n} p_n$. The set of finite sequences over actions in L is denoted by L^*, and the set of derivates of p is defined as $der(p) =_{def} \{p' \mid \exists \sigma \in L^* : p \xrightarrow{\sigma} p'\}$. The set of failure traces of p over L is defined as $f\text{-}traces(p) =_{def} \{\sigma \in (L \cup \mathcal{P}(L))^* \mid p \xrightarrow{\sigma}\}$ where $\mathcal{P}(\cdot)$ denotes the powerset operator. For the notation of transition systems we use some standard process-algebraic operators (cf. LOTOS [6]). For this paper it suffices to use action-prefix $\mu; B$ which can perform action μ and then behave as B, and unguarded choice $\sum \mathcal{B}$ which can behave as any of its members $B \in \mathcal{B}$. We abbreviate $\sum\{B_1, B_2\}$ by $B_1 + B_2$ and $\sum \emptyset$ by **stop**.

In traditional testing theory [1, 3] an LTS abstracts from the initiative of actions. In reality, however, many implementations communicate with their environment via clearly distinguishable input actions (actions that are initiated by the environment and consumed by the implementation, e.g., button push experiments) and output actions (actions that are initiated by the implementation and consumed by the environment, e.g., messages that occur on a display). In testing the distinction between input actions and output actions is crucial to model realistic implementations faithfully. This has triggered research in transition system models where the labelset L is partitioned in a set of input actions L_I and a set of output actions L_U, e.g., input/output automata (IOA, [9]), input/output state machines (IOSM, [11]), input/output labelled transition systems (IOLTS, [4]) and input/output transition systems (IOTS, [14]). These models additionally require that input actions are continuously enabled [2].

A further refinement with respect to distinguishing inputs and outputs was proposed in [5]. There not only a distinction between input actions and output actions is made, but also the interface of the implementation with its environment ("PCO") is explicitly modelled. This is done by partitioning the set of input actions L_I in a set of channels $\mathcal{L}_I =_{def} \{L_I^1, \dots, L_I^n\}$, and the set of output actions L_U in a set of channels $\mathcal{L}_U =_{def} \{L_U^1, \dots, L_U^m\}$. Each channel L_I^j or L_U^k represents a location on the interface of the implementation where the actions in L_I^j or L_U^k may occur, respectively. Moreover, to enlarge the diversity of systems that can be modelled (compared to IOA, IOSM, IOLTS and IOTS) a more liberal condition with respect to the enabling of inputs is imposed: inputs need not always be enabled, but for each input channel input actions must be simultaneously enabled. Such systems are called *multi-input/output transition systems* (MIOTS).

DEFINITION 2 *A multi-input/output transition system (MIOTS) p over partitioning $\mathcal{L}_I$ of L_I and partitioning $\mathcal{L}_U$ of L_U is a transition system with inputs and outputs, $p \in \mathcal{LTS}(L_I \cup L_U)$, such that for all $L_I^j \in \mathcal{L}_I$*

$$\forall p' \in der(p),\ \textit{if}\ \ \exists a \in L_I^j : p' \xrightarrow{a} \quad \textit{then}\ \ \forall b \in L_I^j : p' \xrightarrow{b}$$

The universe of multi-input/output transition systems over $\mathcal{L}_I$ and $\mathcal{L}_U$ is denoted by $\mathcal{MIOTS}(\mathcal{L}_I, \mathcal{L}_U)$.

The formalisms LTS and IOTS are special classes of $\mathcal{MIOTS}(\mathcal{L}_I, \mathcal{L}_U)$ for specific instantiations of the sets $\mathcal{L}_I$ and $\mathcal{L}_U$ [5]. Because many realistic implementations can be modelled as MIOTS we will use MIOTS as the modelling formalism of implementations. The choice to model implementations as MIOTS still leaves freedom to choose the interface of these implementations with their environment by instantiating these MIOTS with the proper partitionings $\mathcal{L}_I$ and $\mathcal{L}_U$. Figure 1 depicts the interface of a multi-input/output transition system.

In [5] an extensional correctness criterion $\leq_{mior}$ has been defined that indicates when an implementation is correct with respect to its specification. This relation can be defined in the same style as equation (1), where the set of observers is taken as the set of *singular observers*. A singular observer acts as a special multi-input/output transition system which supplies inputs and observes outputs, where inputs for the implementation are outputs for the observer and vice versa. Moreover, these observers are equipped with labels to detect *input suspension*, i.e., the non-acceptance of an input action by the implementation (e.g., a button that is pressed but that does not go down), and with labels to detect *output suspension*, i.e., the inability of the implementation to produce an output action (e.g., a display that remains empty). By having the ability to detect input suspension a larger class of implementations can be tested than in, e.g., the testing theory of IOTS [14]. In the extensional

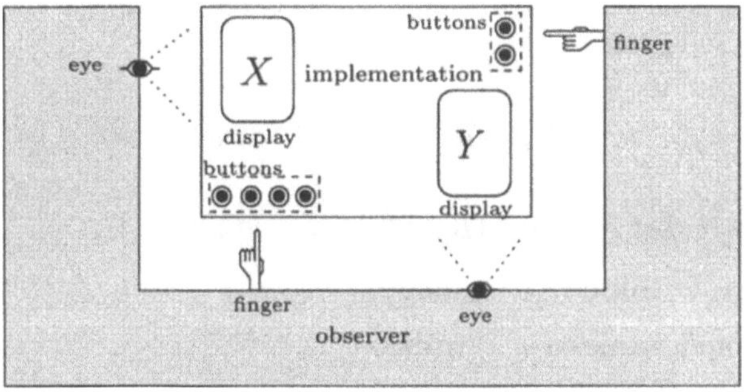

Figure 1 An interface for a multi-input/output transition system

definition of $\leq_{mior}$ an implementation i is related to its specification s if, for every singular observer u, all observations $obs(u, i)$ that u can make of i are included in the set of observations $obs(u, s)$ that u can make of s. We will not present the extensional definition of $\leq_{mior}$ here, but instead we will use an intensional characterization of $\leq_{mior}$ as its definition. For more details we refer to [5].

DEFINITION 3 *$\leq_{mior} \subseteq \mathcal{MIOTS}(\mathcal{L}_I, \mathcal{L}_U) \times \mathcal{LTS}(L_I \cup L_U)$ is defined by*

$$i \leq_{mior} s$$
$$=_{def} \forall \sigma \in (L_I \cup L_U \cup \mathcal{L}_I \cup \mathcal{L}_U)^* : \; out(\, i \textbf{ after } \sigma\,) \subseteq out(\, s \textbf{ after } \sigma\,)$$

where

$$out(\, p \textbf{ after } \sigma\,) =_{def} \quad \{x \in L_U \mid \exists p' : p \xrightarrow{\sigma} p' \xrightarrow{x}\} \qquad (i)$$
$$\cup\{L_I^j \mid 1 \leq j \leq n, \exists p' : p \xrightarrow{\sigma} p' \text{ and } p' \xrightarrow{L_I^j}\} \qquad (ii)$$
$$\cup\{L_U^k \mid 1 \leq k \leq m, \exists p' : p \xrightarrow{\sigma} p' \text{ and } p' \xrightarrow{L_U^K}\} \qquad (iii)$$

The relation $\leq_{mior}$ states that an implementation is $\leq_{mior}$-incorrect if (i) the implementation produces an output, which cannot be produced by the specification after the same trace, or (ii) the implementation has an input suspension at some input channel L_I^j where the specification has none, or (iii) the implementation has an output suspension at some output channel L_U^k where the specification has none.

3 TEST GENERATION FOR MIOTS

Checking for $\leq_{mior}$ requires checking $out(\, i \textbf{ after } \sigma\,) \subseteq out(\, s \textbf{ after } \sigma\,)$ for all $\sigma \in (L_I \cup L_U \cup \mathcal{L}_I \cup \mathcal{L}_U)^*$. Since this is too time consuming in practice, the relation $\textbf{mioco}_{\mathcal{F}}$ restricts $\leq_{mior}$ by checking this condition for all failure traces in $\mathcal{F}$ for particularly chosen $\mathcal{F}$ (cf. the conformance relation **conf** in [1]).

DEFINITION 4 *The relation* $\mathbf{mioco}_{\mathcal{F}} \subseteq \mathcal{MIOTS}(\mathcal{L}_I, \mathcal{L}_U) \times \mathcal{LTS}(L_I \cup L_U)$, *where* $\mathcal{F} \subseteq (L_I \cup L_U \cup \mathcal{L}_I \cup \mathcal{L}_U)^*$, *is defined by*

$$i \ \mathbf{mioco}_{\mathcal{F}} \ s \ =_{def} \ \forall \sigma \in \mathcal{F} : \ out(\, i \ \mathbf{after} \ \sigma \,) \subseteq out(\, s \ \mathbf{after} \ \sigma \,)$$

PROPOSITION 1 *Let* $\mathcal{F}_1, \mathcal{F}_2 \subseteq (L_I \cup L_U \cup \mathcal{L}_I \cup \mathcal{L}_U)^*$

1. $\mathbf{mioco}_{\mathcal{F}_1 \cup \mathcal{F}_2} = \mathbf{mioco}_{\mathcal{F}_1} \cap \mathbf{mioco}_{\mathcal{F}_2}$
2. $\mathcal{F}_1 \subseteq \mathcal{F}_2$ *implies* $\mathbf{mioco}_{\mathcal{F}_1} \supseteq \mathbf{mioco}_{\mathcal{F}_2}$

We use the parameterized relation $\mathbf{mioco}_{\mathcal{F}}$ as the class of correctness criteria for well-chosen $\mathcal{F}$. This is motivated by the fact that for specific instances of $\mathcal{L}_I$, $\mathcal{L}_U$ and $\mathcal{F}$ this relation coincides with well-known implementation relations such as **ioco** and **ioconf** advocated in [14]. The set $\mathcal{F}$ can be considered as a set of test purposes for which tests must be derived [7]. The selection of such traces could be based on testing heuristics or experience. For complex and critical applications, such as communication protocols, the set $\mathcal{F}$ can be very large.

In [5] a test generation algorithm Π has been presented that is able to generate tests from a specification $s \in \mathcal{LTS}(L_I \cup L_U)$. These tests can decide whether an implementation is $\mathbf{mioco}_{\mathcal{F}}$-correct with respect to its specification or not. Tests are modelled as singular observers and use special labels θ_i^j to detect input suspension at channel L_I^j, and the special labels θ_u^k to detect output suspension at channel L_U^k. Tests are built up recursively by either applying an input action a to channel L_I^j and detecting its acceptance or suspension ($t \ ::= \ a; t + \theta_i^j; t$), or observing an output channel L_U^k and detecting the occurrence of an output or output suspension ($t ::= \sum_{x \in L_U^k \cup \{\theta_u^k\}} x; t$), or ending the test [5]. The end states of a test are labelled with **pass** or **fail** and indicate success or failure of test execution.

Figure 2 depicts test algorithm Π from [5]. The algorithm takes a specification $s \in \mathcal{LTS}(L_I \cup L_U)$ and a set of failure traces $\mathcal{F}$ and computes a test case $\Pi_{\mathcal{F},S}$ by applying one of the steps in the algorithm. The set S keeps track of the possible current states of the specification, and initially contains the initial state of specification s. The set $\mathcal{F}$ contains the failure traces for which the condition $out(\, i \ \mathbf{after} \ \sigma \,) \subseteq out(\, s \ \mathbf{after} \ \sigma \,)$ has to be tested. Both sets S and $\mathcal{F}$ are updated in case the test generation algorithm proceeds recursively. We define $S \ \mathbf{after} \ \sigma \ =_{def} \ \{s' \mid \exists s \in S : s \xrightarrow{\sigma} s'\}$ as the set of states that are reachable from a state in S after σ. The trace $\overline{\sigma}$ denotes the trace σ where all occurrences of refusals L_I^j and L_U^k are replaced by there suspension detection labels θ_i^j and θ_u^k, and vice versa.

Running a test against an implementation means that the experiments prescribed in the test (supplying an input to an input channel, or observing an output from an output channel) are applied to the implementation. In case an input action is supplied, then it is either accepted or rejected, after which the

Input:	set of states S
Input:	set of failure traces $\mathcal{F} \subseteq (L_I \cup L_U \cup \mathcal{L}_I \cup \mathcal{L}_U)^*$
Output:	test case $\Pi_{\mathcal{F},S}$

Initial value: $S = \{s_0\}$, where s_0 is the initial state of s.

Apply one of the following non-deterministic choices recursively.

1. (* terminate the test case if there are no more specified traces in $\mathcal{F}$ *)
   if $\mathcal{F} = \emptyset$ then
   $$\Pi_{\mathcal{F},S} := \textbf{pass}$$
2. (* terminate the test case when a trace $\sigma \in \mathcal{F}$ has been performed *)
   if $\epsilon \in \mathcal{F}$ then take some $L_I^j \in \mathcal{L}_I$, and for some $a \in L_I^j$ (* supply input a *)
   $$\Pi_{\mathcal{F},S} := \begin{cases} a;\textbf{pass} + \theta_i^j;\textbf{fail} & \text{if } S\,\textbf{after}\,L_I^j = \emptyset \\ a;\textbf{pass} + \theta_i^j;\textbf{pass} & \text{if } S\,\textbf{after}\,L_I^j \neq \emptyset \end{cases}$$
3. (* terminate the test case when a trace $\sigma \in \mathcal{F}$ has been performed *)
   if $\epsilon \in \mathcal{F}$ then take some $L_U^k \in \mathcal{L}_U$, then (* observe channel L_U^k *)
   $$\Pi_{\mathcal{F},S} := \sum\{x;\textbf{pass} \mid x \in L_U^k \cup \{\theta_u^k\} \text{ and } S\,\textbf{after}\,\overline{x} \neq \emptyset\} + \sum\{x;\textbf{fail} \mid x \in L_U^k \cup \{\theta_u^k\} \text{ and } S\,\textbf{after}\,\overline{x} = \emptyset\}$$
4. (* supply an input for which you want to test deeper *)
   take some $L_I^j \in \mathcal{L}_I$ and $a \in L_I^j$ such that $\{\sigma \mid a{\cdot}\sigma \in \mathcal{F}\} \neq \emptyset$, then
   $$\Pi_{\mathcal{F},S} := a;\Pi_{\mathcal{F}',S'} + \theta_i^j;\textbf{pass}$$
   where $S' = S\,\textbf{after}\,a$, $\mathcal{F}' = \{\sigma \mid a{\cdot}\sigma \in \mathcal{F}\}$
5. (* supply some input and continue if it is refused *)
   take some $L_I^j \in \mathcal{L}_I$ such that $\{\sigma \mid L_I^j{\cdot}\sigma \in \mathcal{F}\} \neq \emptyset$, then
   $$\Pi_{\mathcal{F},S} := a;\textbf{pass} + \theta_i^j;\Pi_{\mathcal{F}'',S''}$$
   where $a \in L_I^j$, $S'' = S\,\textbf{after}\,L_I^j$, $\mathcal{F}'' = \{\sigma \mid L_I^j{\cdot}\sigma \in \mathcal{F}\}$
6. (* Find a channel L_U^k that produces an output for which to test deeper *)
   take some $L_U^k \in \mathcal{L}_U$ such that $\{\sigma \mid \exists x \in L_U^k \cup \{L_U^k\} : x{\cdot}\sigma \in \mathcal{F}\} \neq \emptyset$, then
   $$\Pi_{\mathcal{F},S} := \sum\{x;\Pi_{\mathcal{F}',S'} \mid x \in L_U^k \cup \{\theta_u^k\} \text{ and } \mathcal{F}' = \{\sigma \mid \overline{x}{\cdot}\sigma \in \mathcal{F}\} \text{ and } S' = S\,\textbf{after}\,\overline{x}\}$$

Figure 2 Test generation algorithm.

test continues accordingly. Similarly, for any output channel either an output action is produced by the implementation and observed by the test, or output at this channel is suspended, and the test continues with its corresponding successor experiment. Consequently, when running a test against an implementation the test will always end up in one of its end states (i.e., a **pass** or a **fail** state).

We say that implementation i passes test t if t can only end up in a **pass** state after running against i. This is denoted by the predicate i *passes* t. The dual is denoted by i *fails* t, meaning that t *may* end up in a **fail** state after running against i. For test suite T we define i *passes* $T =_{def} \forall t \in T : i$ *passes* t, and i *fails* $T =_{def} \neg(i$ *passes* $T)$.

To assess the correctness of implementations by means of testing we have to link the passing and failing of these tests when run against implementations to the correctness of these implementations. For that we use the terms soundness, exhaustiveness and completeness of test suites [8]. A test suite T is *sound* (for specification s with respect to $\mathbf{mioco}_{\mathcal{F}}$) if every correct implementation will always pass this test suite: $i\ \mathbf{mioco}_{\mathcal{F}}\ s \implies i$ *passes* T. Test suite T is *exhaustive* if passing test suite T guarantees correctness: $i\ \mathbf{mioco}_{\mathcal{F}}\ s \Longleftarrow i$ *passes* T. Test suite T is *complete* if it is both sound and exhaustive. A sound test suite is never able to reject correct implementations, and an exhaustive test suite is theoretically able to fail with all incorrect ones (which, in practice, may take an infinite amount of time).

It has been shown [5] that every test that can be generated by algorithm Π for $\mathcal{F}$ and s is sound for s with respect to $\mathbf{mioco}_{\mathcal{F}}$. Moreover, the set of all tests that can be generated by algorithm Π for $\mathcal{F}$ and s, denoted by $\Pi_{\mathcal{F}}(s)$, is complete for s with respect to $\mathbf{mioco}_{\mathcal{F}}$.

PROPOSITION 2 *Let* $\mathcal{F} \subseteq (L_I \cup L_U \cup \mathcal{L}_I \cup \mathcal{L}_U)^*$ *and* $s \in \mathcal{LTS}(L_I \cup L_U)$

1. *Any test case obtained from algorithm* Π *for* s *and* $\mathcal{F}$ *is sound for* s *with respect to* $\mathbf{mioco}_{\mathcal{F}}$.
2. *The set of all test cases that can be obtained from algorithm* Π *for* s *and* $\mathcal{F}$ *is complete for* s *with respect to* $\mathbf{mioco}_{\mathcal{F}}$.

The algorithm may generate tests that are not very efficient in detecting incorrect implementations, however, optimizations are not considered here.

4 LOOSER SPECIFICATIONS

We discuss a technique that can be used to isolate a part of the specification, and use this part to generate tests. This technique is called *loosening of specifications* [8]. We analyse the conditions under which it is valid to isolate such a part of the specification. Sections 5.1 and 5.2 will show how to generate a sound and complete test suite from these parts for implementation relation $\mathbf{mioco}_{\mathcal{F}}$ in a factorized way.

For correctness criterion $\mathbf{mioco}_{\mathcal{F}}$ only the behaviour after failure traces specified in $\mathcal{F}$ has to be investigated (definition 4). For analysing whether the responses to failure traces in $\mathcal{F}$ are valid or not there is no need to investigate the complete specification; responses to experiments that are not specified in $\mathcal{F}$ can be discarded. This argument shows that it may be possible to generate tests from a smaller specification (in size). The question is how to obtain such a smaller specification.

Such a smaller specification can be obtained by having the tester provide the input actions of the failure traces in $\mathcal{F}$ for which correctness has to be checked. Because a tester fully controls the input actions of an implementation but not the output actions, such a tester can "steer" the implementation towards checking a specific failure trace in $\mathcal{F}$ as much as possible by providing the input actions that are necessary to perform this failure trace. Deterministic processes that specify such sequences of input actions are called *selection processes.* Such processes can be seen as test purposes. From a selection process q and specification s a specification $s \parallel_{L_I} q$ is isolated that contains the responses to the input sequences specified in q, but discards all responses to input sequences that are not specified in q. The operator $\parallel_{L_I}$ describes how the part $s \parallel_{L_I} q$ is isolated from s, and its formal definition is given below (cf. LOTOS [6])

DEFINITION 5 *The universe of selection processes $\mathcal{SLTS}(L_I)$ over L_I is*

$$\mathcal{SLTS}(L_I) \quad =_{def} \quad \{p \in \mathcal{LTS}(L_I) \mid p \textit{ is deterministic}\}$$

Let $s \in \mathcal{LTS}(L_I \cup L_U)$ and $q \in \mathcal{SLTS}(L_I)$ then the transition system $s \parallel_{L_I} q \in \mathcal{LTS}(L_I \cup L_U)$ is inductively defined by the following inference rules.

$$\frac{s \xrightarrow{a} s', q \xrightarrow{a} q'}{s \parallel_{L_I} q \xrightarrow{a} s' \parallel_{L_I} q'} \ (a \in L_I) \qquad\qquad \frac{s \xrightarrow{x} s'}{s \parallel_{L_I} q \xrightarrow{x} s' \parallel_{L_I} q} \ (x \in L_U)$$

The operator $\parallel_{L_I}$ forces synchronization on actions in L_I, but allows actions not in L_I (i.e., actions in L_U) to occur independently.

We focus on what conditions need to be imposed on q in order to use $s \parallel_{L_I} q$ instead of s as the specification to generate tests from without running the risk to generate tests that are able to reject implementations that are $\mathbf{mioco}_{\mathcal{F}}$-correct for s, that is, what conditions need to be imposed on q in order to generate test suites from $s \parallel_{L_I} q$ that are sound with respect to $\mathbf{mioco}_{\mathcal{F}}$ for s.

As a first step in analysing these conditions we compare the input refusals and output refusals of s with the ones of $s \parallel_{L_I} q$. Because *all* and *only* output actions that $s \parallel_{L_I} q$ can perform are the ones that s is able to perform, any refusal $X \subseteq L_U$ of s is also a refusal of $s \parallel_{L_I} q$ and vice versa. For refusals of input actions $A \subseteq L_I$ the situation is slightly different. Because s and q need to synchronize on input actions it follows that the inability of s to perform an input action is reflected by the inability of $s \parallel_{L_I} q$ to perform this action, but not vice versa!

PROPOSITION 3 *Let $A \subseteq L_I$ and $X \subseteq L_U$.*

1. $s \xrightarrow{A} s$ *implies* $s \parallel_{L_I} q \xrightarrow{A} s \parallel_{L_I} q$
2. $s \xrightarrow{X} s$ *iff* $s \parallel_{L_I} q \xrightarrow{X} s \parallel_{L_I} q$

Proposition 3 states that $s \,\|\, _{L_I} q$ preserves the refusals $A \subseteq L_I$ and $X \subseteq L_U$ of s. This result can be used to show that $s \,\|\, _{L_I} q$ preserves all failure traces of s in $(L_I \cup L_U \cup \mathcal{L}_I \cup \mathcal{L}_U)^*$ for which q specifies the sequences of input actions to be performed. We use $\sigma \lceil L_I$ to denote the sequence that arises from σ when restricted to actions in L_I.

PROPOSITION 4 *For all* $\sigma \in (L_I \cup L_U \cup \mathcal{L}_I \cup \mathcal{L}_U)^*$

$$s \xrightarrow{\sigma} s' \text{ and } q \xrightarrow{\sigma \lceil L_I} q' \quad \text{implies} \quad s \,\|\, _{L_I} q \xrightarrow{\sigma} s' \,\|\, _{L_I} q'$$

Combining the facts that all output actions of $s' \,\|\, _{L_I} q'$ are produced by s' (definition 5) and that all input refusals and output refusals of s' are preserved by $s' \,\|\, _{L_I} q'$ (proposition 3), it follows, together with proposition 4, that $out(\, s \textbf{ after } \sigma\,)$ is included in $out(\, s \,\|\, _{L_I} q \textbf{ after } \sigma\,)$ for those σ such that $\sigma \lceil L_I$ is specified by q.

PROPOSITION 5 *Let* $\sigma \in (L_I \cup L_U \cup \mathcal{L}_I \cup \mathcal{L}_U)^*$

$$q \xrightarrow{\sigma \lceil L_I} \quad \text{implies} \quad out(\, s \textbf{ after } \sigma\,) \subseteq out(\, s \,\|\, _{L_I} q \textbf{ after } \sigma\,)$$

By choosing suitable q it is possible to "steer" for which failure traces the inclusion $out(\, s \textbf{ after } \sigma\,) \subseteq out(\, s \,\|\, _{L_I} q \textbf{ after } \sigma\,)$ holds. In particular, if q contains the input sequences of the failures traces specified in $\mathcal{F}$, then this inclusion holds (at least) for all $\sigma \in \mathcal{F}$. But then, according to definition 4, any implementation that is $\textbf{mioco}_{\mathcal{F}}$-correct for s is also $\textbf{mioco}_{\mathcal{F}}$-correct for $s \,\|\, _{L_I} q$, or alternatively, any $\textbf{mioco}_{\mathcal{F}}$-incorrect implementation for $s \,\|\, _{L_I} q$ is also $\textbf{mioco}_{\mathcal{F}}$-incorrect for s. We define $\mathcal{F} \lceil L_I$ as the set-wise restriction on failure traces in $\mathcal{F}$: $\mathcal{F} \lceil L_I =_{def} \{\sigma \lceil L_I \mid \sigma \in \mathcal{F}\}$.

PROPOSITION 6 *If* $traces(q) \supseteq \mathcal{F} \lceil L_I$, *then*

$$i \textbf{ mioco}_{\mathcal{F}}\ s \quad \text{implies} \quad i \textbf{ mioco}_{\mathcal{F}}\ (s \,\|\, _{L_I}\, q)$$

The significance of proposition 6 is that in order to obtain a sound test suite that can check whether an implementation is $\textbf{mioco}_{\mathcal{F}}$-correct for "big" specification s, it is possible to generate a sound test suite from the "smaller" specification $s \,\|\, _{L_I} q$ as long as q specifies all input sequences of the failures traces in $\mathcal{F}$. If $traces(q) \supseteq \mathcal{F}$ we call the specification $s \,\|\, _{L_I} q$ the *projected* specification of s on $\textbf{mioco}_{\mathcal{F}}$. The reverse implication of proposition 6 does not hold in general: erroneous implementations with respect to $\textbf{mioco}_{\mathcal{F}}$ and s may pass a sound test suite that is generated from $s \,\|\, _{L_I} q$. This is caused by the fact that, due to the pruning of s using q, additional input refusals may be introduced in $s \,\|\, _{L_I} q$ that were not present in s itself (cf. proposition 3.1).

5 FACTORIZED TEST GENERATION

In this section we describe two techniques to generate tests for $\mathbf{mioco}_{\mathcal{F}}$ in a factorized way. We do this by generating tests from a specification that is projected on the correctness criterion $\mathbf{mioco}_{\{\sigma\}}$ for all $\sigma \in \mathcal{F}$. Section 5.1 discusses how a sound test suite can be obtained in this way, and section 5.2 shows how to obtain a complete test suite.

5.1 Factorized test generation (soundness)

In practice, when generating tests for $\mathbf{mioco}_{\mathcal{F}}$, the set $\mathcal{F}$ may contain a large number of failure traces, and the specification s can be very large (e.g., measured in number of states and transitions). Consequently, the generation of tests $\Pi_{\mathcal{F}}(s)$ directly from $\mathcal{F}$ and s can be a time and space consuming task, and tools may not be able to generate this set. In this subsection we present a technique to cope (at least partially) with this complexity.

In order to reduce the size of the specification s selection processes can be used (see section 4). A special class of selection processes is the class consisting of linear sequences over the set of input actions L_I. Such selection processes are called *sticks.*

DEFINITION 6 *Let* $\sigma, \sigma' \in (L_I \cup L_U \cup \mathcal{L}_I \cup \mathcal{L}_U)^*$, *then* $\mathit{stick}(\sigma)$ *is the transition system that is inductively defined by*

$$\begin{array}{lll} \mathit{stick}(\epsilon) & =_{def} & \mathbf{stop} \\ \mathit{stick}(a\cdot\sigma') & =_{def} & \left\{ \begin{array}{ll} a; \mathit{stick}(\sigma') & \mathit{if}\ a \in L_I \\ \mathit{stick}(\sigma') & \mathit{otherwise} \end{array} \right. \end{array}$$

The universe of all sticks is denoted by $\mathcal{STICK}(L_I)$.

Note that $\mathcal{STICK}(L_I) \subset \mathcal{SLTS}(L_I)$. Figure 4(a) depicts the structure of a stick.

From the generalized version of proposition 1.1 it follows that checking for correctness with respect to $\mathbf{mioco}_{\mathcal{F}}$ can be expressed in terms of checking for $\mathbf{mioco}_{\{\sigma\}}$ for each $\sigma \in \mathcal{F}$, viz.

$$\mathbf{mioco}_{\mathcal{F}} = \bigcap_{\sigma \in \mathcal{F}} \mathbf{mioco}_{\{\sigma\}} \tag{2}$$

Each correctness check with respect to $\mathbf{mioco}_{\{\sigma\}}$ can be performed independently. For this correctness criterion it suffices to take a selection process that is able to perform the sequence $\sigma\lceil L_I$ according to proposition 5. The process $\mathit{stick}(\sigma)$ is such a selection process. Thus, according to proposition 6 we have

$$i\ \mathbf{mioco}_{\{\sigma\}}\ s \quad \text{implies} \quad i\ \mathbf{mioco}_{\{\sigma\}}\ (s \,\|_{L_I} \mathit{stick}(\sigma)) \tag{3}$$

Combining the results of equations (2) and (3) shows that instead of generating tests from s for $\mathbf{mioco}_{\mathcal{F}}$ we can generate tests from $s \parallel_{L_I} stick(\sigma)$ for $\mathbf{mioco}_{\{\sigma\}}$ without running the risk that $\mathbf{mioco}_{\mathcal{F}}$-correct implementations are rejected. For all $\sigma \in \mathcal{F}$ this can be done independently. This leads to the parallelization procedure sketched in figure 3.

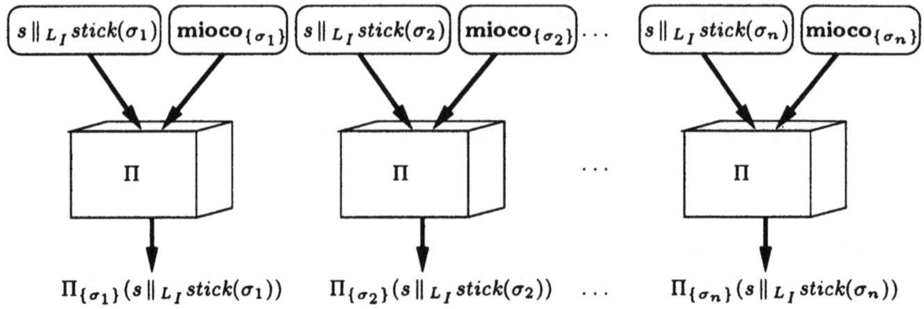

Figure 3 Factorized test suite generation (soundness)

PROPOSITION 7 $i\ \mathbf{mioco}_{\mathcal{F}}\ s$ *implies* $\forall \sigma \in \mathcal{F} : i\ \mathbf{mioco}_{\{\sigma\}}\ (s \parallel_{L_I} stick(\sigma))$

Algorithm Π can be applied to generate tests from $s \parallel_{L_I} stick(\sigma)$ for $\mathbf{mioco}_{\{\sigma\}}$. This gives us a test suite $\Pi_{\{\sigma\}}(s \parallel_{L_I} stick(\sigma))$ for each $\sigma \in \mathcal{F}$. The union of all these tests suites is sound for $\mathbf{mioco}_{\mathcal{F}}$. Now any implementation that fails a test in $\bigcup_{\sigma \in \mathcal{F}} \Pi_{\{\sigma\}}(s \parallel_{L_I} stick(\sigma))$ will also fail test suite $\Pi_{\mathcal{F}}(s)$, and hence is $\mathbf{mioco}_{\mathcal{F}}$-incorrect for s (remember the convention that $\Pi_{\mathcal{F}}(s)$ denotes the set of all tests that are generated by Π from s for $\mathbf{mioco}_{\mathcal{F}}$, and that this set is complete (proposition 2.2)).

COROLLARY 1 i *fails* $\bigcup_{\sigma \in \mathcal{F}} \Pi_{\{\sigma\}}(s \parallel_{L_I} stick(\sigma))$ *implies* i *fails* $\Pi_{\mathcal{F}}(s)$

Instead of generating tests from s for $\mathbf{mioco}_{\mathcal{F}}$ we can generate tests from $s \parallel_{L_I} stick(\sigma)$ for $\mathbf{mioco}_{\{\sigma\}}$. This reduces complexity in several ways. Specification $s \parallel_{L_I} stick(\sigma)$ is in most cases much smaller than s (in number of states and transitions) due to its projection on $\{\sigma\}$, and $\mathbf{mioco}_{\{\sigma\}}$ is less complex, and thus easier to check, than $\mathbf{mioco}_{\mathcal{F}}$. Although the number of test generation activities increases (for each $\sigma \in \mathcal{F}$ a test suite $\Pi_{\{\sigma\}}(s \parallel_{L_I} stick(\sigma))$ is generated) all these test generation activities are simpler and can be done independently.

Note that the reverse of corollary 1 does not hold; an implementation that passes test suite $\bigcup_{\sigma \in \mathcal{F}} \Pi_{\{\sigma\}}(s \parallel_{L_I} stick(\sigma))$ does not have to pass test suite $\Pi_{\mathcal{F}}(s)$. This is a direct consequence of the one-way implication of (3), that is a direct consequence of the one-way implication in proposition 3.1.

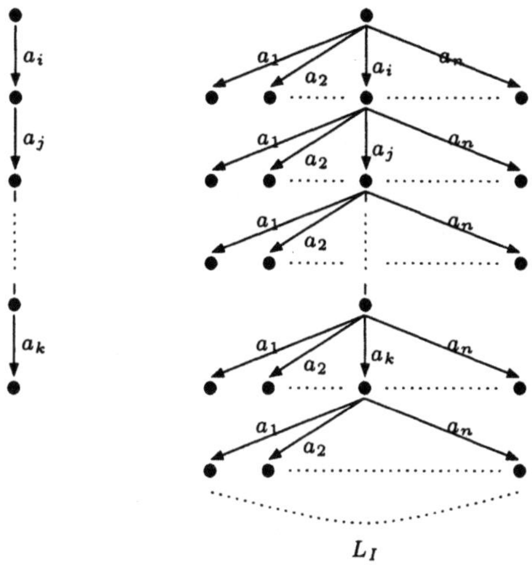

(a) *stick*(σ) (b) *fan*(σ)

Figure 4 *stick*(σ) and *fan*(σ) with $\sigma\lceil L_I = a_i \cdot a_j \cdot \ldots \cdot a_k$

5.2 Factorized test generation (completeness)

The factorized test generation technique sketched in section 5.1 produces a sound but not necessarily complete test suite (corollary 1). In our attempts to develop a test suite that is able to reject as many faulty implementations as possible, we develop in this section a factorized test generation technique which can generate a complete test suite.

For arbitrary selection process q proposition 3.1 states that any input refusal of s is preserved in $s \,\|_{L_I} q$, but not necessarily vice versa: an input refusal of $s \,\|_{L_I} q$ can be caused by s itself, or by pruning of s with q. This also holds if the selection process q is a stick. Consequently, $s \,\|_{L_I} q$ may have input suspension where s has none. This exactly explains the absence of the reverse implication in proposition 6.

In order to prevent the unwanted introduction of input refusals in $s \,\|_{L_I} q$ we have to enforce that every input refusal of $s \,\|_{L_I} q$ is caused by s. This can be done by requiring that q is always able to offer any input action. In that case any input refusal of $s \,\|_{L_I} q$ must be caused by s itself, i.e., $s \xrightarrow{A} s$ iff $s \,\|_{L_I} q \xrightarrow{A} s \,\|_{L_I} q$. Selection processes that are prepared to synchronize on all input actions in states that lie on a particular sequence of input actions are called *fans*. A fan can be seen as a stick where in each state all input actions are offered. Figure 4(b) visualizes the structure of a fan.

DEFINITION 7 *Let $\sigma, \sigma' \in (L_I \cup L_U \cup \mathcal{L}_I \cup \mathcal{L}_U)^*$, then $fan(\sigma)$ is the transition system that is inductively defined by*

$$\begin{array}{lll} fan(\epsilon) & =_{def} & \sum\{a; \mathbf{stop} \mid a \in L_I\} \\ fan(a \cdot \sigma') & =_{def} & \begin{cases} \sum\{b; \mathbf{stop} \mid b \in L_I, b \neq a\} + a; fan(\sigma') & \text{if } a \in L_I \\ fan(\sigma') & \text{otherwise} \end{cases} \end{array}$$

The universe of all fans is denoted by $\mathcal{FAN}(L_I)$.

We now claim that $s \,\|_{L_I} fan(\sigma)$ can be used for complete test generation from s for $\mathbf{mioco}_{\{\sigma\}}$:

$$i \ \mathbf{mioco}_{\{\sigma\}} \ s \quad \text{iff} \quad i \ \mathbf{mioco}_{\{\sigma\}} \ (s \,\|_{L_I} fan(\sigma)) \tag{4}$$

Since $s \,\|_{L_I} fan(\sigma)$ will, in most cases, be smaller than s itself it is profitable to use $s \,\|_{L_I} fan(\sigma)$ for the generation of tests. Together with equation (2) this procedure can be repeated for each $\sigma \in \mathcal{F}$, thereby obtaining a parallelization procedure for the generation of a complete test suite for $\mathbf{mioco}_{\mathcal{F}}$. This procedure can be visualized by replacing all $stick(\sigma_i)$ with $fan(\sigma_i)$ in figure 3.

PROPOSITION 8 $i \ \mathbf{mioco}_{\mathcal{F}} \ s$ *iff* $\forall \sigma \in \mathcal{F} : i \ \mathbf{mioco}_{\{\sigma\}} \ (s \,\|_{L_I} fan(\sigma))$

As test generation algorithm Π is able to generate a test suite $\Pi_{\mathcal{F}}(s)$ from specification s that is complete with respect to $\mathbf{mioco}_{\mathcal{F}}$ (see proposition 2.2), it immediately follows from proposition 8 that an implementation is $\mathbf{mioco}_{\mathcal{F}}$-correct for s in case it passes every test in $\Pi_{\{\sigma\}}(s \,\|_{L_I} fan(\sigma))$ for all $\sigma \in \mathcal{F}$.

COROLLARY 2 i *fails* $\bigcup_{\sigma \in \mathcal{F}} \Pi_{\{\sigma\}}(s \,\|_{L_I} fan(\sigma))$ *iff* i *fails* $\Pi_{\mathcal{F}}(s)$

5.3 Efficient implementation of factorized test generation

Efficient implementation of factorized test generation techniques can be obtained by exploiting the special structure of the set of failure traces $\mathcal{F}$, and by sharing common parts of tests that are generated. In this subsection we briefly discuss (i) reducing the set $\mathcal{F}$ as far as possible without weakening the correctness criterion, and (ii) sharing common prefixes of test generation.

Reduction of $\mathcal{F}$: One way to reduce the set $\mathcal{F}$ of failure traces is to preprocess $\mathcal{F}$ and remove all "equivalent" failure traces that would apriori lead to the generation of identical test cases, that is, to reduce $\mathcal{F}$ to a smaller set $\mathcal{F}'$ without weakening the correctness criterion.

Find the smallest $\mathcal{F}' \subseteq \mathcal{F}$ such that $\{i \mid i \ \mathbf{mioco}_{\mathcal{F}'} \ s\} = \{i \mid i \ \mathbf{mioco}_{\mathcal{F}} \ s\}$

An example of such a reduction is the removal of failure traces that only differ in permutations of failures. Since the algorithm produces tests that check whether $out(\, i \,\mathbf{after}\, \sigma\,) \subseteq out(\, s \,\mathbf{after}\, \sigma\,)$ for each $\sigma \in \mathcal{F}$ and the set of states reachable by failure trace $\sigma_1 \cdot A \cdot X \cdot \sigma_2$ equals the set of states reachable by failure trace $\sigma_1 \cdot X \cdot A \cdot \sigma_2$ (where A and X are refusals), one of these tests is redundant.

Another example of such a reduction has to do with robustness testing. In case the behaviour of an implementation for a failure trace which is not in the specification is checked, then any implementation that accepts this failure trace is considered erroneous. Checking correctness for any longer failure trace would not be senseful, since the implementation was already considered erroneous. Consequently, it suffices to restrict to the smallest prefix of failure traces that occur in $\mathcal{F}$ and not in s.

Sharing common prefixes: For common prefixes of failures traces in $\mathcal{F}$ the application of Π can be done in a shared way, e.g., in case Π has to be applied for failure trace $\sigma \cdot \sigma_1$ and for $\sigma \cdot \sigma_2$, then the application for σ can be shared. So, test generation could be started by a single master test generation process which spawns new test generation processes at points where failure traces in $\mathcal{F}$ bifurcate.

6 CONCLUSIONS

In this paper, factorized test generation techniques were presented which can be used to generate test cases from a specification for implementation relation $\mathbf{mioco}_{\mathcal{F}}$. The factorized techniques consist of two steps. Firstly, the "complex" correctness criterion $\mathbf{mioco}_{\mathcal{F}}$ is split up in several independent and "easy-to-check" correctness criteria $\mathbf{mioco}_{\{\sigma\}}$. Secondly, the specification that is used for test generation for $\mathbf{mioco}_{\{\sigma\}}$ is reduced to a smaller specification than the original one by projecting the original specification on the correctness criterion $\mathbf{mioco}_{\{\sigma\}}$. In this way test generation from a specification for $\mathbf{mioco}_{\mathcal{F}}$ can be done more efficiently, which is necessary to make test generation for realistically-sized applications feasible. Depending on the type of selection process that is used (a stick or a fan) the factorized test generation proves to produce a sound or a complete test suite with respect to $\mathbf{mioco}_{\mathcal{F}}$, respectively.

Related work In the tool TGV [4] tests are generated with respect to a test purpose that is given as a IOLTS automaton. This test purpose acts as a selection process that is used to isolate the relevant part of the specification from which tests are generated. A similar facility is provided by the AUTOLINK tool [13] that supports the (semi-)automatic generation of tests from SDL specifications with respect to test purposes that are specified as Message Sequence Charts. This tool runs in cooperation with the SDL development environment SDT.

Further work The next step to be taken is the implementation of a tool on the basis of the theory presented here. This will require more than the direct translation into actual code of the algorithms presented in this paper. For example, algorithm Π is an abstract, generic algorithm that captures the essential idea of test generation for $\mathbf{mioco}_{\mathcal{F}}$, but which should be optimized. and made more efficient. Moreover, this paper considered factorized test generation from given $\mathcal{F}$. How to select $\mathcal{F}$, the *test selection problem*, was not discussed.

7 REFERENCES

[1] E. Brinksma. A theory for the derivation of tests. In S. Aggarwal et al. (eds.), *PSTV VIII*, pages 63–74. North-Holland, 1988.

[2] E. Brinksma, L. Heerink, and J. Tretmans. Developments in testing transition systems. In M. Kim et al. (eds.), *IWTCS X*, pages 143–166. Chapman & Hall, 1997.

[3] R. De Nicola and M.C.B. Hennessy. Testing equivalences for processes. *Theoretical Computer Science*, 34:83–133, 1984.

[4] J.-C. Fernandez, C. Jard, T. Jéron, and C. Viho. Using on-the-fly verification techniques for the generation of test suites. In R. Alur et al. (eds.), *CAV'96*. LNCS 1102, Springer-Verlag, 1996.

[5] L. Heerink and J. Tretmans. Refusal testing for classes of transition systems with inputs and outputs. In T. Mizuno et al. (eds.), *FORTE X/PSTV XVII*, pages 23–38. Chapman & Hall, 1997.

[6] ISO. *LOTOS*. International Standard IS-8807. Geneve, 1989.

[7] ISO. *Conformance Testing Methodology and Framework*. International Standard IS-9646. Geneve, 1991.

[8] ISO/IEC JTC1/SC21 WG7, ITU-T SG 10/Q.8. *Framework: Formal Methods in Conformance Testing*. Committee Draft CD 13245-1, ITU-T proposed recommendation Z.500. Geneve, 1996.

[9] N.A. Lynch and M.R. Tuttle. An introduction to Input/Output Automata. *CWI Quarterly*, 2(3):219–246, 1989.

[10] R. Milner. *Communication and Concurrency*. Prentice-Hall, 1989.

[11] M. Phalippou. *Relations d'Implantation et Hypothèses de Test sur des Automates à Entrées et Sorties*. PhD thesis, Uni. Bordeaux I, France, 1994.

[12] I. Phillips. Refusal testing. *Theoretical Computer Science*, 50(2):241–284, 1987.

[13] M. Schmitt, B. Koch, J. Grabowski, and D. Hogrefe. – Autolink – a tool for the automatic and semi-automatic test generation. In A. Wolisz et al. (eds.), *Formale Beschreibungstechniken für verteilte Systeme*, number Nr. 315 in GMD-Studien, St. Augustin, 1997.

[14] J. Tretmans. Test generation with inputs, outputs and repetitive quiescence. *Software—Concepts and Tools*, 17(3):103–120, 1996.

Test Generation for Specifications Modeled by Input/Output Automata*

Q. M. Tan† *and A. Petrenko*‡
†*Département d'IRO, Université de Montréal*
C.P. 6128, Succ. Centre-Ville, Montréal, P.Q. H3C 3J7, Canada

‡*CRIM, Centre de Recherche Informatique de Montréal*
1081 McGill College Avenue, Montréal, P.Q. H3A 2N4, Canada

Abstract

In this paper, we consider conformance testing of communication systems modeled by I/O automata. A framework is proposed for testing I/O automata with full fault coverage for implementations with at most m states. The notion of state identification, which was originally is defined in the realm of I/O finite state machines, is applied. Based on this notion, a test derivation algorithm is given for test suites which guarantee fault coverage. This algorithm is an analogue of the so-called FSM-based HSI-method.

Keywords

software testing, network protocols, conformance testing, test derivation, state identification, I/O automata

1 INTRODUCTION

Communication is a process in which a system interacts with its environment. In many real-life applications, communication is performed via ***inputs*** and ***outputs***. For such a system, inputs are initialized by the environment and always enabled in the system for any interaction, while outputs are initialized by the system and cannot be blocked by the environment. In this context, when the system has its outputs and inputs at the same time, a choice mechanism is

*This work was supported by the NSERC grant OGPO194381

needed to decide which action among the outputs and inputs actions is executed; no communication can be blocked unless no outputs from the system and no inputs from the environment at all. Systems with such input/output interactions are called *I/O automata.*

The choice mechanism of I/O interactions can be implemented in a simple way such that a system and the environment are interacting in an alternative order. The alternative input/output interactions are usually assumed for so-called input/output finite state machines (FSMs). More general interactions with no distinction between inputs and outputs are called rendezvous. The rendezvous interactions are usually assumed for so-called labeled transition systems (LTSs).

The I/O automata is one of the important formalisms in the realm of conformance testing of communication systems, in particular, network protocols. There has been much work done on testing labeled transition systems [2, 9, 10] and finite state machines [3, 7, 4, 5]. However, for I/O automata, few studies [8, 6, 11, 12] have been done. It is argued that the input/output model is closer to the reality than the rendezvous model, while the alternative input/output model is only a limited case of I/O automata. Therefore, it is of both practical and theoretical interests to study conformance testing based on I/O automata.

A testing scenario for I/O automata was given in [8]. Conformance testing based on I/O automata was discussed in [6]. A test generation method was proposed in [11] for a quiescent trace preorder with unspecified behavior concerned. In [12], the refusal testing notion was further applied for test derivation. These existing methods usually use an exhaustive testing approach in order to prove the correctness of an implementation against its specification with respect to a particular conformance relation. Apparently, this approach is often impractical since it may involve a test suite of a big size or infinite behavior. Moreover, no fault coverage measure can be given for conformity of an implementation with its specification.

In this paper, we deal with test derivation for specifications modeled by I/O automata with guaranteed fault coverage for a well-defined fault model, which is defined as a set of mutants of a given specification with at most m states, where m is a known integer. I/O automata considered are transition-deterministic. We give in Section 2 the basic definitions and notations of I/O automata. In Section 3, a framework is proposed for testing I/O automata. In Section 4, a test derivation method, with the notion of state identification which was originally defined in the realm of FSMs, is given.

2 BASIC DEFINITIONS AND NOTATIONS

The starting point for conformance testing is a specification in some (formal) notation, an implementation given in the form of a black box, and a set of conformance requirements the implementation should satisfy. In this paper, I/O

notation	meaning
$(X \cup Y)^*$	set of sequences over $X \cup Y$; σ or $a_1 \ldots a_n$ denotes such a sequence
$p{-}a_1 \ldots a_n{\rightarrow}q$	there exist p_k, $1 \leq k < n$, such that $p{-}a_1{\rightarrow}p_1 \ldots p_{n-1} {-}a_n{\rightarrow}q$
$p{-}\sigma{\rightarrow}$	there exists q such that $p{-}\sigma{\rightarrow}q$
$p{\not-}\sigma{\rightarrow}$	no q exists such that $p{-}\sigma{\rightarrow}q$
$out(p)$	$out(p) = \{a \in Y \mid p{-}a{\rightarrow}\}$
p **after** σ	p **after** $\sigma = \{q \in S \mid p{-}\sigma{\rightarrow}q\}$; S **after** $\sigma = s_0$ **after** σ
$Tr(p)$	$Tr(p) = \{\sigma \in (X \cup Y)^* \mid p{-}\sigma{\rightarrow}\}$; $Tr(S) = Tr(s_0)$

Table 1 Notations for I/O automata

automata are considered for specifications; implementations are also assumed to be described in the same model as its specification; the conformance requirements is supposed to be defined by a specific conformance relation–*trace equivalence*.

Definition 1 (*I/O automaton (IOA)*): An I/O automaton is a 4-tuple $< S, X, Y, \Delta, s_0 >$, where:
- S is a finite non-empty set of states, $s_0 \in S$, is the initial state.
- X is a finite non-empty set of inputs.
- Y is a finite non-empty set of outputs.
- $\Delta \subseteq S \times (X \cup Y) \times S$ is a transition set, where $\forall p \in S$ and $\forall a \in X, \exists q \in S$ such that $(p, a, q) \in \Delta$.

An element $(p, a, q) \in \Delta$ is usually denoted by $p - a \rightarrow q$, and called a *transition*. If $a \in X$, it is called an *input transition*; otherwise $a \in Y$ and it is called an *output transition*. $p{-}a{\rightarrow}q$ also means that action a is enabled in state p. Note that I/O automata we give here are always input-enabled, i.e. completely specified.

An I/O automaton is said to be *transition-deterministic* if there exist no $p{-}a{\rightarrow}q_1$, $p{-}a{\rightarrow}q_2 \in \Delta$ where $q_1 \neq q_2$. An I/O automaton is also said to be *output-deterministic* if there exist no states enabling two or more different outputs, i.e. there are no $p{-}y_1{\rightarrow}q_1$, $p{-}y_2{\rightarrow}q_2 \in \Delta$ where $y_1, y_2 \in Y$ and $y_1 \neq y_2$. In this paper, we only address the transition-deterministic IOAs. Note that the transition-deterministic I/O automata can still be non-deterministic, because when there may be two or more outputs enabled in a state, which output is executed is not determined.

Given two sequences σ_1, σ_2, we use $\sigma_1.\sigma_2$ to represent the concatenation of σ_1 and σ_2. Given two sets of sequences V_1, V_2, we use $V_1@V_2$ to represent the concatenation of V_1 and V_2, that is, $V_1@V_2 = \{\sigma_1.\sigma_2 \mid \sigma_1 \in V_1 \wedge \sigma_2 \in V_2\}$. We also use $pref(\sigma)$ to denote a set of prefixes of σ and $pref(V)$ to denote a set of prefixes of all sequences in V. We also use V^m to represent the concatenation of set V of sequences with itself for m times. Formally, $V^m = V@V^{m-1}$ and $V^0 = \{\varepsilon\}$.

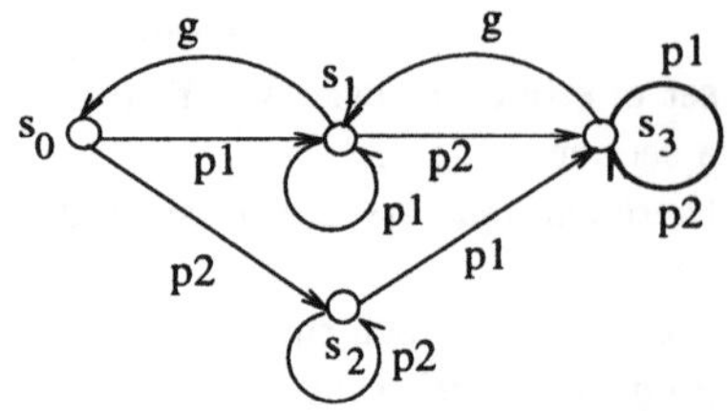

Figure 1 An I/O automaton

Let $\mathsf{S} =< S, X, Y, \Delta, s_0 >$ be an I/O automaton and $p, q \in S$, the conventional notations shown in Table 1 are used in this paper. Furthermore for the convenience of the presentation, we always use $\mathsf{I}, \mathsf{P}, \mathsf{S}, \ldots$ to represent I/O automata; $I, P, Q, \ldots$ for sets of states; $a, b, c, \ldots$ for inputs or output; $x, x_1, \ldots$ for inputs; $y, y_1, \ldots$ for outputs; $i, p, q, s, \ldots$ for states. The set $Tr(p)$ is called the set of *traces* for p.

An IOA can also be represented by a directed graph where nodes are states and labeled edges are transitions. An IOA graph, which describes a slide window of size 2, is shown in Figure 1. In this IOA, where $p1$ and $p2$ are the input actions, g is the output action. In the initial state s_0, the two buffers in the window are empty, in which $p1$ puts a package with the sequence number i to the current buffer and $p2$ puts a package with the sequence number $i+1$ to the next buffer. The arrival of packages might not be in order, but g must get a package from the window in turn. Once a package is removed from the current buffer, the buffer becomes the next empty buffer (i.e. the pointer j to the current buffer now is $mod_2(j+1)$). In the state s_3 the window is full.

Definition 2 (*Trace equivalence*): The trace equivalence relation between two states i and s, written $i = s$, holds iff $Tr(i) = Tr(s)$.
Given two IOAs I and S, we say that I is trace-equivalent to S, written $\mathsf{I} = \mathsf{S}$, iff $i_0 = s_0$.

The trace equivalence relation requires that an implementation has the same traces of interactions as its specification.

3 CONFORMANCE TESTING

Conformance testing is a finite set of experiments, in which a set of test cases, derived from a specification according to the given conformance relation, is applied by a tester or experimenter to the implementation under test (IUT), such that from the results of the execution of the test cases, it can be concluded whether or not the implementation conforms to the specification.

Definition 3 (*Test cases and test suite*): Given an IOA $\mathsf{S} =< S, X, Y, \Delta, s_0 >$, a test case for S is a 3-tuple $< \rho, \Omega, \ell >$ where

- $\rho \in (X \cup Y)^*@Y$;
- $\Omega = (pref(\rho)\backslash\{\rho\})@Y$ is a set of all possible observations while executing this test case.
- $\ell : \Omega \longrightarrow \{\textbf{pass}, \textbf{fail}, \textbf{inconclusive}\}$ is an acceptance function.

A test suite for S is a finite set of test cases for S.

Before a test case T, where $\rho = a_1.a_2 \ldots a_n$, is applied, the IUT M is in the initial state m_0. Starting with a_1, a_i is applied in turn until a_n, if no exception occurs. Let currently M be in state p and meanwhile a_i be applied, the I/O interaction between T and M can be defined as shown in the following table.

I/O	Available Transitions	Execution	Test Process
$a_i \in X$	$p{-}a_i{\rightarrow}q$, no $p{-}y{\rightarrow}$	a_i	proceed
$a_i \in X$	$p{-}a_i{\rightarrow}q$, $p{-}y{\rightarrow}$	a_i	proceed
$a_i \in X$	$p{-}a_i{\rightarrow}q$, $p{-}y{\rightarrow}$	y	exception
$a_i \in Y$	$p{-}a_i{\rightarrow}q$	a_i	proceed
$a_i \in Y$	$p{-}y{\rightarrow}q$, $(y \neq a_i)$	y	exception
$a_i \in Y$	no $p{-}y{\rightarrow}q$		exception(deadlock)

Once an exception occurs at a_i, the test execution is terminated, and the exception function $\ell(a_i)$ tells the result of this test run. The deadlock is a specific exception, which may be detected by a time mechanism. A test case should be designed to be "sound", i.e. no deadlock or **fail** exception should happen if a test suite is applied to a valid implementation. A reliable reset also is assumed to exist always in any IUT, which brings the IUT from any state to the initial state such that a next test run can be executed.

In this paper, the implementations are modeled by I/O automata, which may not be output-deterministic. In order to test nondeterministic implementations, one usually makes a so-called *complete-testing assumption:* it is possible, by applying a given test case to the implementation a finite number of times, to exercise all possible execution paths of the implementation which are traversed by the test case. Without such an assumption, no test suite can guarantee full fault coverage (in terms of conformance relations) for nondeterministic implementations. Therefore, the test application should include several test runs and lead to a complete set of observations $Obs_{(\mathrm{T,M})} = \Omega \cap Tr(m_0)$.

Based on $Obs_{(\mathrm{T,M})}$, the success or failure of testing needs to be concluded. The way a verdict is drawn from $Obs_{(\mathrm{T,M})}$ is the *verdict assignment* for T. Intuitively, the success should mean that no unexpected behavior is observed and the test purpose has been achieved. If we define *the test purpose* of T, written $Pur(\mathrm{T})$, to be $Pur(\mathrm{T}) = \{\sigma \in \Omega \mid \ell(\sigma) = \textbf{pass}\}$ then the conclusion can be drawn as follows.

Definition 4 (*Verdict assignment*): Given an IUT M, a test case T, let $Obs_{fail} = \{\sigma \in Obs_{(T,M)} \mid \ell(\sigma) = \textbf{fail}\}$ and $Obs_{pass} = \{\sigma \in Obs_{(T,M)} \mid \ell(\sigma) = \textbf{pass}\}$, then
M passes T if $Obs_{fail} = \emptyset \wedge Obs_{pass} = Pur(\mathsf{T})$
M fails T otherwise.

The goal of conformance testing is to gain confidence in an implementation under test concerning its conformance with the specification. Increased confidence is normally obtained through time and effort spent in testing the implementation, which, however, are limited by practical and economical considerations. In order to have a more precise measure of the effectiveness of testing, a fault model and fault coverage criteria [1] are introduced. Like in the FSM realm, we also take the mutation approach [1] to define the fault model in the IOA realm to be a set $\mathcal{F}(m)$ of all transition-deterministic IOAs with at most m states and the same alphabets. Based on $\mathcal{F}(m)$, a test suite with complete fault coverage for an IOA specification can be defined as follows.

Definition 5 (*Complete test suite*): Given an IOA S and a fault model $\mathcal{F}$, a test suite *TS* for S is said to be complete over $\mathcal{F}(m)$, if for any M in $\mathcal{F}(m)$, M = S iff M passes T for each T in *TS*.

A complete test suite guarantees that for any implementation M in $\mathcal{F}(m)$, if M passes all test cases, it is a *conforming* or *valid* implementation of the given specification, and any faulty implementation in the fault model is detected by failing at least one test case.

In the context of trace equivalence, a conforming implementation should have the same traces as a specification. Therefore each test case specifies a sequence of inputs and outputs, which is either valid or invalid trace for the specification, to verify that an implementation has implemented the valid one and not the invalid one. If such a sequence is implemented in the implementation, then there must exist a test run such that the sequence is observed. If the sequence is a valid trace, a **pass** verdict should be assigned to this test run, which implies that the sequence in the test case should be labeled with **pass**; no conclusion could be made if a test run completes before the end of the sequence, so all the proper prefixes of the sequence should be labeled with **inconclusive**. On the other hand, if the sequence is an invalid trace, a **fail** verdict should be assigned to this test run, which implies the sequence in the test case should be labeled with **fail**. Based on this reasoning, we conclude that all test cases must be of the following form:

Definition 6 (*Acceptance function*): Given an IOA specification S and T $=< \rho, \Omega, \ell >$, the acceptance function $\ell(\sigma)$, $\sigma \in \Omega$, satisfies:

$$\ell(\sigma) = \begin{cases} \textbf{pass} & (\sigma \in Tr(s_0) \cap pref(\rho)) \wedge \\ & (\sigma.a \in pref(\rho) \rightarrow \sigma.a \notin Tr(s_0)) \\ \textbf{fail} & \sigma \notin Tr(s_0) \\ \textbf{inconclusive} & \text{otherwise.} \end{cases}$$

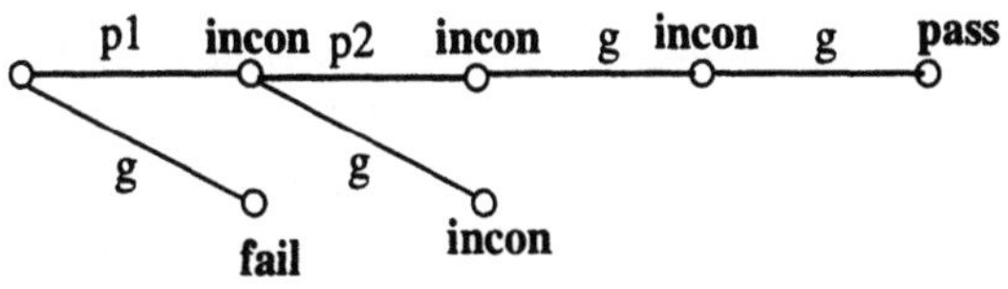

Figure 2 Test case for I/O sequence $p1.p2.g.g$

Obviously, if $\ell(\sigma) =$ **fail**, all the sequences in Ω which have σ as a prefix are labeled with **fail**. On the other hand, if $\ell(\sigma) =$ **pass**, all the sequences in Ω which are prefixes of σ are labeled with **inconclusive** and all the sequences in Ω which have σ as a proper prefix cannot be labeled with **pass** any more.

Theorem 1 *Given a test case* T *for IOA* S*, for any IOA* I*, if* $\mathsf{I} = \mathsf{S}$*, then* I *passes* T.

A test case with I/O sequence $p1.p2.g.g$ for the IOA given in Figure 1 is shown in Figure 2. Note that in the figure Ω is represented as a tree.

4 TEST GENERATION

We intend to apply the notion of state identification, which was originally defined in the realm of FSMs, for the IOA-based test derivation. In this section, we discuss how the notion is adapted to the realm of IOAs. Based on this, a test derivation method with full fault coverage is presented for an IOA specification.

4.1 State Identification in Specifications

Similar to the case of FSMs, in order to identify states in a given IOA specification, first the specification is required to have certain testability properties, one of which is the so-called reducibility.

Definition 7 (*Distinguishable states of an IOA*): Two states m and s of an IOA are said to be distinguishable if there exists $\sigma \in (X \cup Y)^*$, such that $m' = m$ **after** $\sigma, s' = s$ **after** σ and $out(m') \neq out(s')$.

Obviously, two distinguishable states are not trace equivalent. This means that for two distinguishable states, there exists a sequence of inputs and outputs, such that it is a trace for one of the two states, but it is not a trace for the other.

Definition 8 (*Reduced IOA*): An IOA is said to *reduced* if all its states are pairwise distinguishable.

Here we assume that the given IOA specification S is in the reduced form and has n states $s_0, s_1, \ldots s_{n-1}$.

There are several state identification notions that may be borrowed from the FSM-based testing. Here we only present the so-called *harmonized state identifiers* [5] for the IOA-based testing. The other notions of state identification could be defined similarly, as we did for test derivation based on labeled transition systems in [9].

Definition 9 (*harmonized state identifiers* (HSI)): $< H_0, H_1, \ldots, H_{n-1} >$ is said to be a set of harmonized state identifiers for an IOA S, if

- $H_i \subseteq (X \cup Y)^*@Y$ for $i = 0, 1, \ldots, n-1$;
- For each pair of different states $s_i \neq s_j$, there exists $\sigma \in pref(H_i) \cap pref(H_j)$ such that $\sigma \in Tr(s_i) \oplus Tr(s_j)$, where the operator $\oplus$ is denoted to be $A \oplus B = (A \backslash B) \cup (B \backslash A)$.

H_i is said to be a harmonized state identifier for state s_i. The harmonized state identifier for s_i captures the following property: for any other state s_j, there exists a sequence σ_i in $Pref(H_i)$ that distinguishes s_i from s_j and σ_i is also in $Pref(H_j)$.

Harmonized state identifiers always exist for any reduced IOA. As an example, for the IOA in Figure 1, we can choose the harmonized state identifiers $H_0 = \{g, p1.g.g\}, H_1 = \{g.g\}, H_2 = \{g, p1.g.g\}, H_3 = \{g.g\}$. Considering H_0: g is used to distinguish s_0 from s_1 and s_3, so g is also in $pref(H_1)$ and $pref(H_3)$; $p1.g.g$ is used to distinguish s_0 from s_2, so H_2 has $p1.g.g$.

4.2 State identification in implementations

Given an IOA specification S with n states and an IOA implementation I with m states, similar to the FSM-based testing, there are the two phases for the IOA-based testing. In the first phase, the state identification facility is applied to I to check if it can also properly identify the states in I. Once I passes the first phase, we can in the second phase test whether each transition and its tail state are correctly implemented.

In order to perform the first testing phase, proper transfer sequences are needed to bring I from the initial state to those particular states in I to which H_i should be applied. Moreover, it should be guaranteed that all the sequences in H_i are applied to the same particular state in I. Since a reliable reset is assumed, we can guarantee this in the following way: after a sequence in H_i is applied, the implementation I is reset to the initial state, and brought to the same particular state by the same transfer sequence, and then another sequence in H_i is applied. This process is repeated until all the sequences are applied.

Accordingly, let Q be a state cover for S, i.e. for each state s_i of S, there exists exactly one sequence σ such that $s_0 \text{-}\sigma\text{\to} s_i$, we can use $< N_0, N_1, \ldots N_{n-1} >$ to cover all states of I, where

$$N_i = \{\sigma \in Q@((X \cup Y)^0 \cup (X \cup Y)^1 \cup \ldots \cup (X \cup Y)^{m-n}) \mid s_0 \text{-}\sigma\text{\to} s_i\} \quad (1)$$

and construct a set of sequences to be executed by I from the initial state in the first testing phase as follows:

$$TS_1 = \bigcup_{i=0}^{n} N_i@H_i \tag{2}$$

Using TS_1, we can make test cases for the IOA S for the first testing phase according to Definitions 3 and 6. We call any I/O sequence that is used to make a test case for an IOA a *test sequence.*

After I passes all test cases made by TS_1, the states of I are partitioned into n groups, each of which is mapped by a state of S, written $f(s_k)$, $0 \leq k \leq n-1$. For all states i_j in the same group $f(s_k)$, if $\sigma \in H_k$ then $\sigma \in Tr(i_j)$, and for two states $i_k \in f(s_k)$ and $i_j \in f(s_j)$, there exists $\sigma \in H_i$ such that $\sigma \in Tr(i_i) \oplus Tr(i_j)$.

In the second phase of testing, for testing a given transition $s_i - a \rightarrow s_j$ in S, it is necessary to first bring I into each state $i_k \in f(s_i)$, then apply a at this state and see if a can be executed; moreover, let I be in i_l after a is executed, it is necessary to check that $i_l \in f(s_j)$ which should be verified by H_j. If any undefined output transition out of s_i has been implemented in I, an exception might occur, i.e. an unexpected output is observed. If a defined output transition out of s_i has not been implemented in I, a deadlock exception might occur. In the case of any **inconclusive** exception, testing is repeated.

Obviously, N_i may be used to bring I to any state $i_k \in f(s_i)$. Using this state cover, we can obtain a transition cover $< E_0, E_1, \ldots E_{n-1} >$ for the implementation, where

$$E_i = \{\sigma \in \bigcup_{k=0}^{n-1} (N_k@(X \cup Y)) \mid s_0 - \sigma \rightarrow s_i\} \tag{3}$$

Next, H_i is used to verify the tail states of E_i. Excluding the transitions that have been tested in the first testing phase, we can construct the set of test sequences for the second testing phase as follows:

$$TS_2 = \bigcup_{i=0}^{n-1} (E_i \backslash N_i)@H_i \tag{4}$$

We conclude that the set of test sequences is expressed as follow, by combining the two sets of test sequences for the first and second testing phases:

$$TS' = TS_1 \cup TS_2 = (\bigcup_{i=0}^{n-1} N_i@H_i) \cup (\bigcup_{i=0}^{n-1} (E_i \backslash N_i)@H_i) = \bigcup_{i=0}^{n-1} E_i@H_i \tag{5}$$

We have seen that the above process of checking experiments for IOAs is an analogue to the checking experiments for FSMs [10]. It is expected that a test suite which is derived from the IOA S based on the above process is complete for the fault model $\mathcal{F}(m)$. Below we formally present the test derivation algorithm.

4.3 Test selection algorithm

Following the testing process given in the above subsection, we have the following test generation algorithm.

Algorithm 1 *Test Generation Algorithm:*

Input: *An IOA* S *with* n *states in the reduced form and an upper bound* m *on the number of the states of IOA implementations.* $(n \leq m.)$

Output: *A test suite TS.*

Step 1: *Find a tuple of harmonized state identifiers* $\{H_0, H_1, \ldots, H_{n-1}\}$ *from* S.

Step 2: *Construct a minimal set* $Q \subseteq (X \cup Y)^*$ *such that* $\forall s_i \in S\ \exists \sigma \in V\ (s_0 - \sigma \rightarrow s_i)$.

Step 3: *Construct the sets* $< E_0, E_1, \ldots E_{n-1} >$ *such that* $E_i = \{\sigma \in Q@((X \cup Y)^0 \cup (X \cup Y)^1 \cup \ldots \cup (X \cup Y)^{m-n+1}) \mid s_0 - \sigma \rightarrow s_i\}$.

Step 4: *Construct the set of test sequences* $TS' = \bigcup_{i=0}^{n-1} E_i@H_i$

Step 5: *Construct a test case* T *for each* $\rho \in TS'$:

- *compute* Ω *according to Definition 3,*
- *label each* $\sigma \in \Omega$ *according to Definition 6.*

All the resulting test cases constitute the test suite TS.

The above method is an analogue of the FSM-based HSI-method [5]. The following theorem guarantees that a test suite obtained with the above method has complete fault coverage for the fault model $\mathcal{F}(m)$.

Theorem 2 *Given an IOA specification* S *in the reduced form and the fault model* $\mathcal{F}(m)$, *a test suite obtained with Algorithm 1 is complete for* S.

If we do not consider the optimization of state cover and state identification, for given n and m, the complexity of this algorithm is effective. The size of a test suite, i.e. the sum of the lengths of all I/O sequences obtained with the above test generation algorithm, is less than $n^3|X \cup Y|^{m-n+1}$.

For the other alternatives of state identification, such as W-set [3, 4] and UIO-sequences [7], the corresponding test generation methods could be devised similarly for IOAs. In [9, 10], we presented a list of the test generation methods based on the different notions of state identification for labeled transition systems.

4.4 Example

We use the IOA in Figure 1 which describes a slide window specification of size 2 as an example to the application of the test generation algorithm.

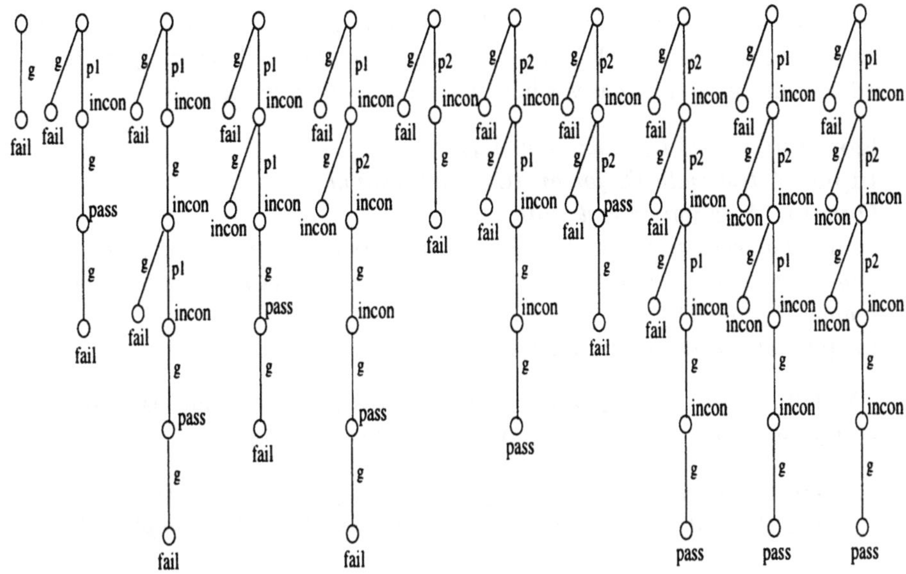

Figure 3 Test suite for a slide window specification

We assume the fault model to be $\mathcal{F}(4)$, i.e. implementations considered are transition-deterministic IOAs with at most 4 states and with the alphabet $X = \{p1, p2\}$ and $Y = \{g\}$. This IOA specification is reduced, and it has HSI identifiers

$$H_0 = \{g, p1.g.g\}, H_1 = \{g.g\}, H_2 = \{g, p1.g.g\}, H_3 = \{g.g\}.$$

If we chose $Q = \{\varepsilon, p1, p2, p1.p2\}$ as a state cover for the specification, then we have

$$\begin{aligned} E_0 &= \{\varepsilon, p1.g\}, & E_1 &= \{p1, p1.p1, p1.p2.g\} \\ E_2 &= \{p2, p2.p2\}, & E_3 &= \{p1.p2, p2.p1, p1.p2.p1, p1.p2.p2\} \end{aligned}$$

and further a set of test sequences

$$\begin{aligned} TS' = \; & \{g, p1.g.g, p1.g.p1.g.g, p1.p1.g.g, p1.p2.g.g.g, p2.g, p2.p1.g.g, p2.p2.g, \\ & p2.p2.p1.g.g, p1.p2.p1.g.g, p1.p2.p2.g.g\}. \end{aligned}$$

Note that in TS' we remove those sequences which are prefixes of the others.

Using TS', we can constitute a test suite for the slide window specification, as shown in Figure 3. This test suite is complete for $\mathcal{F}(4)$.

4.5 Test optimization

From the example in the above subsection, we can see that in the resulting test suite, even though we have removed those test sequences which are prefixes of other test sequences, there are still some test cases which are implied by others. For example, the test case with the test sequence g is implied by any other test cases, because its possible observation set Ω is contained in the possible observation set Ω' of another test case. The following algorithm is used to remove the test cases which are implied by other test cases.

Algorithm 2 *Test Optimization Algorithm*
Input: *A test suite TS for an IOA specification.*
Output: *A optimized test suite.*
Step1 *For each test case* $\mathsf{T} \in TS$*, let* ρ *be its test sequence, if there exists another test case* TS' *with the possible observation set* Ω' *such that* $\rho \in \Omega'$*, remove* T *from TS.*
Step2 *Return the resulting TS as the output.*

The following theorem guarantees that, if the given test suite is complete, then the test suite obtained with Algorithm 2 is also complete.

Theorem 3 *Given a test suite TS for an IOA specification* S*, if TS is complete then the test suite obtained with Algorithm 2 is also complete for* S.

For the test suite derived in the example in the above subsection, after the test optimization algorithm is applied, the resulting test suite has the set of test sequences:

$$TS' = \{p1.g.p1.g.g, p1.p1.g.g, p1.p2.g.g.g, p2.p1.g.g, p2.p2.p1.g.g, p1.p2.p1.g.g, p1.p2.p2.g.g\}.$$

The four test cases with the test sequnces $g, p1.g.g, p2.g$ and $p2.p2.g$ are removed by the algorithm.

5 CONCLUSION

This paper deals with conformance testing of communication systems which are specified in the formalism of I/O automata. The I/O automaton model is a more powerful model than the I/O finite state machine model (FSM), which is extensively used in conformance testing of communication systems. A framework has been introduced for testing transition-deterministic I/O automata for trace equivalence. In this framework, test cases, test execution, verdict assignment and fault model are defined. The fault model is a set of mutants of a given specification with at most m states, where m is a known integer. The notion of state identification, called harmonized state identifiers (HSI), which was originally defined in FSMs, has been introduced for test derivation for I/O automata. Based on the state identification technique, a test generation method with complete fault coverage, which is an analogue of the FSM-based HSI-method, has been presented.

Although we only introduced one notion of state identification, the other notions of state identification, such as a W-set and UIO-sequences, could also be defined similarly, and thus a set of useful and competing test derivation methods with fault coverage could be developed for testing specifications modeled by I/O automata.

6 REFERENCES

[1] G. v. Bochmann and A. Petrenko. (1994) Protocol testing: Review of methods and relevance for software testing. In *ACM 1994 Intern. Symp. on Software Testing and Analysis*, 109–124.

[2] E. Brinksma. (1988) A theory for the derivation of tests. In *IFIP PSTV VIII*, 63–74.

[3] T. S. Chow. (1978) Testing software design modeled by finite-state machines. *IEEE Transactions on Software Engineering*, SE-4(3):178–187.

[4] S. Fujiwara et al. (1991) Test selection based on finite state models. *IEEE Transactions on Software Engineering*, SE-17(6):591–603.

[5] A. Petrenko. (1991) Checking experiments with protocol machines. In *IFIP 4th IWPTS*, 83–94.

[6] M. Phalippou. (1993) Executable testers. In *IFIP 6th IWPTS*, 35–50.

[7] K. Sabnani and A. T. Dahbura. (1988) A protocol test generation procedure. *Computer Networks and ISDN Systems*, 15(4):285–297.

[8] R. Segala. (1993) Quiescence, fairness, testing, and the notion of implementation. In *CONCUR'93*.

[9] Q. M. Tan, A. Petrenko, and G. v. Bochmann. (1997) Checking experiments with labeled transition systems for trace equivalence. In *IFIP 10th IWTCS*.

[10] Q. M. Tan. (1998) On conformance testing of systems communicating by rendezvous. Ph.D. thesis, University of Montreal, Canada.

[11] J. Tretmans. (1995) Testing labelled transition systems with inputs and outputs. In *IFIP 8th IWPTS*.

[12] L. Heerink, and J. Tretmans. (1997) Refusal testing for classes of transition systems with inputs and outputs. In *FORTE X/PSTV XVI*.

7 BIOGRAPHY

Qiang-Ming Tan received the B.S. degree and the M.S degree in computer science from Chongqing University, Chongqing, China, in 1982 and 1984, respectively. Since 1993, he has been with the Université de Montréal, PQ, Canada for the Ph.D. degree in conformance testing on communication protocols. From 1984 to 1992, he was a lecturer in the Department of Computer Science of Chongqing University. Now he is also working with Claremont Technology Inc. Canada.

Alexandre Petrenko Alexandre Petrenko received the Dipl. degree in electrical and computer engineering from Riga Polytechnic Institute and the Ph.D.

in computer science from the Institute of Electronics and Computer Science, Riga, USSR. In 1996, he has joined CRIM, Centre de Recherche informatique de Montréal, Canada. He is also an adjunct professor of the Université de Montréal, where he was a visiting professor/researcher from 1992 to 1996. From 1982 to 1992, he was the head of a research department of the Institute of Electronics and Computer Science in Riga. From 1979 to 1982, he was with the Networking Task Force of the International Institute for Applied Systems Analysis (IIASA), Vienna, Austria. His current research interests include high-speed networks, communication software engineering, formal methods, conformance testing, and testability.

APPENDIX

We here only give the proof of Theorem 2. Given an IOA specification S and an IOA implementation M, we assume the following:
(1) All states of S and M are reachable from the initial state s_0 and m_0, respectively.
(2) $\overline{\mathsf{S}}$ is the reduced form of S and has at most n states with $n > 1$.
(3) $\overline{\mathsf{M}}$ is the reduced form of M and has at most m states with $m \geq n$, i.e. $\overline{\mathsf{M}} \in \mathcal{F}(m)$.
(4) $\overline{s}_i, \overline{s}_j, \overline{s}_k, \overline{s}_l$ and $\overline{m}_i, \overline{m}_j, \overline{m}_k, \overline{m}_l$ represent the states of $\overline{\mathsf{S}}$ and $\overline{\mathsf{M}}$, respectively.
(5) A tuple of harmonized state identifiers $\{H_0, H_1, \ldots, H_{n-1}\}$.
(6) Q is a state cover for $\overline{\mathsf{S}}$.
(7) TS is a test suite obtained with Algorithm 1 and TS' is the set of test sequences for TS.

Definition 10 *V–equivalence.* Given a set $V \subseteq \Sigma^*$, the V–equivalence relation between two states p and q, written $p =_V q$, holds if and only if for all $\sigma \in pref(V)$, $\sigma \in Tr(p) \Leftrightarrow \sigma \in Tr(q)$.
Given two IOAs S and M with initial states s_0 and m_0 respectively, we say that M is V-equivalent to S, written S $=_V$ M, if only if $s_0 =_V m_0$.

notation	meaning
$[\overline{s}_i, \overline{m}_i]{-a\rightarrow}[\overline{s}_j, \overline{m}_j]$	For $a \in \Sigma$, $\overline{s}_i {-a\rightarrow} \overline{s}_j$ and $\overline{m}_i {-a\rightarrow} \overline{m}_j$
$[\overline{s}_i, \overline{m}_i]{-\sigma\rightarrow}[\overline{s}_j, \overline{m}_j]$	For $\sigma \in \Sigma^*$, $\overline{s}_i {-\sigma\rightarrow} \overline{s}_j$ and $\overline{m}_i {-\sigma\rightarrow} \overline{m}_j$
$[\overline{s}_i, \overline{m}_i]$ **after** V	given a pair of states $[\overline{s}_i, \overline{m}_i] \in \overline{S} \times \overline{M}$, and a set $V \subseteq \Sigma^*$ $[\overline{s}_i, \overline{m}_i]$ **after** $V = \{[\overline{s}_j, \overline{m}_j] \mid \exists \sigma \in pref(V)\ ([\overline{s}_i, \overline{m}_i]{-\sigma\rightarrow}[\overline{s}_j, \overline{m}_j])\}$
D	$D = [\overline{s}_0, \overline{m}_0]$ **after** Σ^*
D_r	$D_r = \{[\overline{s}_i, \overline{m}_j] \in D \mid \overline{s}_i =_{H_i} \overline{m}_j\}$
$\overline{\Sigma}^k$	$\overline{\Sigma}^k = \bigcup_{i=0}^{k} \Sigma^i$

Lemma 1 *For* $V \subseteq \Sigma^*$, *assume* $|[\overline{s}_0, \overline{m}_0]$ **after** $V| \geq k$. *If* $|D| > k$, *then*

$|[\overline{s}_0, \overline{m}_0]$ **after** $V.\overline{\Sigma}^1| \geq k+1$; *if* $|D| \leq k$, *then* $[\overline{s}_0, \overline{m}_0]$ **after** $V@\overline{\Sigma}^1 =$ $[\overline{s}_0, \overline{m}_0]$ **after** V.

Proof:
(I) To prove that the lemma holds when $|D| > k$.
The lemma holds when $|[\overline{s}_0, \overline{m}_0]$ **after** $V| > k$. Consider $|[\overline{s}_0, \overline{m}_0]$ **after** $V| = k$.

(1)	$|D| > k$ and $|[\overline{s}_0, \overline{m}_0]$ **after** $V| = k$	hypothesis
(2)	$[\overline{s}_0, \overline{m}_0]$ **after** $V \subseteq D$	definition of D
(3)	$\exists[\overline{s}_k, \overline{m}_k] \in D \backslash [\overline{s}_0, \overline{m}_0]$ **after** V	(1),(2)
	$\exists[\overline{s}_i, \overline{m}_i] \in [\overline{s}_0, \overline{m}_0]$ **after** V	(1)
	$\exists \sigma \in pref(V)\ \exists \sigma.a \in \Sigma^*$	
	$([\overline{s}_0, \overline{m}_0] - \sigma \rightarrow [\overline{s}_i, \overline{m}_i] - a \rightarrow [\overline{s}_k, \overline{m}_k])$	(2)
(4)	$[\overline{s}_k, \overline{m}_k] \in [\overline{s}_0, \overline{m}_0]$ **after** $V@\overline{\Sigma}^1 \backslash [\overline{s}_0, \overline{m}_0]$ **after** V	(3)
(5)	$[\overline{s}_0, \overline{m}_0]$ **after** $V@\overline{\Sigma}^1 \geq k+1$	(4).

(II) To prove that the lemma holds when $|D| \leq k$.

(1)	$|D| \leq k$ and $|[\overline{s}_0, \overline{m}_0]$ **after** $V| = k$	hypothesis
(2)	$[\overline{s}_0, \overline{m}_0]$ **after** $V \subseteq D$	definition of D
(3)	$[\overline{s}_0, \overline{m}_0]$ **after** $V@\overline{\Sigma}^1 = [\overline{s}_0, \overline{m}_0]$ **after** V	(1),(2).

Lemma 2 *Assume* $\overline{s}_0 =_Q \overline{m}_0$. *If* $|D| > m$, *then* $|[\overline{s}_0, \overline{m}_0]$ **after** $Q@\overline{\Sigma}^{m-n}| \geq m$; *and if* $|D| \leq m$, *then* $[\overline{s}_0, \overline{m}_0]$ **after** $Q@\overline{\Sigma}^{m-n} = D$.

Proof:
(I) To prove that the lemma holds when $|D| > m$.

(1)	$\overline{s}_0 =_Q \overline{m}_0$ and $|D| > m$	hypothesis
(2)	$|[\overline{s}_0, \overline{m}_0]$ **after** $Q| \geq n$	initially connected $\overline{S}$, (1)
(3)	$|[\overline{s}_0, \overline{m}_0]$ **after** $Q@\overline{\Sigma}^{m-n}| \geq m$	Lemma 1, (1),(2).

(II) It is evident from Lemma 1 when $|D| \leq m$.

Lemma 3 *If* $\overline{s}_i =_{H_i} \overline{m}_k$ *and* $\overline{s}_j =_{H_j} \overline{m}_k$, *then* $i = j$.

Proof:

(1)	For $V \subseteq \Sigma^*, \overline{s}_i =_V \overline{m}_k \Leftrightarrow \overline{s}_i =_{pref(V)} \overline{m}_k$	evident
(2)	$\overline{s}_i =_{H_i} \overline{m}_k$ and $\overline{s}_j =_{H_j} \overline{m}_k$	hypothesis
(3)	$\overline{s}_i =_{pref(H_i)} \overline{m}_k$ and $\overline{s}_j =_{pref(H_j)} \overline{m}_k$	(1),(2)
(4)	$i \neq j$	assumption
(5)	$\exists \sigma \in (Tr(\overline{s}_i) \oplus Tr(\overline{s}_j)) \cap pref(H_i) \cap pref(H_j)$	definition of H_i, (4)
(6)	let $\sigma \in Tr(\overline{s}_i)$, then $\sigma \in Tr(\overline{m}_k)$	(3)
(7)	$\sigma \in Tr(\overline{s}_j)$	(3),(6)
(8)	$i = j$	(6),(7)$\not\Leftrightarrow Tr(\overline{s}_i) \oplus Tr(\overline{s}_j)$.

Lemma 4 $|D_r| \leq m$.

Proof:

(1)	$\lvert\overline{M}\rvert \le m$	hypothesis
(2)	$\lvert D_r\rvert > m$	assumption
(3)	$\exists[\overline{s}_i, \overline{m}_k], [\overline{s}_j, \overline{m}_k](i \ne j, \overline{s}_i =_{H_i} \overline{m}_k \wedge \overline{s}_j =_{H_j} \overline{m}_k)$	(1),(2)
(4)	$\lvert D_r\rvert \le m$	(3)$\not\Leftrightarrow$Lemma 3 (definition of H_i).

Lemma 5 *If* $\overline{s}_0 =_{TS'} \overline{m}_0$*, then* $[\overline{s}_0, \overline{m}_0]$ **after** $Q@\overline{\Sigma}^{m-n} = D_r = D$.

Proof:

(I) To prove that the lemma holds when $\lvert D\rvert \le m$.

(1)	$\lvert D\rvert \le m$	hypothesis
(2)	$\overline{s}_0 =_{TS'} \overline{m}_0$	hypothesis
(3)	$\overline{s}_0 =_Q \overline{m}_0$	(2)
(4)	$[\overline{s}_0, \overline{m}_0]$ **after** $Q@\overline{\Sigma}^{m-n} = D$	(1),(3),Lemma 2
(5)	$\forall[\overline{s}_i, \overline{m}_j] \in [\overline{s}_0, \overline{m}_0]$ **after** $Q@\overline{\Sigma}^{m-n}(\overline{s}_i =_{H_i} \overline{m}_j)$	(2)
(6)	$D = Dr$	(4),(5),definition of D_r.

(II) To prove that the lemma holds when $\lvert D\rvert > m$.

(1)	$\lvert D\rvert > m$	assumption
(2)	$\overline{s}_0 =_{TS'} \overline{m}_0$	hypothesis
(3)	$[\overline{s}_0, \overline{m}_0]$ **after** $Q@\overline{\Sigma}^{m-n+1} \subseteq D$	definition of D
(4)	$\forall[\overline{s}_i, \overline{m}_j] \in [\overline{s}_0, \overline{m}_0]$ **after** $Q@\overline{\Sigma}^{m-n+1}$ $(\overline{s}_i =_{H_i} \overline{m}_j)$	(2)
(5)	$[\overline{s}_0, \overline{m}_0]$ **after** $Q@\overline{\Sigma}^{m-n+1} \subseteq D_r$	(3),(4),definition of D_r
(6)	$\lvert[\overline{s}_0, \overline{m}_0]$ **after** $Q@\overline{\Sigma}^{m-n+1}\rvert \ge m+1$	(1),(2),Lemma 2,Lemma 1
(7)	$\lvert D_r\rvert \ge m+1$	(3),(4)
(8)	$\lvert D\rvert \le m$	(5)$\not\Leftrightarrow$Lemma 4
(9)	$[\overline{s}_0, \overline{m}_0]$ **after** $Q@\overline{\Sigma}^{m-n} = D_r = D$	(6),Lemma 2.

Lemma 6 *If* $\overline{m}_0 =_{TS'} \overline{s}_0$*, then* $\overline{m}_0 =_{Tr(\overline{s}_0)} \overline{s}_0$.

Proof:

(1)	$\overline{m}_0 =_{TS'} \overline{s}_0$	hypothesis
(2)	$\forall[\overline{s}_i, \overline{m}_i] \in D\ \exists\sigma \in Q@\overline{\Sigma}^{m-n}\ ([\overline{s}_0, \overline{m}_0]{-}\sigma{\rightarrow}[\overline{s}_i, \overline{m}_i])$	(1),Lemma 5
(3)	$\overline{m}_i =_{out(\overline{s}_i)} \overline{s}_i$	(1)
(4)	$not(\overline{m}_0 =_{Tr(\overline{s}_0)} \overline{s}_0)$	assumption
(5)	$\exists[\overline{s}_i, \overline{m}_i] \in D\ \exists y \in out(\overline{s}_i)\ not(\overline{m}_i =_{\{y\}} \overline{s}_i)$	(4)
(6)	$\overline{m}_0 =_{Tr(\overline{s}_0)} \overline{s}_0$	(5)$\not\Leftrightarrow$(3).

Lemma 7 *If* $\overline{m}_0 =_{TS'} \overline{s}_0$ *and* $\overline{m}_0 \ne \overline{s}_0$*, then there exist* $\mathsf{T} \in TS$ *and* $\sigma \in \Omega$ *such that* $\ell(\sigma) =$ **fail** *and* $\sigma \in Tr(\overline{m}_0)$.

Proof:

(1)	$\overline{s}_0 =_{TS'} \overline{m}_0$	hypothesis
(2)	$\forall[\overline{s}_i, \overline{m}_i] \in D\ \exists\sigma \in Q@\overline{\Sigma}^{m-n}\ ([\overline{s}_0, \overline{m}_0]{-}\sigma{\rightarrow}[\overline{s}_i, \overline{m}_i])$	(1),Lemma 5
(3)	$\overline{m}_i =_{out(\overline{s}_i)} \overline{s}_i$	(1)

(4)	$\overline{m}_0 \neq \overline{s}_0$	hypothesis
(5)	$\exists y \in out(\overline{m}_i) \backslash out(\overline{s}_i)\ \exists [\overline{s}_i, \overline{m}_i] \in D\ \exists \sigma' \in prep(TS')$ $([\overline{s}_0, \overline{m}_0] {-}\sigma'{\rightarrow} [\overline{s}_i, \overline{m}_i] \wedge \sigma'.y \in Tr(\overline{m}_0))$	(1),(3),(4)
(6)	$\exists \mathsf{T} \in TS\ (\sigma' \in \rho \wedge \sigma'.y \in \Omega)$	(5),Definition 3
(7)	$\ell(\sigma'.y) = \mathbf{fail}$	(6), Definition 6

Lemma 8 $\overline{\mathsf{M}}$ *passes TS iff* $\overline{m}_0 = \overline{s}_0$.

Proof:

(1)	$\overline{m}_0 = \overline{s}_0 \Longrightarrow \overline{\mathsf{M}}$ passes TS	Theorem 1
(2)	$\overline{\mathsf{M}}$ passes TS	hypothesis
(3)	$\overline{m}_0 \neq \overline{s}_0$	assumption
(4)	1) $not(\overline{m}_0 =_{TS'} \overline{s}_0)$ or 2) $\overline{m}_0 =_{TS'} \overline{s}_0 \wedge \overline{m}_0 \neq \overline{s}_0$	(3)
(5)	$not(\overline{m}_0 =_{TS'} \overline{s}_0)$	(4) 1)
(6)	$\exists \sigma \in TS'\ (\sigma \in Tr(\overline{m}_0) \backslash Tr(\overline{s}_0)$ or $\sigma \in Tr(\overline{s}_0) \backslash Tr(\overline{m}_0))$	Lemma 6
(7)	$\exists \mathsf{T} \in TS\ (\sigma = \rho)$	T made by σ
(8)	$\ell(\sigma) = \mathbf{fail} \wedge \sigma \in Obs_{(\mathrm{T},\overline{\mathrm{M}})}$ or $\ell(\sigma) = \mathbf{pass} \wedge \sigma \notin Obs_{(\mathrm{T},\overline{\mathrm{M}})}$	Definition 6, (6)
(9)	$\overline{m}_0 =_{TS'} \overline{s}_0 \wedge \overline{m}_0 \neq \overline{s}_0$	(4) 2)
(10)	$\exists \mathsf{T} \in TS \wedge \sigma \in \Omega\ (\ell(\sigma) = \mathbf{fail} \wedge \sigma \in Obs_{(\mathrm{T},\overline{\mathrm{M}})})$	Lemma 7
(11)	M fails T, i.e. M fails TS	Definition 4,(8),(10)
(12)	$\overline{m}_0 = \overline{s}_0$	(11)$\not\Leftrightarrow$(2).

Theorem 2 M passes TS iff $m_0 = s_0$.

Proof:

(1)	$s_0 = \overline{s}_0$, $m_0 = \overline{m}_0$	reduced forms of S and M
(2)	$\overline{\mathsf{M}}$ passes TS iff $\overline{m}_0 = \overline{s}_0$	Lemma 8
(3)	M passes TS iff $\overline{\mathsf{M}}$ passes TS	(1)
(4)	$m_0 = s_0$ iff $\overline{m}_0 = \overline{s}_0$	(1)
(5)	M passes TS iff $m_0 = s_0$	(3),(4).

Distinguishing tests for nondeterministic finite state machines

S. Yu. Boroday
Institute of Applied Mathematics and Mechanics,
National Academy of Sciences of Ukraine,
74, R.Luxemburg street, Donetsk, 340114, Ukraine
Tel: (038) 622 510 144
E-mail: math@iamm.ac.donetsk.ua

Abstract

In this paper, testing of deterministic implementations of nondeterministic specification FSMs is considered. Given two nondeterministic FSMs, a black box deterministic FSM is known to be a correct implementation of at least one them. We want to derive a test that determines whether this black box is a correct implementation of the first NDFSM. No upper bound on the number of states of the black box is known. The necessary and sufficient conditions for test existence are found. A method for constructing a conditional test of a minimal length is proposed. Upper bounds of multiplicity, length and overall length close to minimal are obtained.

Keywords

Finite state machine, nondeterministic finite state machine, distinguishing test, conformance testing.

1 INTRODUCTION

FSM based languages are widely used for protocol specification. Protocol conformance testing is often formalised as a problem of verifying the equivalence of a deterministic implementation to a given deterministic specification machine.

Since most existing protocols allow alternatives and options, a nondeterministic model of specifications seems to be a more versatile model for describing protocols [Petrenko, Yevtushenko, Bochmann]. An implementation machine in this paradigm is still deterministic. The implementation FSM conforms to a given nondeterministic specification FSM (NDFSM) if the implementation machine produces an output sequence that can also be produced by the specification machine in response to every input sequence. Unlike testing of deterministic FSMs, testing of NDFSMs has not been studied sufficiently well.

Conformance testing of deterministic implementations against NDFSMs is considered in [Petrenko, Yevtushenko, Lebedev, Das]. An upper bound on the implementation's state number is assumed. In [Petrenko, Yevtushenko, Bochmann] tests distinguishing states of NDFSM were also studied. Based on these tests, a method for deriving test suites from a given NDFSM complete within the given number of states is proposed in that paper.

In [Lukjanov], the following problem is considered. A black box DFSM is known to be a correct implementation of at least one of the two NDFSMs *A* and *B*. The experimenter wants to test a black box DFSM to determine whether or not the DFSM is a correct implementation of *A*. No upper bound on the number of states of the black box is given. The black box can be a correct implementation of one of them only, or it can correctly implement each of them at the same time. A test that determines whether a black box is a correct implementation of the NDFSM *A* is called a distinguishing test. In [Lukjanov], sufficient conditions for test existence and a test derivation method for some cases of specification machines are obtained.

In this paper, we address the above problem and obtain necessary and sufficient conditions. A method for deriving a conditional test with a minimal length and multiplicity is proposed. Upper bounds on length, multiplicity and overall length are obtained. These bounds are close to minimal.

2 PRELIMINARIES

Definition 1. *A nondeterministic finite state machine (NDFSM)* is a quintuple $A=(S,X,Y,F,s_0)$ where S, X, Y are finite and nonempty sets of states, input and output symbols, respectively, s_0 is an initial state, $F\colon X{\times}Y \to 2^{(S\times Y)}$ is a behaviour function.

Let sp denote a set of states in which an NDFSM can move from state s upon input word $p\in X^*$. $s(p,q)$ denotes the set of states, in which the NDFSM can move from s upon input word $p \in X^*$ with output word $q \in Y^*$. $\lambda\,(s,p)$ denotes a set of all output words, which the NDFSM can produce from the state s upon input p. λ_A denotes the set of all input output words $p,\lambda_A(s_0,\ p)$ and is called the FSMs behaviour. A deterministic FSM R is a correct implementation of the NDFSM A if $\lambda_R \subseteq \lambda_A$. If $|s(x,y)|\leq 1$ for all $x\in X$, $y\in Y$, $s\in S$, the NDFSM is observable. If $s(x,y)=\varnothing$

for some $x \in X$, $y \in Y$, $s \in S$, the NDFSM is called partially defined, otherwise it is called completely defined. It is known that for each NDFSM an observable NDFSM with the same behaviour exists. Deterministic acceptors (Rabin Scott automata) in the alphabet $X \times Y$ can be considered as a (partially defined) NDFSM with the input alphabet X, output alphabet Y and a set of final states. So notations extend on acceptors (Rabin Scott automata) in the alphabet $X \times Y$. $L(Q)$ denotes the set of words which the automaton Q accepts.

Let us consider the notion of a testing acceptor — a special case of a conditional test. An acyclic acceptor Q in the alphabet $X \times Y$ is called a single testing acceptor if for every acceptor's nonfinal state s there exists only one x_s such that $sx_s \neq \varnothing$. The acceptor is called acyclic if $s(p,q)=s$ implies p,q are the empty words. In [Petrenko, Yevtushenko, Bochmann], a final state of the testing acceptor is called a 'fail' state. A testing acceptor corresponds to an algorithm that conducts a single adaptive experiment with the black box as follows.

1. Declare the initial state s_0 as a current state.
2. From the current state s submit an input signal x_s such that $sx_s \neq \varnothing$ to the black box.
3. Read the output signal y produced by the black box.
4. If $s(x_s, y) \neq \varnothing$, then assume state $s(x_s, y)$ as a current state and go to Step 2
5. If the current state is final, then return the result 'fail', else 'pass'.

The number of inputs x_s submitted during this test is called the length of the test.

Since the acceptor is acyclic, testing is always finite and the result is defined. It is obvious that the result is 'pass' iff $L(Q) \cap \lambda_R = \varnothing$.

Let Q_i be a maximal single testing acceptor which is a subautomaton of Q. This means that a single testing acceptor Q_i can be obtained from Q by 'erasing' some (or none) states and transitions and every Q' such that if Q_i is a subautomaton of Q' and Q' is a subautomaton Q then $Q' = Q_i$. Let $\{Q_1, \ldots, Q_N\}$ be the set of all such acceptors. This set is obviously finite. Define an arbitrary order on acceptors.

The following test procedure corresponds to a testing acceptor Q.

1. Order acceptors in a sequence $Q_1, \ldots, Q_N$.
2. $i := 1$, L be an empty set.
3. If there are $px, qy \in L$, $y' \in Y$ such that px, qy' is a prefix of a word from $L(Q_i)$ but px, qy is not then go to 6.
4. Reset the black box to the initial state.
5. Perform test Q_i. If the result is 'fail', go to 9.
6. $i := i + 1$.
7. if $i \leq n$ go to 1.
8. The result is 'pass'. Stop.
9. The result is 'fail'. Stop.

Step 3 is included to prevent redundant testing when the result of the test Q_i with black box is obviously 'pass'.

The maximal number of resets during testing is called the multiplicity of the test Q with the given black box. The maximal multiplicity of the test Q with an arbitrary FSM is called the multiplicity of the test Q. The maximal length of test Q_i during testing is called the length of the test Q. The maximal length of the test with an arbitrary FSM is called the length of the test Q. The number of all inputs offered to the black box is called the overall length of the test.

The order of acceptors $Q_1,...,Q_N$ can affect the test multiplicity and length in a general case.

It is obvious that the result is 'pass' iff $L(Q)\cap\lambda_R=\varnothing$. So more sophisticated algorithms for checking the intersection of $L(Q)$ and the black box behaviour could be developed to reduce the overall length of a test.

A testing acceptor Q is called a test distinguishing the NDFSM A from the NDFSM B if the result of the test Q with every correct implementation of A is 'pass' and the result of the test with a black box which is a correct implementation of B but is not a correct implementation of A is 'fail'. Initialised, completely defined implementation and specification FSMs only are considered in this paper.

The existence of a test distinguishing A from B does not imply the existence of a test distinguishing B from A. But if both tests exist we can test whether the given black box is a correct implementation of A, B or both of them.

In the next section, we consider the problems of the existence and derivation of NDFSM distinguishing tests.

3 TEST EXISTENCE

It is known that a distinguishing test exists not for all NDFSMs. The following theorem gives necessary and sufficient conditions for the test's existence:

Theorem 1. A test distinguishing an NDFSM A from B exists if and only if $[\lambda_B\setminus \lambda_A]\cap\lambda_{]A\cap B[}$ is a finite set.

Here $[W]$ denotes the closure of the set of words W under prefix (if $pp'\in W$ then $p\in[W]$). $]A\cap B[$ is the largest completely defined NDFSM whose behaviour is included in A and B behaviours. The FSM $]A\cap B[$ is the largest if the behaviour of any NDFSM C is included in the behaviours of A and B then NDFSM C behaviour is also included in the behaviour of $]A\cap B[$.

The conditions of Theorem 1 are constructive. Let us make some auxiliary constructions based on the state pair graph in order to prove the theorem.

1. Build an acceptor (the graph of A and B state pairs) in the alphabet $X\times Y$ with the state set $S^A\times S^B\cup\{f\}$, transitions are defined as follows:

$$(s^A,s^B)(x,y) = \begin{cases} (t^A,t^B), & \text{if } s^A(x,y)=t^A \text{ and } s^B(x,y)=t^B;\\ f, & \text{if } s^A(x,y)=\varnothing, \text{ but } s^B(x,y)\neq\varnothing. \end{cases}$$

2. i:=0, V_0:=$\{f\}$.
3. i:=i+1 and V_i:= V_{i-1}.
4. For $s^A\in S^A$, $s^B\in S^B$, do:

if $(s^A,s^B)\notin V_i$ and $(s^A,s^B)x \subseteq V_{i-1}$ for a certain x, then do $V_i := V_i \cup (s^A,s^B)$ and mark all transitions $((s^A,s^B),x,y)$ of the constructed automaton, where $y\in Y$.

5. If $V_i \neq \varnothing$, go to 3.
6. $V:=V_i$.
7. Delete for nonmarked transitions outgoing from all states from V, the obtained automaton with the final state f is denoted $D(A,B)$.
8. Delete all states from which f is not reachable.
9. Delete all marked transitions outgoing from V. The resulting automaton with V as the set of final states is denoted $KER(A,B)$.

In step 1, the acceptor which accepts the opening of the set of word $\lambda_B \setminus \lambda_A$ under prefix was constructed. The opening under prefix means the set of all words from $\lambda_B \setminus \lambda_A$ whose own prefix can not be found in $\lambda_B \setminus \lambda_A$. On step 2–5 all r-distinguishable state pairs are included in V. As shown in [Petrenko, Evtushenko, Bochmann] s_A and s_B are r-distinguishable if and only if there exist no correct implementation of B which is a correct implementation of A.

According to [Petrenko, Yevtushenko, Bochmann] and [Lukjanov], a testing acceptor $D(A,B)$ with the initial state $(s^A,s^B)\in V_i$ is a simple test distinguishing A and B.

The set $[\lambda_B \setminus \lambda_A] \cap \lambda_{]A\cap B[}$ in Theorem 1 is the set of prefixes of $L(KER)$. Since final states of KER do not have cycles, $[\lambda_B \setminus \lambda_A] \cap \lambda_{]A\cap B[}$ and $L(KER)$ are either both finite or both infinite at the same time. This implies the following theorem that is equivalent to Theorem 1.

Theorem 2. A distinguishing test for NDFSMs A and B exists if and only if $L(KER)$ is a finite set, i.e. KER is acyclic.

To prove the statement, we use the following proposition.

Proposition 1. Let $L\subseteq \lambda_A$ and $L\subseteq \lambda_C$ where A is an NDFSM, C is a DFSM and both FSMs are completely defined. Then L is included in the behaviour of a correct implementation of A.

Proof of Theorem 1. Conditions are obviously sufficient. We prove that they are necessary. Assume that $[\lambda_B \setminus \lambda_A] \cap \lambda_{]A\cap B[}$ is infinite and Q is a testing acceptor distinguishing A from B. Since the language is regular, there exist such words v, u, w in the alphabet $X\times Y$ such that $vu^* \subseteq \lambda_{]A\cap B[}$ but $vu^*w \subseteq \lambda_B \setminus \lambda_A$. Due to Proposition 1, there exists a correct implementation DFSM C of $]A\cap B[$, $vu^* \subseteq \lambda_C$. The DFSM C is a correct implementation of both NDFSMs A and B. Perform the test Q with C. Let the length of the test Q with the DFSM C be equal to k.

It is obvious that $\lambda_C \setminus vu^*(X\times Y)^* \cup vu^kw \subseteq \lambda_B$.

Let H be an implementation DFSM of B which includes $\lambda_C \setminus vu^*(X\times Y)^* \cup vu^kw$ in its behaviour as H. Such an implementation exists according to Proposition 1. H is not a correct implementation of A because $vu^*w \subseteq \lambda_B \setminus \lambda_A$. Thus, results of the test Q on H and C are different. Therefore, H and C have different outputs for at least one input word of length k. This contradicts the construction of H and C. The theorem has been proven.

4 TEST GENERATION

The following lemma can be proven similarly to Theorem 1.

Lemma 1. Let Q be a testing acceptor distinguishing A from B. Then every word from $L(KER)$ is a prefix of a word from $L(Q)$.

$D(A,B)$ with the initial state (s^A,s^B) from V is a minimal testing acceptor distinguishing A and B. Classes of correct implementations of states s^A, s^B do not intersect. From these facts and Lemma 1, the following theorem is obtained.

Theorem 3. If a test distinguishing NDFSM A from B exists then $D(A,B)$ is a distinguishing test of minimal length and multiplicity.

Theorem 3 obviously defines a test generation procedure. Lemmea 1 implies that the length and multiplicity of the test D in the worst case do not depend on the assumed order of simple tests $Q_1,\ldots,Q_N$.

Several testing acceptors with various overall lengths of corresponding tests can be constructed. To find an algorithm for constructing tests of a minimal overall length we needs a more complex conditional test model than a testing acceptor. But the latter can yet be useful.

By erasing outputs in the testing acceptor graph we also can obtain an unconditional test.

It is obvious that the test length, multiplicity and overall length do not exceed nm, $|X|^{nm}$, $nm|X|^{nm}$ respectively, where n and m are state numbers of NDFSMs. These upper bounds have the same order as minimal.

5 ACKNOWLEDGEMENTS

The author thanks Alexandre Petrenko and Igor Grunsky for discussions and helpful suggestions. The author would like also to thank them and anonymous reviewers for comments that helped improve the presentation of the paper.

6 REFERENCES

Boroday, S. Yu. (1995) Experiments in classes of NDFSM implementations. Proceeding of II Ukrainian conference on automatic control 'Automatika 95', Lvov, p 101.

Lukjanov B. D. (1995) On distinguishing and checking experiments with nondeterministic automata, *Kibernetika i systemny analiz,* **5**. Plenum Publishing Corporation. New York. pp. 56-66.

Petrenko, A., Yevtushenko, N., Lebedev, A., Das, A. (1993) Nondeterministic state machines for protocol testing. *IFIP Transactions Protocol Test Systems*

VI (the Proceedings of IFIP TC6 Sixth international Workshop on Protocol Test Systems 1993) North-Holland, 1994, pp.193-208.

Petrenko, A., Yevtushenko, N., Bochmann G. (1996) Testing deterministic implementations from nondeterministic FSM specifications. *IFIP Testing of Communicating Systems (the Proceedings of IFIP TC6 9th International Workshop on Protocol Test Systems 1996)* Chapman & Hall, 1996, pp125-140.

Starke, P.H. (1972) Abstract automata. North-Holland/American Elsevier, 419p.

7 BIOGRAPHY

Boroday Sergey is an engineer of the Institute of Applied Mathematics and Mechanics (Donetsk, Ukraine), the National Academy of Sciences of Ukraine. He received his Ph.D. degree in Math from the Saratov State University (Russia) in 1997. His main research interests are in automata theory and software, hardware and protocol testing.

PART FOUR

Testing in Context

Evaluation of test coverage for embedded system testing

Jinsong Zhu, Son T. Vuong
Department of Computer Science
University of British Columbia
Vancouver, B.C., Canada V6T 1Z4

Samuel T. Chanson
Department of Computer Science
Hong Kong University of Science and Technology
Clear Water Bay, Kowloon, Hong Kong

Abstract

In this paper we propose an original approach to the evaluation of test suites for embedded system testing, where the implementation under test (IUT) is embedded in a composite system as a component module. We define a coverage measure based on the identification of the IUT within the test context with respect to observational equivalence at the composite system level. The problem of limited IUT controllability and observability caused by the test context is handled when computing the coverage. The approach is purely functional and only assumes a general fault model where the number of states in the IUT is upper bounded. A tool has been developed and an example is given to illustrate and validate the approach.

Keywords

Protocol testing, embedded test, test in context, fault coverage

1 INTRODUCTION

With the development of distributed component computing, it has become increasingly important to test components that are embedded in a composite system. Often this has to be done at the composite system level due to the difficulty in isolating an embedded component that is fully integrated into the system. All components other than the one to be tested constitute the *test context* [7]. In such an environment, testing is performed by applying test suites at the system level in order to exercise the behavior of the target component since the interfaces of the embedded module are not directly accessible by the tester. The correctness of the component can only be inferred from the global system behavior, *i.e.,* system input and output events. Internal events between components act as constraints on the global behavior and are not directly controllable or observable. This type of testing is known as *embedded testing* as in the International Standards Organization (ISO) conformance testing framework for communication protocols [7].

Most work in the area of protocol testing has mainly been on single component testing where each component of the system is tested in isolation. Effectiveness of such tests can be evaluated in various ways, ranging from fault simulation [4, 15], structural analysis [17], to faulty machine identification [16, 18].

Recently some work aiming at testing embedded components (also called testing in context) has appeared [13, 14, 12, 9]. The work so far has mainly been on defining fault models and generating effective test suites with respect to given fault models. However, the problem of evaluating the coverage of a given test suite for embedded component testing has not been addressed. The objective of this paper is to propose a solution to this problem.

The traditional fault simulation method as used in fault coverage evaluation for isolated machines [4, 15] can be adapted for use in embedded systems by defining an appropriate fault model as in [13]. However, in many complex systems detailed fault classes are unknown or are difficult to be established. It is simply impossible to explicitly construct all faulty machines in a realistic situation as pointed out in [13]. We therefore propose a new approach to the evaluation of test coverage for embedded system testing that is based on faulty machine identification by given external behavior, *i.e.,* a test suite. No detailed fault classes are needed. In the approach, the components of a system are modeled as two communicating finite state machines (FSMs), one being the test context and the other being the IUT. The test context is assumed to be faultless. We identify the IUT component (an FSM) using both reachability computation and constraint based search, the latter has been effectively used in our earlier work on the identification of single FSMs [18]. The test coverage of the test suite is measured by the number of FSMs that are not conforming to the component specification. The approach increases the computing efficiency in two ways: 1) The global identification is avoided; 2) The constraint based search is optimized by domain reduction and backjumping. The approach can be readily extended to enhance the quality of test suites by generating additional test cases

The rest of the paper is organized as follows. Section 2 gives an overview of the general principles of embedded system testing and test architectures. Section 3 describes our approach of coverage measure in embedded testing. The next two sections provide a more formal description of the algorithm followed by an illustrative example, respectively. We close with a discussion of the proposed approach and some future work.

2 EMBEDDED TESTING ARCHITECTURE

We model a composite system as a set of communicating FSMs. Figure 1 shows a test architecture that depicts such an environment.

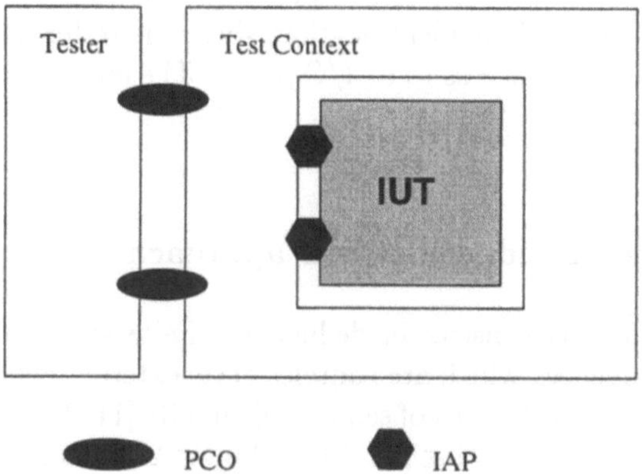

Figure 1 Test architecture for embedded component testing

Major components of the architecture are (see [10] for details):

- a tester;
- an implementation under test (IUT);
- test context;
- points of control and observation (PCO);
- implementation access points (IAP).

The tester contains a test engine that executes test suites and determines whether the IUT (the targeted component, also called Module Under Test, or MUT for brevity) conforms to its specification by observing system responses at the PCOs. The test context is a set of FSMs that are not being tested; instead they act as an environment in which the IUT runs, and which interacts with the IUT through the IAPs. The IAPs are not directly controllable or observable by the tester.

An example of embedded module is the Service Specific Connection Oriented Protocol (SSCOP) used in the B-ISDN ATM Adaption Layer (AAL) [8]. SSCOP is embedded in the AAL, on top of the Common Part Convergence Sublayer (CP-AAL) and below a Service Specific Coordination Function (SSCF) module. Testing of SSCOP needs to be an embedded testing if it is only accessible via the CP-AAL and SSCF.

3 TEST COVERAGE MEASURE

As mentioned before, our coverage measure is based on machine identification. The number of faulty machines that cannot be detected by a test suite reflects the effectiveness of the test suite. We will first give an overview of the measure as applied to single machine identification where the tester has full control of the IUT (*i.e.*, has direct access to its I/O ports). The measure is then extended to embedded component testing.

3.1 The machine identification approach

Machine identification consists in deducing a finite state machine from its input/output sequences which are samples of its external behavior. The issue was initially studied in the area of sequential circuits [11, 2, 5] and later used in communication protocol testing [16, 18, 3]. Figure 2 shows the different sets of machines that can be defined according to this method: S is the specification machine, SCM (the area of the inner-most circle) is the set of machines that conform to S, and TCM is the set of machines that are compatible with the test suite, *i.e.*, they are perceived as correct by the test suite. The shaded area thus represent those machines that do not conform to the specification but cannot be detected by the test suite. It is a measure of the *inefficiency* of the test suite. The coverage, or the *goodness* of the test suite, can be defined as the inverse of the size of the shaded area.

Definition 1 (Coverage measure) *Let T be a given test suite, S the specification, and n the upper bound of the number of states in any implementation of S. Let K_{TCM} be the number of T-conforming machines for S, and K_{SCM} is the number of S-conforming machines. The test coverage of T is defined as:*

$$Cov(S,T,n) = \frac{1}{K_{TCM} - K_{SCM} + 1}.$$

The "1" in the denominator is necessary since it is possible that $K_{TCM} = K_{SCM}$ which should be interpreted as complete coverage instead of infinite coverage, which does not make good mathematical sense. The definition also

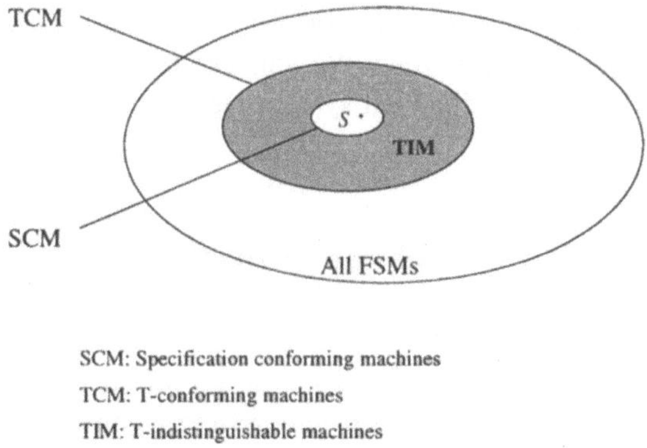

Figure 2 Machine identification with respect to a test suite

makes sure that if a test suite subsumes another one, it must have a higher coverage value.

To compute Cov, we need to:

- calculate K_{TCM}, i.e., to identify the number of machines that accept the test suite, and
- determine how many of these T-conforming machines actually conform to the specification.

To calculate K_{TCM}, identification and construction of the T-conforming machines are necessary. Traditional methods such as [11, 2] may be employed. Assumptions are usually made to make the problem solvable, First of all, the space of all FSMs must be limited since otherwise K_{TCM} may be infinite, causing a zero fault coverage for any test suite. A widely used assumption is the "upper-bound state" assumption where faults in the IUT can only increase the number of states to a predefined boundary. The upper bound should also be relatively small to make it manageable in practice, since the space of FSMs grows exponentially with the number of states. Secondly, in order to construct the T-conforming machines in a deterministic way, the IUT must be modeled as a deterministic finite state machine, using either the Moore or Mealy model (the latter is commonly used in communication protocol modeling).

The identification approach has the major advantages of generality and independence of detailed fault models such as output faults, head and tail state faults, state transfer faults, and their combinations. However, the identification process typically introduces high computational cost as the identification problem is known to be NP-complete with respect to the number of states [6]. In [18], heuristics developed in artificial intelligence have been employed in an effort to reduce the computing time and it has been shown to be effective for some moderate-size protocols [18]. This technique should be even more feasible at higher levels of abstraction for test design where the number of protocol states would be smaller. We will use this AI based method in our coverage calculation for embedded system testing.

3.2 Identification of embedded machines

In embedded system testing, the coverage could have been calculated by identifying the whole global system. However, this approach would not be interesting because it may run out of steam in handling a complex composite system and also violates our motto of "divide and conquer". Besides, when we assume the test context is faultless (which is perfectly reasonable in embedded system testing), it is simply unnecessary to identify the global system because a lot of energy would be wasted in identifying possible faults in the context.

To calculate the coverage in embedded system testing, we assume that only the embedded component (MUT) is faulty, the test context being thoroughly tested or for our purpose being simply faultless. The MUT is identified by the test suite applied to the whole system (see below). This approach avoids the identification of the global system and allows us to focus on the MUT for a better coverage.

After identification of the MUT, comparison must be made to determine whether it conforms to its specification. As explained in [13, 14], this comparison must be done at the composite system level, *i.e.*, the conformance of MUT must be defined by the global behavior after composition with its test context. This is because with limited controllability and observability of the MUT, we can only determine the conformance by its globally observable behavior. If an MUT does not conform to its specification at the component level, but the global behavior of the MUT in composition with the context conforms to the composition of its specification and the context, we can only conclude that the MUT works correctly *in that context*. This is often called *equivalence in context* [13]. The observational equivalence relation [1] applies in this situation and is used for determining the equivalence in context.

Besides the equivalence issue, we also need to impose the so-called *I/O ordering constraint* [13]. This constraint requires that the next input is submitted to the composite system only after it has produced an output in response to the previous input. This allows the system to have a finite number of steady states. Also, in order for the composition machine (the MUT and the context) to be valid, the product of MUT and context should not contain livelocks.

Since a test suite may consist of many test cases each starting from the initial system state, a special input *reset* is used to bring the system to its initial state. This initial state is a combination of the initial states of the context machine and the MUT. Whenever *reset* is encountered, both the context and MUT are set to its initial state, and no output (or a *null* output) will be produced. The faults in the IUT will not affect this capability.

Let M_I be the embedded MUT module to be identified given its specification S, the test context C, and the test suite TS. We consider TS as composed of a linear sequence of test steps (I/O pairs). If there are many test cases, they will simply be concatenated with *reset/null* to form the test suite.

Our approach in identifying M_I is to first build a test tree for M_I, which represents all sampled behaviors of M_I as excited by the test suite. This test tree is then collapsed for the identification of M_I using the approach proposed in [18]. We follow the following steps to build the test trees:

1. Apply a test step x/y (an I/O pair) in the test suite to the system. Since C is known, reachability computation can be performed until the input reaches M_I (we do not exclude the case where an input is directly applied to an embedded module by bypassing C; this is done simply by skipping the reachability computation part). For example, an input x applied to the system via C may produce an internal event in C as an input to M_I. A special case in this step is the *reset* signal. No reachability computation is needed; we simply reset the context machine and move to the root of the current test tree.
2. When an input reaches M_I, we start to construct a test tree for M_I. At this point, the output caused by that input from C may be any possible event that is allowed in M_I's output alphabet. Try each possible event according to the next step.
3. For each possible event o_p, continue the reachability computation of Step 1, until a global output is produced. If, during the reachability computation, an output of the context is an input back to M_I, Step 2 will be called. This will lead to a recursive execution of Steps 2 and 3. When a global output is finally produced, it is compared with the output of the test step. If they are different, a contradiction is found and hence o_p is ruled out. Repeat this step for the next possibility until a global output compatible with the test step is observed or no possibility is compatible. If a compatible output is found, the test tree goes one step farther with a label of i/o, where i is the internal input to M_I excited by the global input, and o is the compatible output. If no compatible output can be found, backtrack to the previous test step. The current test tree is also "rewound" accordingly.
4. If all test steps have been processed, we have obtained a test tree for M_I. At this point, we backtrack to the previous test step, with C and the test tree being set to their respective previous states, and continue searching for the next output possibility with respect to these states. This step is repeated until all output possibilities have been exhausted, resulting in all test trees for M_I.

If the test suite was correctly generated for S, we will end up with at least one test tree for M_I; otherwise the algorithm will produce no test trees which would indicate an incorrect given test suite. Figure 3 gives a simplified illustration of this process. Suppose a test step consists of x/y. The global input x causes a transition x/a in C, in which a serves as an internal input to M_I, which outputs y. An edge labeled a/y in M_I's test tree is then obtained. The dotted line from M_I to C indicates a possibility where M_I produces an output to C in response to its input. In general, a finite number of events may be exchanged between C and M_I before a global output is produced. Due to the assumption that there is no livelock between C and M_I, a global input will always cause a global output after its propagation within the system. However, this assumption may not hold for a faulty IUT when we explore all output possibilities of M_I. In such a case, we will terminate the algorithm after a predefined number of interactions between C and M_I, which should be large enough to signify a livelock. More sophisticated mechanism in detecting a livelock on-the-fly requires further investigation.

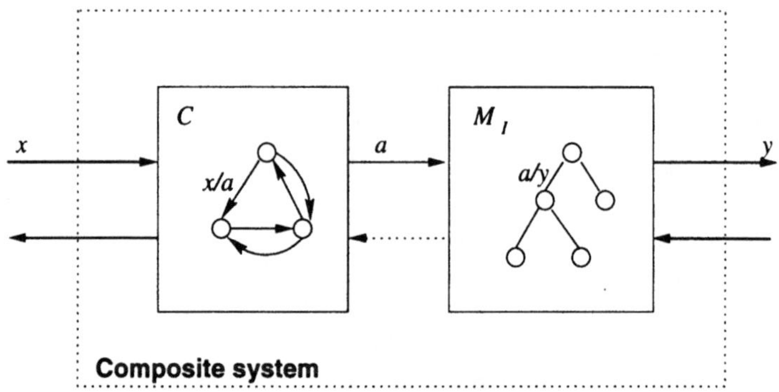

Figure 3 Test tree construction for embedded component

With the test trees constructed, the identification of M_I can be solved as a single machine identification problem using our tool COV as presented in [18]. Each test tree is fed to COV to build the corresponding M_I(s).

In order to determine whether an M_I is a conforming implementation, the final step is to compute the composition of C and M_I, denoted as $C \circ M_I$ [13]. It can be obtained from the product of C and M_I ($C \times M_I$) by hiding the internal actions since observational equivalence is used in our testing. The assumption that $C \times M_I$ has no livelock assures the existence of $C \circ M_I$. The composition is then compared with $C \circ S$ for observational equivalence. Those equivalent in the global behaviors are counted as K_{SCM}. This completes our calculation of K_{TCM} and K_{SCM}.

4 ALGORITHM

This section presents the above algorithm in more formal notations. Our formal model is the familiar Mealy machine frequently used in communication protocol modeling. A deterministic FSM is represented by a quintuple $M = \langle Q, X, Y, \delta, \lambda \rangle$, where Q, X, Y are the internal states, input alphabet and output alphabet respectively. $\delta : Q \times X \rightarrow Q$ is the next state function and $\lambda : Q \times X \rightarrow Y$ is the output function. The functions δ and λ can be extended to handle an input sequence $\sigma = x_1 x_2 ... x_k$ in the usual way: $\delta(q_1, \sigma)$ is the final state after σ is applied to state q_1, and $\lambda(q_1, \sigma)$ denotes the corresponding output sequence. That is, $\lambda(q_1, \sigma) = y_1 y_2 \cdots y_k$ where $y_i = \lambda(q_i, x_i)$ and $q_{i+1} = \delta(q_i, x_i)$ for $i = 1, ..., k$, and $\delta(q_1, \sigma) = q_{k+1}$.

A test case of length k is an element of $(X \times Y)^*$ and is denoted by $\tau_k = \langle x_1/y_1, ..., x_k/y_k \rangle$, which satisfies $\lambda(q_1, x_1 x_2 \cdots x_k) = y_1 y_2 \cdots y_k$. The special test step *reset/null* is given by $r/-$ for short. If a test suite contains a set of test cases each starting from the initial state, it can be considered as a linear test sequence formed by concatenating the test cases with $r/-$. In the following algorithm, we will not make distinction between a test suite and the test sequence obtained by concatenating the test cases in the test suite.

The inputs to our algorithm are the initial state of the MUT to be identified, the initial state of the context, and the test suite. In other words, we build our test tree starting from the initial state of the system. The output is the set of test trees for the MUT, obtained by applying the test suite to the system in the given test context. These trees are then used to infer the MUT as per the single machine identification algorithm [18]. The final composition of the inferred machines and the context is done to single out conforming solutions.

Figure 4 gives the algorithm for constructing the test tree. It is a recursive procedure that processes the test suite one test step each time. The exceptional case where a livelock may exist is not shown.

The first line checks if the whole test suite has been processed. If so the test tree has been constructed and is saved, and the algorithm backtracks to search for other solutions. Line 2 handles the special case of the reset signal, resulting in the initialization of the context as well as the current test tree. The current tree node points to the root. Line 5 applies reachability computation (RC) from the global input until the MUT module M_I is reached, which means an internal input for M_I is produced from the global input. In case where a global output is generated before M_I is reached, *i.e.*, this test step does not really test any part of M_I, the procedure backtracks to the next test step without adding any edge to the test tree. Otherwise, the algorithm proceeds to line 7 to search for an output symbol for a potential edge in the test tree. This is done by analyzing all possible outputs for M_I with regard to the present internal input. In Line 8, for each output possibility, if it is a system output and it is consistent with the output given in the test step, we have found a consistent edge for M_I. If it is a context input instead, the algorithm goes

Algorithm: Test tree construction for M_I.

$\sigma = \langle \sigma_1, ..., \sigma_n \rangle$: the test suite, where $\sigma_i = x_i/y_i$ is the ith test step;

v_i: represents tree node i;

q_i: represents state i of the context;

TT: the test tree being constructed.

boolean TestTree(σ_i, v_i, q_i)

{

1. **if** (σ_i = NULL) { save TT; **return**(FALSE);}

2. **if** ($\sigma_i = r/-$) { $v_i \leftarrow v_0; q_i \leftarrow q_0$; }

3. **else** {

4. $in \leftarrow x_i$;

5. RC from in to M_I; Context reaches q_j with input z;

6. **if** (($\lambda(q_j, z) = a_i$) is input to M_I) {

7. **for** $b_i \in Y$ **do** {

8. **if** ($b_i == y_i$) { add edge $v_i \stackrel{a_i/b_i}{\longrightarrow} v_j$ to TT; **break**; }

9. **else if** (b_i is input to C) { $in \leftarrow b_i$; **goto** Line 5 };

10. }

11. **if** (no compatible y found for all b_i) **return**(FALSE);

12. }

13. **else if** (a_i is system output **and** $a_i \neq y_i$) **return** (FALSE);

14. }

15. **return** TestTree(σ_{i+1}, v_j, q_j);

}

Figure 4 Test tree construction algorithm

back to Line 5 for further reachability computation. If all failed, we go on to try the next output b_i. If no compatible b_i can be found, which indicates an inconsistent b_i in a previous test step, the algorithm backtracks. Line 15 repeats the processing for the next test step.

5 TOOL AND EXAMPLE

To validate the approach, an experimental tool called ECOV (Figure 5) has been implemented in C on a Linux 2.0 Pentium machine. It is a tool set that consists of tools for test tree construction (TTREE), machine identification by the test tree (COV), and a composition algorithm (COMP) that checks the observational equivalence in context. In the tool set, COV was developed in [18] and is integrated into ECOV via I/O files. ECOV takes as input a test suite, a test context, and the MUT specification, and produces a coverage value and non-conforming machines which may be used for further analysis and test suite enhancement.

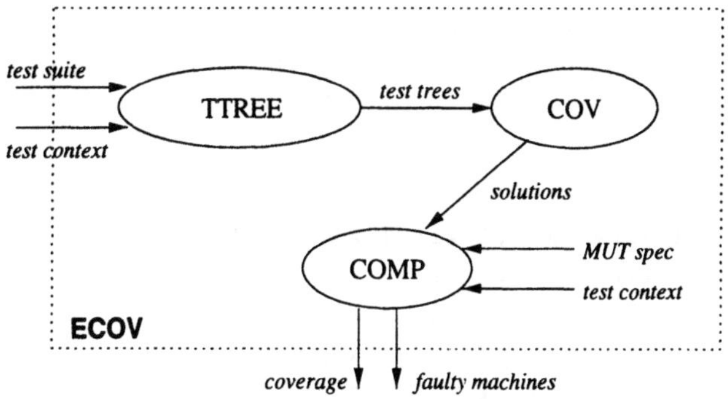

Figure 5 ECOV tool set

To illustrate the approach, we use the simple example in Figure 6 which is taken from [13]. The test context C takes inputs x_1, x_2 from the system environment and internal input z_1, z_2 from the component under test, and generates system outputs y_1, y_2. C also generates internal outputs u_1, u_2 as inputs to S. The composed machine $S \circ C$ is generated by determinizing the product machine $S \times C$ under the I/O ordering constraint.

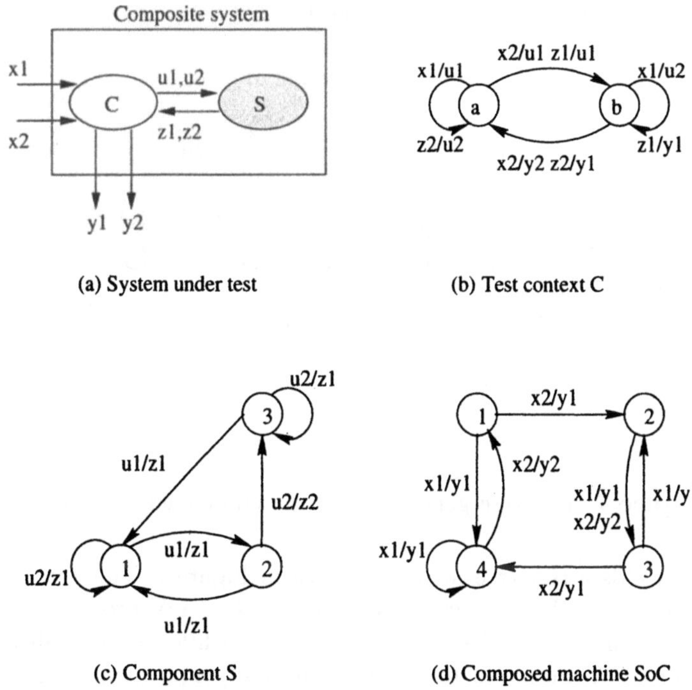

Figure 6 Example composite system

We use the following test suite from [12] for the component S. This is a complete test suite with respect to the fault model where faults in S's implementation do not increase the number of states:

```
T = {
    r/- x1/y1 x1/y1 x1/y1 x1/y1 x1/y1 x2/y2
    r/- x1/y1 x1/y1 x1/y1 x2/y2
    r/- x1/y1 x1/y1 x2/y2 x2/y1 x1/y1 x2/y1
    r/- x1/y1 x1/y1 x2/y2 x2/y1 x2/y2
    r/- x1/y1 x2/y2 x1/y1 x1/y1 x2/y2
    r/- x1/y1 x2/y2 x1/y1 x2/y2
    r/- x2/y1 x2/y2 x1/y1 x1/y1 x2/y1 x1/y1 x2/y2
    r/- x2/y1 x2/y2 x1/y1 x1/y1 x2/y1 x2/y2
    r/- x2/y1 x1/y1 x1/y1 x1/y1 x2/y1
    r/- x2/y1 x1/y1 x1/y1 x2/y2
    r/- x2/y1 x1/y1 x2/y1 x1/y1 x1/y1 x2/y2
    r/- x2/y1 x1/y1 x2/y1 x2/y2 x2/y1
    r/- x2/y1 x2/y2 x1/y1 x2/y2
}
```

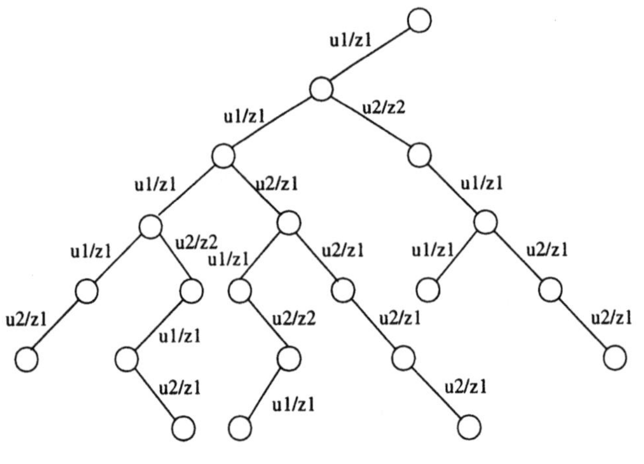

Figure 7 Test tree generated for the embedded component

Our tool yields a test tree for S as shown in Figure 7. This tree is further fed into our tool COV to be collapsed into FSMs. COV produces 19 FSMs, one of which conforms to S at the module level, while the rest, although do not conform to S, are found to be observationally equivalent with S within context C, *i.e.*, the composition of those FSMs conform to the composition of S and C. Therefore, there are no non-conforming solutions at the composite system level, which leads to the conclusion that T is indeed complete with respect to the embedded testing architecture.

Figure 8 shows two of the generated FSMs that do not conform with S per se but are conforming at the system level. The first one has only two states, yet it is considered conforming when tested in context. This allows us to discover the smallest FSM that implements S without causing system behavior discrepancy. The second one differs from S wildly but still exhibits observational equivalence at the system level.

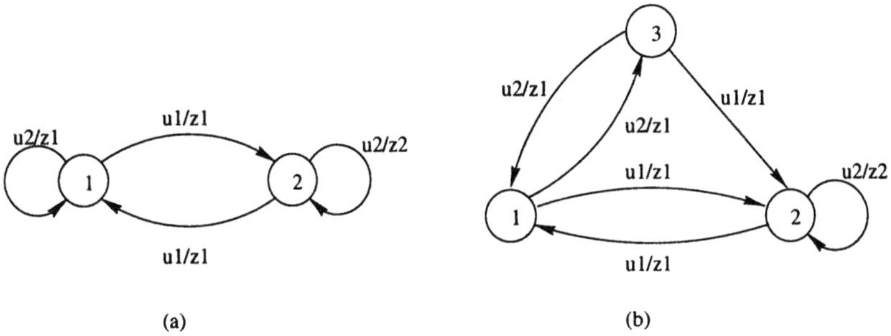

Figure 8 Two FSMs observationally equivalent with S

6 DISCUSSION

So far we have focused on the case where the test context is modeled as a single finite state machine. However, the approach itself does not restrict us only to that situation.

Consider the more general case in which the composite system is composed of multiple modules, and only one of the modules is to be tested as an embedded module (while all other modules are considered faultless). It is obviously unnecessary to compose all the faultless modules to produce a test context of single machine, since some modules may not reach the MUT at all.

Our approach can be easily adapted to this multi-module case while avoiding the costly composition of all text context modules. The structure of the algorithm in Figure 4 does not have to be altered; the only part that is affected is the reachability computation part. When a reachability computation from system input to MUT input is needed, the single module test context is simply replaced by the multi-module one with internal connection between the modules being taken into account. These internal connections are used as constraints on possible interactions between modules, which allows the reachability computation to be performed only for the modules that can reach to the MUT. This should increase the computing efficiency in many cases than if all context modules are simply composed. The backward reachability computation from MUT output to system output can be performed similarly. The algorithm generates test trees for the MUT just as in the single module case.

Another issue worth discussion is the generation of test suite with complete fault coverage based on the machine identification approach. Our approach supports this by adding extra test cases that distinguish the T-indistinguishable machines (TIMs) (Figure 2). In the normal test environment where the MUT is directly accessible (non-embedded case), this can be done by generating distinguishing test sequences for the TIMs. Algorithms exist that solve this problem.

In embedded system testing, similar approach can be adopted for generating the distinguishing sequences. However, these test sequences must be "externalized" to the system input and output in order for the test sequence to be executable. Fortunately, Petrenko et al have already developed a method for the externalization process [12]. This should allow us to generate complete test suite after coverage evaluation.

7 CONCLUSIONS

We have presented our work on fault coverage evaluation for embedded system testing. The approach is based on the faulty machine identification method that has been successfully employed in direct single module (as opposed to embedded module) testing. By generating test trees for the embedded module, faulty machines can be constructed and therefore fault coverage can be computed. The determination of conformance is done at the composite system level due to the limited observability and controllability imposed by the test context.

The approach effectively extends the single machine identification method into the area of embedded module testing. It inherits the advantages of the method such as its generality and independence of detailed fault classes and their combinations (only the most general fault model is needed where the number of states in an IUT is upper bounded). As a coverage evaluation tool, it not only verifies a given test suite as a complementary of test generation, but can also be used to augment a test suite (by generating additional test steps that detect the faulty machines). The main drawback is higher computational complexity in collapsing the test trees.

We have developed a set of tools that accomplishes the coverage evaluation. The tools have been applied to a simple example as presented in Section 5. The example serves to demonstrate and validate our approach. However, as future work, real protocols such as SSCOP should be used to test the effectiveness of the approach. Moreover, the tool set should be integrated in a more coherent way, since currently the tools are linked via files. For example, the test trees generated by TTREE (Figure 5) are written to a file which is then read by COV. It would certainly be more efficient to directly collapse a test tree once it is generated.

8 ACKNOWLEDGMENTS

The authors wish to thank the anonymous reviewers for their very helpful comments.

9 REFERENCES

[1] S. Abramsky. Observation equivalence as a testing equivalence. *Theoretical Computer Science*, 53:225–241, 1987.

[2] A.W. Biermann and J.A. Feldman. On the synthesis of finite-state machines from samples of their behavior. *IEEE Transactions on Computers*, 21(6), June 1972.

[3] G.v. Bochmann, A. Petrenko, and M. Yao. Fault coverage of tests based on finite state models. In *Protocol Test Systems VII*, Japan, 1994.

[4] A. Dahbura and K. Sabnani. An experience in estimating fault coverage of a protocol test. In *Proc. IEEE INFOCOM*, 1988.

[5] E.M. Gold. System identification via state characterization. *Automatica*, 8:621–636, 1972.

[6] E.M. Gold. Complexity of automaton identification from given data. *Information and Control*, 37:302–320, 1978.

[7] ISO/IEC 9646. *Information Technology - Open Systems Interconnection - Conformance testing methodology and framework*. March 1991.

[8] ITU-T Recommendation Q.2110. *B-ISDN ATM Adaption Layer - Service Specific Connection Oriented Protocol (SSCOP)*. ITU Telecommunication Standards Sector, Geneva, 1994.

[9] Luiz Paula Lima Jr. and Ana R. Cavalli. A pragmatic approach to generating test sequence for embedded systems. In *IFIP 10th Int. Workshop on Testing of Communicating Systems*, Cheju Island, Korea, September 1997.

[10] JTC1/SC21/WG1/Project 54.1. *Framework: Formal Methods in Conformance Testing*. February 1995.

[11] J. Kella. Sequential machine identification. *IEEE Transactions on Computers*, 20(3), March 1971.

[12] A.F. Petrenko and N. Yevtushenko. Fault detection in embedded components. In *IFIP 10th Int. Workshop on Testing of Communicating Systems*, Cheju Island, Korea, September 1997.

[13] A.F. Petrenko, N. Yevtushenko, and G.v. Bochmann. Fault models for testing in context. In *Proc. IFIP FORTE IX/PSTV XVI*, Kaiserslautern, Germany, October 1996.

[14] A.F. Petrenko, N. Yevtushenko, G.v. Bochmann, and R. Dssouli. Testing in context: framework and test derivation. *Computer Communications*, 19:1236–1249, 1996.

[15] D.P. Sidhu and T.-K. Leung. Formal methods for protocol testing: A detailed study. *IEEE Transactions on Software Engineering*, April

1989.
[16] S.T. Vuong and K.C. Ko. A novel approach to protocol test sequence generation. In *Proc. GLOBECOM'90*, December 1990.
[17] M. Yao, A. Petrenko, and G.v. Bochmann. A structural analysis approach to the evaluation of fault coverage for protocol conformance testing. In D. Hogrefe and S. Leue, editors, *FORTE'94*, 1994.
[18] J. Zhu and S.T. Chanson. Toward evaluating fault coverage of protocol test sequences. In *Proc. IFIP 14th Int. Symp. on Protocol Specification, Testing, and Verification*, Vancouver, Canada, June 1994.

10 BIOGRAPHY

Mr. Jinsong Zhu received his B.Sc and M.Sc in Computer Science from Tsinghua University, China in 1985 and 1987 respectively. He was a faculty member at Tsinghua University from 1987 to 1991. From 1995 to 1996, he was a Senior Software Engineer at the Infonet Software Solutions, Inc., Canada. He is currently working towards his Ph.D. degree at the Computer Science Department of the University of British Columbia. His research interests include protocol testing theory, high speed networks, mobile computing, and distributed object computing technology.

Dr. Son T. Vuong received his Ph.D. in computer science from the University of Waterloo, Canada. He had some industrial experience on image processing at CCRS and design of packet-switched networks at BNR (Nortel). He joined the faculty of the Department of Computer Science at the University of Waterloo for two years before moving to the University of British Columbia in 1983, where he was a founder of the Distributed Systems Research Group and is presently an Associate Professor. Dr. Vuong's areas of research interest include protocol engineering, distributed multimedia systems, high speed networks and mobile computing. Dr. Vuong served on several conference program committees and was the (co)chair and local organizer of numerous international conferences: ICDCS'95, PSTV'94, FORTE'89, IWPTS'88.

Dr. Samuel Chanson received his Ph.D. degree from the University of California, Berkeley in 1975. He was a faculty member at Purdue University for two years before joining the University of British Columbia where he became a full professor and director of its Distributed Systems Research Group. In 1993 he joined the Hong Kong University of Science & Technology as Professor and Associate Head of the Computer Science Department. He has chaired and served on the program committees of many international conferences on distributed systems and computer communications, including IEEE ICDCS, IFIP PSTV and IWTCS.

Dr. Chanson's research interests include protocols, high speed networks, multimedia communication and load balancing in workstation clusters. He has published more than 120 technical papers in the above areas.

Test suite minimization for testing in context

Nina Yevtushenko
Tomsk State University
36, Lenin str., Tomsk, 634050, Russia
Yevtushenko.RFF@elefot.tsu.tomsk.su

Ana Cavalli, Luiz Lima Jr.[1]
National Institute of Telecommunications
9, rue Charles Fourier, 91011 Evry Cedex, France
{Ana.Cavalli,lima}@hugo.int-evry.fr

Abstract

In this paper, the problem of test suite minimization for testing in context is studied and some results are proposed. The test architecture considered is general enough to allow the context and the component to have external inputs and outputs. Using a given fault model for testing in context, conditions are provided to detect all the redundant transitions that do not need to be tested. A complete test suite for the fault model is obtained and a method to select a minimal subset of this test suite with the same fault detection power is then proposed. Furthermore, the proposed method can be used to reduce a test suite produced by human experts while preserving its fault detection power.

Keywords

Test derivation, test in context, embedded testing, test of communicating FSMs.

1 INTRODUCTION

As the complexity of communication protocols increases, there is a strong need for systematic methods for test derivation with guaranteed fault coverage based on formal description techniques. Formal description techniques for communication protocols usually use a system of communicating finite state machines (FSMs) as their

1. Work funded by CNPq.

underlying model. The 'black-box' testing of a system can be performed by constructing the composite machine that describes a behaviour of a system at points that are accessible for testing and then applying test suite derivation methods for FSMs [Vasilevsky 73][Chow 78][Vuong 89][Fujiwara 91][Petrenko 96-1][Yannakakis 95]. However, this approach provides very long tests due to a huge number of states of the composite machine.

If only some components of the system need testing, the problem of testing an embedded component machine (or 'gray-box' testing [IS-9646 91] or testing in context [Petrenko 96-2]) appears. In this case, a specification system is represented as a composition of two FSMs. One of these machines, called *component*, is the composite FSM of all machines which need testing. The other machine, called *context*, is the composite machine of all other machines that are assumed to be faulty-free. If the set of Points of Control and Observation (PCOs) of a system contains all the access points to the component (i.e. there are no internal signals exchanged between the component and the context), then testing in context reduces to 'black-box' testing in isolation and a number of methods can be called for deriving a test suite with guaranteed fault coverage. Otherwise, 'black-box' testing at the PCOs will (unnecessarily) test the context as well, that is assumed to be faulty-free. The issue has been studied in a number of papers [Lima 97][Petrenko 97][Petrenko 96-2][Huang 96][Lee 96][Kim 95][PT 96]. Nevertheless, as far as we know, the application of embedded testing methods in precise fields has not been greatly explored [Lima 98].

Not all transitions of the system come from the combination of transitions of the context and the component. Some concern uniquely the context and thus, do not need testing. Various methods for selecting such transitions have been proposed. Based on the so-called fault function, the authors of paper [Petrenko 96-2] suggest a method for translation of component's transfer and output faults into faults of the composite FSM of the specification and call a method elaborated in [Petrenko 92] for complete test suite derivation. Another approach is based on a coverage of component's transitions by performing a random walk of the specification composite FSM [Lee 96] but fault detection power of the method is unknown. Paper [Lima 97] is devoted to determining a certain part of the composite machine comprising transitions should be tested in the case when any fault of the component does not increase the number of states of the specification system. In the paper, transitions which the component does not affect are called redundant transitions while other transitions are called suspicious transitions. If the context is faulty-free then an external sequence traversing only redundant transitions is superfluous and can be deleted from the test suite without loss of its completeness. A method for determining such transitions is proposed, and sufficient conditions for deleting superfluous test cases are established. The authors notice, nevertheless, that a more rigorous analysis is necessary to elaborate a method for checking suspicious transitions. This paper continues this work.

The paper is structured as follows. After the preliminaries of Section 2, Section 3 explains how to deal with transitions concerning only the faultless context (redundant transitions) in the machine representing the system's global behaviour. In Section 4, given a set of external input sequences (from Section 3), we propose a method for derivation of a regular language describing a part of the component's behaviour that can be tested with the set and a complete test suite minimization technique when the obtained regular set is finite. We conclude with a discussion on the future work in Section 5.

2 PRELIMINARIES

2.1 Finite State Machines

A Finite State Machine (FSM) (often simply called 'a machine' throughout this paper) is an initialized completely specified deterministic Mealy machine which can be formally defined as follows. A FSM A is a 6-tuple $(S, X, Y, \delta, \lambda, s_0)$ where S is a finite set of states with s_0 as the initial state; X - a finite nonempty set of input symbols; Y - a finite nonempty set of output symbols, δ and λ are the next state and output functions δ: $S \times X \rightarrow S$ and λ: $S \times X \rightarrow Y$. In usual way, functions δ and λ are extended to functions on the set $S \times X^*$ where X^* is the set of all finite input sequences including the empty sequence ε. The FSM A is said to be *connected* if each state of A is reachable from the initial state, i.e. for each state $s \in S$ there exists an input sequence $\alpha \in X^*$ such that $\delta(s_0,\alpha) = s$.

Two states s_i and s_j are said to be *distinguishable* states of A if there exists an input sequence $\alpha \in X^*$ such that the FSM A produces different output sequences at the states s_i and s_j to the α; otherwise, states s_i and s_j are said to be *equivalent*. An FSM with pairwise distinguishable states is called a *minimal* FSM.

Given sequences $\alpha = x_1...x_k \in X^*$ and $\beta = y_1...y_k \in Y^*$, the sequence $x_1y_1...x_ky_k$ is called a trace of the FSM A if $\beta=\lambda(s_0,\alpha)$. For a sequence σ over alphabets X and Y, $X \cap Y = \varnothing$, the X-projection of σ is obtained by deleting from σ all symbols of the set Y.

Given an FSM $B = (T, X, Y, \psi, \varphi, t_0)$, B is said to be *equivalent to* A, written $B \cong A$, if for any sequence $\alpha \in X^*$, it holds that $\lambda(s_0,\alpha) = \varphi(t_0,\alpha)$, otherwise B is distinguishable from A, written $B \not\cong A$. In other words, B is equivalent to A if A and B exhibit the same behaviour under all input sequences. Protocol conformance testing is often formalized as FSMs equivalence problem.

2.2 Fault models

A fault model is a triple $<A, \sim, \Re>$ [Petrenko 96-2], where A is a reference FSM, $\Re$, usually called a *fault domain*, is the set of all (possibly faulty) implementation FSMs defined over the same input alphabet as the reference machine, and ~ is a conformance relation. If the reference FSM A is a deterministic minimal FSM with n states, $\Re_n$ is a set of all deterministic FSMs over the same input alphabet as A with at most n states and ~ is the equivalence relation $\cong$ then the fault model $<A, \cong, \Re_n>$ is a classical 'black-box' model. A complete test suite TS w.r.t. the fault model $<A, \cong, \Re_n>$ is a finite set of finite input sequences of the reference FSM A such that for any

FSM $B\in\Re_n$, if $B\neq A$ then there exists input sequence $\alpha\in TS$ such that A and B produce different output sequences to α. There exists a number of methods for complete test suite derivation w.r.t. this model. Below we sketch the **Wp-method** [Fujiwara 91] for a complete test derivation.

Let $A = (S, X, Y, \delta, \lambda, s_0)$ be a minimal connected FSM with n states. A subset $V\subseteq X^*$ comprising the empty sequence ε is called *a state cover set* if for each state $s\in S$, there exists a sequence $\alpha\in V$ such that $\delta(s_0,\alpha)=s$. We further denote VX the set obtained by concatenating each sequence $\alpha\in V$ with each input $x\in X$. The set VX is usually called *a transition cover set* since each transition of the FSM A is traversed with an appropriate sequence of the set VX.

Given state $s\in S$ of the FSM A, a finite set W_s of finite input sequences is called *a state identifier* of the state s if for each state $p\in S$, $p\neq s$, there exists an input sequence $\alpha\in W_s$ such that $\lambda(p,\alpha)\neq\lambda(s,\alpha)$. An input sequence α is said to be a *distinguishing sequence* of the FSM A if the output responses of the FSM A at different states to α are different. When the FSM A has a distinguishing sequence α then the set $\{\alpha\}$ is a state identifier of any state of A. When a state identifier W_s is fixed for each state $s\in S$ a procedure for derivation of a complete test suite w.r.t. the fault model $<A, \cong, \Re_n>$ comprises two phases.

In the former phase, the part T_1 of the test suite is derived. The T_1 checks whether an implementation FSM has exactly n states and each state identifier preserves its feature. This part of a test suite is obtained by concatenating each sequence $\alpha\in V$ with each sequence of the union of state identifiers W_s over all states $s\in S$. If the implementation and reference FSMs have one and the same output sequence to each sequence of T_1 then any state of the reference FSM has a corresponding state in the implementation FSM with the same state identifier.

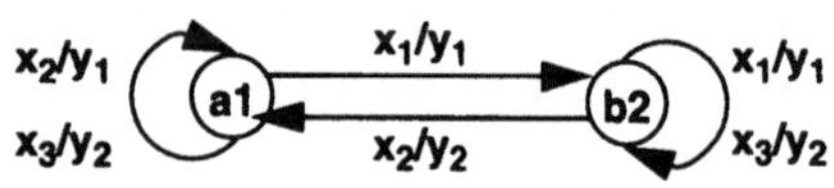

Figure 1 FSM *RS*.

The part T_2 of a test suite derived in the latter phase, checks whether every transition of the implementation FSM is correctly implemented. The T_2 is obtained by concatenating each sequence $\alpha\in VX$ with each sequence of the state identifier of the state $\delta(s,\alpha)$ where α takes the FSM A from the initial state. Merging the T_1 and T_2 we obtain a complete test suite $TS = T_1\cup T_2$ w.r.t. the fault model $<A, \cong, \Re_n>$. For the sake of simplicity, each sequence that is a prefix of another sequence can be deleted from the test suite TS.

Example 1. Consider an FSM *RS* in Figure 1 (*a1* is the initial state). An input sequence x_2 is a distinguishing sequence of the *RS*. We select the state cover set $V = \{\varepsilon, x_1\}$. Then $T_1 = \{x_2, x_1x_2\}$ while the transition cover set $VX = \{x_1, x_2, x_3, x_1x_1, x_1x_2, x_1x_3\}$. Concatenating each sequence of VX with x_2 we obtain $T_2 = \{x_1x_2, x_2x_2, x_3x_2, x_1x_1x_2, x_1x_2x_2, x_1x_3x_2\}$. After deleting from $T_1\cup T_2$ all prefixes of other sequences the set $TS = \{x_2x_2, x_3x_2, x_1x_1x_2, x_1x_2x_2, x_1x_3x_2\}$ is obtained that is a complete test suite w.r.t. the fault model $<RS, \cong, \Re_2>$. The tree of the test suite is shown in Figure 2. The test suite comprises five test cases of total length 18. ❑

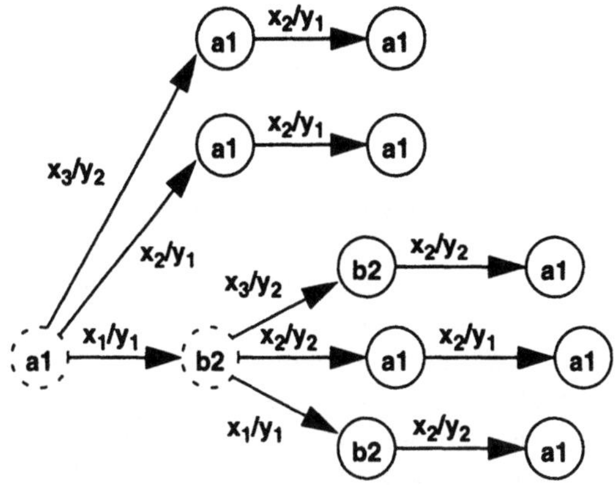

Figure 2 The tree of a complete test suite w.r.t. the fault model <RS, ≡, $\Re_2$>. States that should be identified are depicted in dashed lines.

2.3 Composition of FSMs

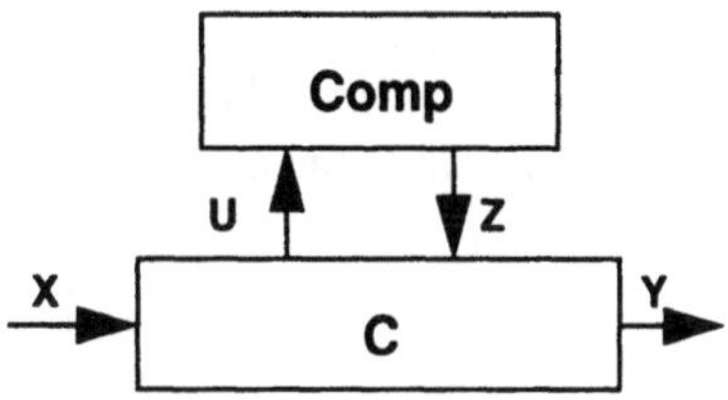

Figure 3 Composition of two FSMs: the context *C* and the component *Comp*.

We consider a composition as a system of two component FSMs, as shown in Figure 3. For the sake of simplicity, we consider pairwise disjoint sets *X*, *Y*, *U* and *Z*. The system at hand has a single message in transit, i.e. the environment submits the next external input to the system only when it has produced an external output to the previous input. This accounts for synchronous communication between the environment and the system. Notice, however, that asynchronous communication through bounded input queues can be simulated by explicitly introducing new contexts representing the behaviour of the bounded queue in the system [Tretmans 92][Phalippou 94]. The input queues must be bounded, so to assure that the corresponding FSM exists (i.e. it is really finite).

Under these assumptions, when there are no live-locks in the composition, we can derive the composite machine of FSMs *C* and *Comp* using, for instance, the algorithm described in [Lima 97] in order to avoid generating the Cartesian product *C*×*Comp* (which is often huge).

Example 2. Consider the two machines *C* and *Comp* shown in Figure 4a and Figure 4b. Considering states *a* and *1* the initial states, the composite machine *RS* = *C*•*Comp* is the FSM of Figure 1. ❑

We now demonstrate that the composition of Figure 3 is general enough to discuss the problems related to testing a component machine. We claim that the gen-

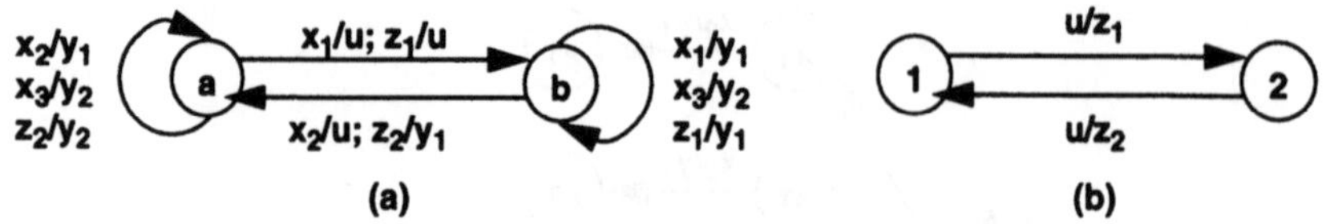

Figure 4 The context C(a) and the component *Comp* (b).

eral model of Figure 5 can be transformed into that in Figure 3 with the component and composite FSM being isomorphic to those of the original composition.

If the external actions X' and Y' are redirected through the context along with the replacement of the external alphabets X' and Y' by the internal alphabets U' and Z' (Figure 6), we obtain an equivalent (isomorphic) system due to one-to-one correspondence. For each input $x \in X'$, the context C at any state produces a corresponding internal output $u \in U'$ while for each internal input $z \in Z'$ the context produces the corresponding $y \in Y'$. *Comp* is isomorphic to *EComp* w.r.t. the one-to-one correspondence between X' and U' and Y' and Z' while the composite FSMs of the compositions coincide.

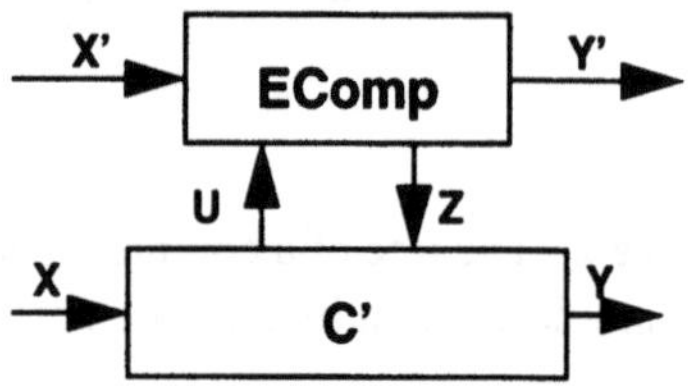

Figure 5 General composition of two FSMs. The context C' and the component *EComp* have external inputs and outputs.

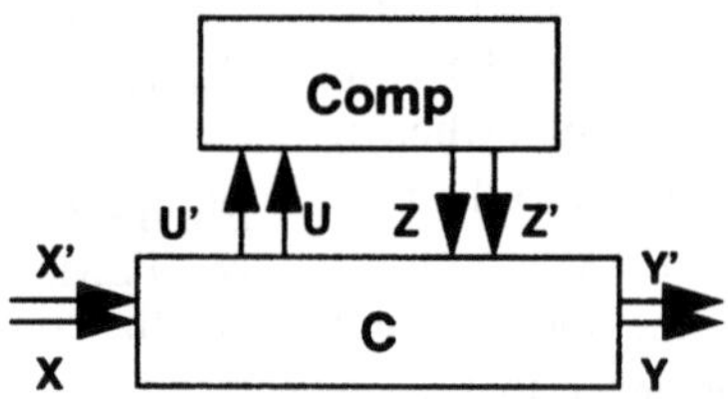

Figure 6 The transformed composition of two FSMs, the context C and the component *Comp*.

2.4 Fault model for testing in context

Many compound systems are formally specified as a system of interacting FSMs. If only some components of a system need testing we face the problem of testing an embedded component or testing in context. In this case, a system under test can be represented as the composition of two FSMs. As stated in Section 1, one of these machines, called the *component*, is the composite FSM of all component machines which need testing while another machine, called the *context*, is the composite machine of all other component machines that are assumed to be faulty-free. As demonstrated above, we can use the composition in Figure 3 to discuss problems related to testing in context.

We further suppose that the behaviour of a specification system is described by a deterministic FSM represented in the tester by a test suite. The tester executes test cases against an implementation system usually called a *System Under Test* (*SUT*), and checks the results on receipt of a response from the SUT. In this paper, we suppose that the SUT's behaviour can be represented by a deterministic composite machine. If the output response of the SUT to a proper test case differs from that of the specification system, then the tester produces the verdict '**fail**'; in this case, the SUT is a nonconforming implementation of the specification system. Otherwise, the tester produces the verdict '**pass**' to the test case. If the verdict '**pass**' is produced to every test case of the test suite then we conclude that the SUT is a conforming implementation of the specification system.

Let a specification system be given as the composition of two interacting FSMs: the context *C* and the component *Comp*. Let this composition be represented by the composite machine *RS* with *n* states. An implementation system is also the composition of two interacting machines: the context *C* (a faultless implementation of the specification context *C*) and an implementation component machine *Imp*. Let the implementation system's behaviour be represented by the composite machine *IM* with at most *n* states. If $\Re_{n,C}$ is the set of all composite machines of such implementation systems, then we are required to derive a complete test suite w.r.t. the fault model $<RS, \cong, \Re_{n,C}>$, i.e. a set of input sequences of *RS* such that for each implementation system of the domain that exhibits a different external behaviour from that of *RS*, there exists at least one test case when the tester produces the verdict '**fail**'.

As mentioned above, regular methods exist for the derivation of a complete test suite w.r.t. the fault model $<RS, \cong, \Re_n>$. The test suite is also complete w.r.t. $<RS, \cong, \Re_{n,C}>$ because $\Re_{n,C} \subseteq \Re_n$. The test suite checks the component as well, but it also checks (unnecessarily) the context that is assumed to be faulty-free; thus, it can be reduced in particular cases. In the paper [Lima 97], some sufficient conditions for reducing the test suite are established. These conditions are briefly summarized in the following section (Section 3.1).

3 DEALING WITH REDUNDANT TRANSITIONS

3.1 Reducing a test suite

Each transition of the composite machine *RS* comes from a combination of transitions of the two modules, the context and the component, but there may be transitions in the composite machine which are not affected by the component. In paper [Lima 97], such transitions are called *redundant* transitions; otherwise, a transition of the *RS* is called *suspicious*. If the context is assumed to be faulty-free then external sequences traversing only redundant transitions are superfluous and can be deleted from the test suite without loss of its completeness. A method for determining such transitions is proposed and sufficient conditions for deleting superfluous test cases are established. These conditions can be formalized as follows.

Proposition 3.1. Given the reference composite machine *RS*, let β be an external input sequence only traversing redundant transitions of the *RS*. The output sequence of an implementation system to β coincides with that of FSM *RS*.

Corollary 3.2. Given a complete test suite *TS* w.r.t. the fault model <*RS*, ≅, $\Re_{n,C}$>, let β∈ *TS* be a sequence only traversing redundant transitions of the *RS*. Then *TS*\{β} is also a complete test suite w.r.t. the fault model.

Example 3. Consider sequences x_2x_2 and x_3x_2 of the complete test suite *TS* w.r.t. the fault model <*RS*, ≅, $\Re_2$> obtained in Section 2.2. The sequences only traverse redundant transitions $(a1,x_2)$, $(a1,x_2)$ and $(a1,x_3)$, $(a1,x_2)$ of the composite FSM *RS*. Thus, $TS = \{x_1x_1x_2, x_1x_2x_2, x_1x_3x_2\}$ is a complete test suite w.r.t. the fault model <*RS*, ≅, $\Re_{2,C}$>. It has three sequences of total length 12. All sequences of the set *TS* traverse a suspicious transition. ❑

Nevertheless, it was pointed out that a more rigorous analysis is necessary in order to elaborate a method for checking suspicious transitions.

3.2 Redundant transitions

It may well happen that a redundant transition of the specification system becomes suspicious due to an unexpected interaction in the implementation system that brings the context to a wrong state. Consider, for instance, the system of two FSMs: context *D* (Figure 7a) and component *Spec* (Figure 7b).

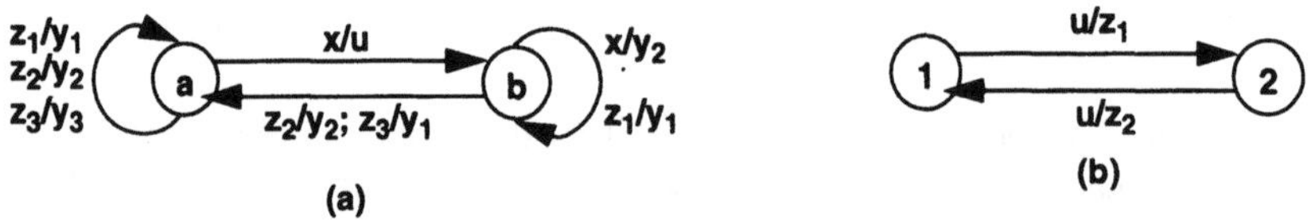

Figure 7 The context *D* (a) and the component *Spec* (b).

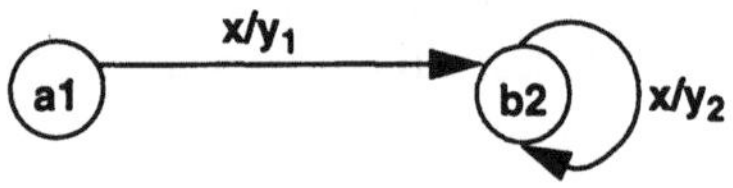

Figure 8 The composite FSM *D•Spec*.

The composite FSM is shown in Figure 8.

We use input sequence *x* to reach state *b2* from the initial state *a1* and the same sequence *x* as a distinguishing sequence. Thus, $T_1 = \{x, xx\}$ coincides with $VX = \{x, xx\}$ and $T_2 = \{xx, xxx\}$, i.e. $TS = \{xxx\}$ is a complete test suite w.r.t. the fault model <*D•Spec*, ≅, $\Re_2$>. Note that the transition from the state *b2* under input *x* is a redundant transition. If we do not check this transition, then *TS'* = $\{xx\}$. However, *TS'* is not complete w.r.t. the fault model <*D•Spec*, ≅, $\Re_{2,C}$>.

Assume that the implementation component FSM *Imp* depicted in Figure 9a replaces *Spec*. The composite FSM *D•Imp* is shown in Figure 9b. By direct inspection, one can assure that the input sequence *xxx* is the shortest sequence distinguishing nonequivalent FSMs *D•Spec* and *D•Imp*.

The reason is that an implementation component machine, being an arbitrary FSM over alphabets *U* and *Z*, can induce unexpected internal interactions in the implementation system. These unexpected internal interactions bring the implemen-

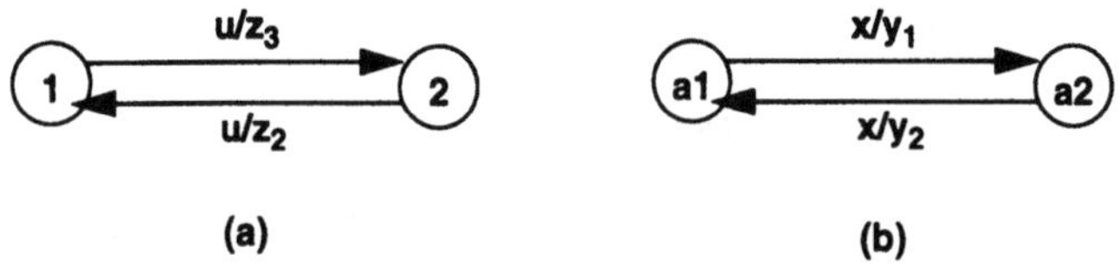

Figure 9 The implementation component FSM *Imp* (a) and the implementation composite FSM *D•Imp* (b).

tation system to a wrong state while producing the expected output. These kinds of faults are known as *latent faults* [Petrenko 96-2] and can only be detected when longer external test cases are submitted. Therefore, some effort must be put into characterizing such cases, so to improve the proposed solution.

Given an external input sequence $\alpha \in X^*$ of the FSM *RS* and an internal trace over alphabets *U* and *Z*, the trace is said to be unexpected w.r.t. α if the trace may be induced in an implementation system not coinciding with that of the specification system when α is submitted. Unexpected internal traces represent alternative paths that a system may take due to faults in the implementation component machine.

Using the concept of unexpected internal traces in the previous example, we observe that a latent fault appears because:

1. there exist unexpected internal traces uz_2 and uz_3 w.r.t. the external input *x*; and
2. one of these unexpected internal traces, namely uz_3, produces the expected external output y_1.

Generally speaking, given a redundant transition of the FSM *RS* at state *s*, the latent fault may occur under the following conditions:

1. the external input sequence $\alpha \in VX$ that takes the *RS* from its initial state to state *s* traverses suspicious transitions; and
2. at least one of the unexpected internal traces w.r.t. α results in the expected output sequence.

Based on this observation we conclude that a redundant transition becomes suspicious if conditions 1 and 2 above apply, and then the tail state of such transition must be checked as well.

On the other hand, if for some test case of a given test suite, none of the unexpected internal traces w.r.t. the case result is the expected external output sequence, then the output sequence of an implementation to the test case coincides with that of the specification system. In this case, the test case is superfluous and can be deleted from the test suite without loss of its completeness. When at least one of the unexpected internal cases w.r.t. the test case results in an unexpected external output sequence, other test case should be examined to recognize those that are superfluous.

4 TEST SUITE MINIMIZATION

4.1 Detectable internal traces

The example of Section 3.2 clearly demonstrates that the fault detection power of an external input sequence α in the given context can be characterized by the set of all unexpected internal traces such that, if an implementation component machine comprises such a trace, then the specification and implementation systems produce different output sequences to α. We call such internal traces *detectable with the external input sequence* α. If an implementation component FSM has no traces detectable with α, then the specification and implementation systems produce the same output sequence to α. The formal definition of such internal traces is given in [Petrenko 97] where they are called forbidden internal traces w.r.t. α. As the set of traces of an FSM is a prefix-closed set, an extension of a trace which is detectable with an external input sequence is also detectable with this sequence. Conversely, a set of internal traces that are detectable with an external input sequence includes that for its prefix.

Given a set T of external input sequences, an internal trace is said to be *detectable with the set* T if it is detectable with at least one sequence of the set. Due to the definition of detectable traces, the following statement holds.

Proposition 4.1. Given the composite FSM RS of the specification system with the context C and a complete test suite TS w.r.t. fault model $<RS, \equiv, \Re_{n,C}>$, let P be a subset of TS. If all internal traces detectable with TS are detectable with P then P also is a complete test suite w.r.t. the fault model.

Due to Proposition 4.1, to compare the fault detection power of two test suites we need a procedure for characterizing internal traces that are detectable with a given external input sequence. We represent such traces as sequences recognized by a designated final state '**fail**' of an appropriate acceptor or recognizer [Hopkroft 79]. Our first step is to describe all possible traces that may be induced in an implementation system when the external sequence is submitted. Because a component FSM is completely embedded within the context (see Section 2.3), these traces can be described as traces of the context LTS LC obtained by unfolding each atomic transition *input/output* in the context FSM C. The context LTS LC for the context C in Figure 4a is shown in Figure 10.

Given an external input $x \in X$, we construct the acceptor $LC(x)$. States of the acceptor are states of LC with the initial state s_0. The initial and final states of the acceptor are special states that cannot be merged with other states with the same names called *intermediate states*, while it is possible to merge two such intermediate states. There is a transition labelled with x from the initial state to state s if x takes the LC from the initial state s_0 to the state s. For two intermediate states p and r, there is a transition labelled with action $a \in U \cup Z$ if there is a transition labelled with a from state p to state r in the LTS LC. There is a transition labelled with $y \in Y$ from the intermediate state p to the final state s if there exists an outgoing transition at the state p to state s labelled with y in the LTS LC. The connected part of the acceptor comprising the initial state is denoted $LC(x)$.

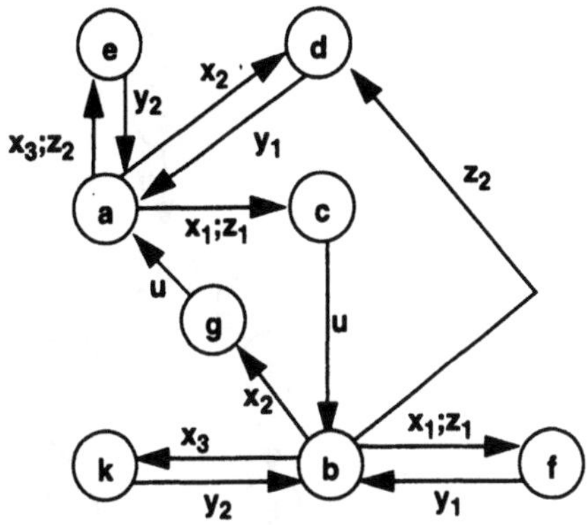

Figure 10 The context LTS LC representing all the traces of the context C.

Given a final state s of the acceptor $LC(x)$, the set of all sequences labelling the paths from the initial to the final state s is a regular set recognized by the final state s [Hopkroft 79]. Due to construction of the acceptor, the regular expression specifying the set can be written as xfy where f comprises only internal actions of the set $U \cup Z$. Thus, the set of the $(U \cup Z)$-projections of the regular set recognized by a final state also is a regular set specified by a proper regular expression.

To construct the acceptor $LC(x_1...x_k)$ we construct at each terminal state s of the acceptor $LC(x_1)$ the acceptor $LC(x_2)$ with s as the initial state. The construction of the acceptor $LC(x_1...x_k)$ implies the following statement.

Proposition 4.2. Given the acceptor $LC(x_1...x_k)$, let IS be the implementation system of a component over alphabets U and Z within context C. **1)** If σ is a trace of the system when the external input sequence $x_1...x_k$ is submitted, then the acceptor $LC(x_1...x_k)$ has a path from the initial state to a final state labelled with the sequence σ. **2)** If an $(U \cup Z)$-projection of sequence σ labelled a path from the initial to a final state is a trace of the implementation component, then the output sequence of the implementation system to $x_1...x_k$ is the Y-projection of the sequence σ.

Example 4. Consider an implementation system with the context of Figure 4a and an input sequence $x_1x_2x_2$. When the external input x_1 is applied to the context at the initial state a the context produces the internal output u and enters the state b. Its next action depends on an output produced by a component machine to the input u. If the component produces z_1 then the context remains at the state b producing y_1 while the context enters the state a producing y_1 if the component has the output response z_2 to u. In both cases, the context produces an external output y_1 to x_1 and the next external input x_2 can be applied to the context. Since we are interested in all the traces that may be induced when the external input sequence is submitted to the context we do not merge states with the same names separated with external actions. The procedure of constructing the acceptor $LC(x_1x_2x_2)$ is illustrated in Figure 11. ❑

The next step is to transform the acceptor, so that the set of all internal traces detectable with the given external input sequence would be represented by the set of $(U \cup Z)$-projections of the sequences recognized by a designated state '**fail**.'

Procedure 4.1. Derivation of the regular expression for forbidden traces w.r.t. a given external input sequence.

Inputs: The composite machine *RS* of the specification system with a context *C* and an external input sequence α.
Output: A regular expression E_α for internal traces detectable with α.
Step 1. Derive the context LTS *LC* unfolding each atomic transition *input/output* in the context FSM *C* and construct the acceptor $LC(\alpha)$.
Step 2. For each path of the acceptor $LC(\alpha)$ from the initial state to a final state such that the $(X \cup Y)$- projection of the sequence labelled the path is not a trace of the specification composite FSM *RS*, replace the last state of the path with a designated deadlock state **'fail.'**
Step 3. If for some transition labelled with external input $x \in X$ or with an internal action $z \in Z$, all the subsequent paths have a final state **'fail'** then replace the final state of the transition with the **'fail'** state.
Step 4. Derive a regular expression E_α for $(U \cup Z)$- projections of sequences labelling all paths from the initial state to **FAIL** state.

The regular language obtained by Procedure 4.1, is called the *characterization fault detection set* $D(\alpha)$ *of the sequence* α which completely characterizes the set of nonconforming implementation systems that can be detected with the external input sequence α.

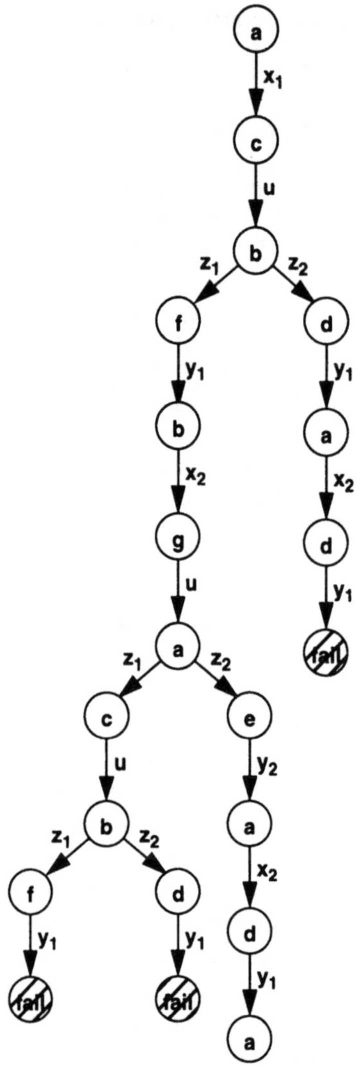

Figure 11 The fragment of the acceptor $LC(x_1x_2x_2)$. The external projections of executed traces are underlined.

Proposition 4.3. Given a context *C*, a component specification *Comp*, a component implementation *Imp*, the composite machines *RS* (system specification) and *IS* (system implementation), and an external input sequence α, let E_α be a regular expression obtained by the Procedure 4.1. The external sequence α distinguishes FSMs *RS* from *IS* iff the language specified by the expression E_α comprises a trace of the component *Imp* of the implementation system.

Proof. First part. Let the language specified by the regular expression *E* comprise a trace β/γ of a component *Imp*. By construction of E_α, there is a path in the acceptor (obtained at Step 2 of Procedure 4.1) from the initial state to the **'fail'** state labelled with a sequence such that: 1) its $(U \cup Z)$- projection $\beta\beta'/\gamma\gamma'$ is a trace of the implementation component *Imp*; 2) its *X*- projection is α; and 3) its *Y*- projection σ is different from the *RS*'s output sequence to α. The **'fail'** state replaces an appropriate final state of the acceptor $LC(\alpha)$. Due to Proposition 4.2 (Part 2), the implementation system produces the output sequence σ when α is submitted to the system.

Second part. We now assume that the *IS* has the output sequence σ to α that is different from that of the *RS*. Then the acceptor obtained at Step 2 of Procedure 4.1 comprises the path from the initial state to the **'fail'** state labelled with a trace induced in the implementation system when α is submitted to the system (Proposition 4.2, Part 1). Therefore, the *X*- projection of the sequence labelling the path is α, the *Y*- projection is σ and the $(U \cup Z)$- projection is a trace β/γ of the component *Imp*. By construction, the language specified by regular expression *E* comprises a prefix to β/γ, i.e. a trace of the component *Imp*. ❑

Corollary 4.4. Given an internal trace τ and an external input sequence α, τ is detectable with α iff the set $D(\alpha)$ contains a prefix of τ.

Given the set *T* of external input sequences, the union of the characterization fault detection sets over all sequences in the set *T* is called the *characterization fault detection set D(T) of T.*

Example 5. Consider a sequence $x_1x_2x_2$ of the complete test suite $TS = \{x_1x_1x_2, x_1x_2x_2, x_1x_3x_2\}$ w.r.t. the fault model $<RS, \cong, \Re_{2,C}>$ obtained in Section 3 after deleting test cases traversing only redundant transitions. This sequence traverses suspicious transitions. We need now to determine the characterization fault detection set $D(x_1x_2)$ for its prefix x_1x_2.

Consider in the acceptor the path labelled with the sequence $x_1uz_2y_1x_2y_1$ from the initial state. A system executes this sequence of actions when the context is combined with an implementation component machine producing at the initial state the output z_2 to the input u. Once the component of a system produces the output z_2 to the input u, the system will produce the output sequence y_1y_1 to the external input sequence x_1x_2. This output does not coincide with the expected output y_1y_2 of the reference composite FSM (Figure 1) to the input x_1x_2. Thus, the final state of the path is replaced by a **'fail'** state.

Consider now the paths labelled with sequences $x_1uz_1y_1x_2uz_1uz_1y_1$ and $x_1uz_1y_1x_2uz_1uz_2y_1$. A system executes these sequences of actions when the context is combined with an implementation component machine producing at the initial state the output sequence $z_1z_1z_1$ or $z_1z_1z_2$ to the input sequence *uuu*. Thus, since the implementation component of a system at hand produces the output sequence $z_1z_1z_1$ (or $z_1z_1z_2$) to the input sequence *uuu*, the system produces an output sequence (y_1y_1) to the input sequence x_1x_2 that is different from the expected output sequence y_1y_2 of the reference FSM *RS* (Figure 1). The prefix $x_1uz_1y_1x_2uz_1$ of the sequences takes the acceptor from the initial state to a state where all the subsequent paths lead to a **'fail'** state. In other words, if an implementation component machine has a trace uz_1uz_1, then the system will always produce the unexpected output sequence y_1y_1 to the input sequence x_1x_2, regardless of the output of the implementation component to the next input u. That is why we replace the state where $x_1uz_1y_1x_2uz_1$ takes the acceptor to from the initial state by a '**fail**' state. Therefore, $D(x_1x_2) = \{uz_1uz_1, uz_2\}$. By direct inspection of Figure 11, one can assure the set $D(x_1x_2x_2)$ coincides with $D(x_1x_2)$, i.e. the test case $x_1x_2x_2$ can be replaced by x_1x_2 without loss of completeness of the TS w.r.t. the fault model $<RS, \cong, \Re_{2,C}>$.

On the other hand, if the component of an implementation system has no traces of the set $D(x_1x_2)$, then even unexpected internal interactions in an implementation system result in the expected external output sequence y_1y_2 of the reference FSM RS to the external sequence x_1x_2. In a similar way, we determine the sets $D(x_1x_1x_2) = \{uz_2uz_2, uz_2uz_1uz_1, uz_1uz_1\}$ and $D(x_1x_3x_2) = \{uz_2, uz_1uz_1\}$. Thus, the fault characterization set $D(TS) = \{uz_2uz_2, uz_2uz_1uz_1, uz_2, uz_1uz_1\}$ comprises a prefix of any internal trace detectable with the test suite TS. ❑

Once the component of an implementation system has a trace belonging to $D(TS)$, there exists a test case in the TS to which the implementation and reference systems produce different output sequences. Conversely, if the component of an implementation system has no traces of the set $D(TS)$ then the implementation and reference systems produce the same output sequence to each test case of the TS. In other words, if the composite machine of the implementation system has at most two states then it is equivalent to the RS. That is to say that all components having no traces of the set $D(TS)$ that combined with the context possess the composite machine with at most two states are equivalent to the specification component *Comp* in the context C. At the PCOs it is impossible to recognize which of them serves as the component of the system at hand.

4.2 Minimizing a test suite

Removing redundant transitions

Using the results of Section 3 and Section 4.1, we can reduce the procedure of derivation of a complete test suite w.r.t. the fault model $<A, \cong, \Re_n>$ (as explained in Section 2.2) to the procedure to generate a complete test suite w.r.t. the fault model $<A, \cong, \Re_{n,C}>$.

Procedure 4.2. Deriving a complete test suite w.r.t. the fault model $<A, \cong, \Re_{n,C}>$.

Input: The composite FSM RS of a specification system with n states that is minimal and connected; the state cover set of RS; a transition cover set; and the set of state identifiers for all the states of RS.

Output: A complete test suite w.r.t. the fault model $<A, \cong, \Re_{n,C}>$.

Step 1. Concatenate each sequence of the state cover set with each sequence of the union of the state identifiers over all states of the FSM RS. Denote T_1 the set obtained. Let Q' be the subset of states of the context such that, for any state $q \in Q'$, there exists a sequence of the state cover set traversing only redundant transitions taking the context from the initial state to state q.

Step 2. Concatenate each sequence of the transition cover set traversing either a suspicious transition or taking the FSM *RS* from the initial state to state (q,t), ($q \notin Q'$) with each sequence of the identifier for the state to which the sequence takes the FSM *RS* from the initial state. Denote T_2 the set obtained and *TS* the union of T_1 and T_2. Each sequence that is a prefix of another sequence can be deleted from the *TS*.

Comparing the above procedure with that in Section 2.2 we do not include in T_2 any subset obtained by concatenating a sequence of the transition cover set traversing only redundant transitions and taking the FSM *RS* from the initial state to the state (q,t), where $q \in Q'$, with an appropriate state identifier, i.e. we do not test an unnecessary transition [Lima 97].

Proposition 4.5. Given the composite FSM *RS* of the specification system with *n* states, the set *TS* derived by Procedure 4.2 is a complete test suite w.r.t. the fault model $<RS, \cong, \Re_{n,C}>$.

Proof. Let α be a sequence of the state cover set traversing only redundant transitions and taking the FSM *RS* from the initial state to state (q,t), where $q \in Q'$. Moreover, let β be a sequence of the transition cover set traversing only redundant transitions and taking the FSM *RS* from the initial state to the state (q,t'). By construction, the set T_1 contains all sequences of the state identifier $W_{qt'}$ for state (q,t'). Each internal trace detectable with the sequence $\beta\gamma$, $\gamma \in W_{qt'}$ is detectable with $\alpha\gamma$ because the parts of both acceptors $LC(\beta\gamma)$ and $LC(\alpha\gamma)$ comprising internal traces start at the same initial state *q* of the context. Thus, the subset containing the concatenation of the subsequence β with each sequence of the set is superfluous in the complete test suite w.r.t. the fault model $<RS, \cong, \Re_{n,C}>$.

Deleting superfluous test cases

Suppose now that we derive a complete test suite w.r.t. the fault model $<RS, \cong, \Re_{n,C}>$ using Procedure 4.2 (i.e. not including into the test suite a sequence traversing only redundant transitions - Corollary 3.2). There may exist a proper subset of the test suite such that any internal trace detectable with the test suite is detectable with the subset, i.e. some sequences of the test suite may still be superfluous. Due to Proposition 4.3, to compare a fault detection power of two external input sequences, it is sufficient to compare the corresponding regular sets (the comparison of arbitrary regular sets is out of the scope of this paper). Let us assume that the characterization fault detection set *D*(*TS*) is finite for a given complete test suite *TS* w.r.t. the fault model $<RS, \cong, \Re_{n,C}>$, as it often occurs in practical situations. In this case, the problem of determining a minimal subset of a complete test suite *TS*, that is also complete w.r.t. $<RS, \cong, \Re_{n,C}>$, is reduced to derivation of a minimal column coverage *P* of the boolean matrix *B*, i.e. a minimal row subset of *P* such that for each column there exists at least one row of *P* comprising 1s in this column. The rows of the matrix *B* correspond to sequences of the test suite *TS* while the columns correspond to internal traces of the characterization fault detection set *D*(*TS*). Item b_{ij} is '1' iff for the sequence $\alpha_i \in TS$, the characterization fault detection set $D(\alpha_i)$ contains a prefix of the internal trace $(\beta/\gamma)_j \in D(TS)$.

Example 6. The boolean matrix *B* for our working example is shown in Table 1. By direct inspection, one can assure that set $D(x_1x_2)$ contains the prefixes of all traces of the set *D*(*TS*), i.e. all internal traces detectable with the test suite $TS = \{x_1x_1x_2, x_1x_2, x_1x_3x_2\}$ (obtained in Section 4.1) are detectable with the external input sequence x_1x_2. Therefore, the subset $\{x_1x_2\}$ is a complete test suite w.r.t. the fault model $<RS, \cong, \Re_{2,C}>$ being sufficient to detect an implementation system that possesses the composite machine with at most two states and is not equivalent to the reference composite FSM *RS* in Figure 1 when the context is faulty-free. One can compare this test suite (of total length 2) with a complete test suite w.r.t. the fault model $<RS, \cong, \Re_2>$ obtained in Section 2.2 that comprises five test cases of total length 18. ❑

In a number of practical situations it is nearly impossible to derive the composite machine of the overall system due to its huge number of states. In this case, a test suite often is derived by a test engineer who is a high-level expert in the area. The obtained test suite checks important features of the protocol's implementations but it is difficult to estimate its fault coverage in the formal way. The above approach can be used to reduce the given test preserving its fault detection power.

Let a specification system be a composition of the context *C* and the component *Comp* possessing the composite machine *RS* and let $\Re$ be a finite set of FSMs over the same input alphabet as *RS*. We denote $\Re_C$ the subset of $\Re$ comprising each machine of the set $\Re$ that is the composite machine of some implementation system with the same context. Given a set *T* of test cases that is complete w.r.t. the fault model $<RS, \cong, \Re_C>$, a subset *P* of the set *T* is said to have the same *fault detection power in the given context* if it is also a complete test suite w.r.t. that fault model.

TABLE 1. The boolean matrix B

	u z_2 u z_2	u z_2 u z_1 u z_1	u z_1 u z_1	u z_2
$x_1x_1x_2$	1	1	1	0
x_1x_2	1	1	1	1
$x_1x_3x_2$	1	0	1	0

Given the context *C* and the set *T* of external test cases, we can use Procedure 4.1 to determine the characterization fault detection *D(T)* and derive its subset *P* with the same fault detection power as a minimal coverage of an appropriate boolean matrix when *D(T)* is finite.

5 FUTURE WORK AND CONCLUSION

Our future work is basically related to the generalization of the proposed approach. It can be easily generalized to the case when some transitions of the specification context and component are undefined while their implementations are assumed to be completely specified. In undefined situations, the implementation can produce an error message, for example, or have a loop labelled with the **NULL** output. In this case, conformance testing for protocols can be formalized as FSMs quasi-equivalence problem where a complete FSM B is said to be quasi-equivalent to (possibly partial) FSM A if A and B exhibit the same behaviour under all input sequences where a behaviour of A is specified. We can call methods for a complete test suite derivation w.r.t. the fault model $<A, \cong_q, \Re_n>$ [Petrenko 96-1][Yannakakis 95] (where $\cong_q$ is a quasi-equivalence relation) and use the method proposed in this paper to minimize the obtained test suite.

We also would like to generalize this approach to a system of communicating nondeterministic FSMs. As the authors of the paper [Lima 97] notice, Proposition 3.1 and Corollary 3.2 hold in this case. But now it is insufficient to keep in a test suite external sequences detecting each trace detectable w.r.t. the test suite. The subset of remaining test cases should also check whether the set of output responses of an IUT to an external input sequence α contains each output sequence of the reference composite FSM to α. The problem is closely related to the problem of equation solving [Merlin 83][Kim 72][Parrow 89][Watanabe 93], where we are required to derive the largest specification of the sub-module that combined with the given context is equivalent to the specification's composite FSM. To the best of our knowledge the problem is not solved to the general case; solutions only exist for particular cases.

In this paper we presented a test derivation strategy for testing in context. The system studied was composed of two communicating FSMs, the context and component, and the test architecture was generic, i.e. the context and the component may have external inputs and outputs. Using the given fault model for testing in context, conditions were provided to detect redundant transitions that did not need to be tested.

Furthermore, given an external input sequence, a regular set was derived such that a nonconforming implementation system could be detected with this input sequence if and only if its component comprised a trace of the set. Based on this approach a method was also presented to construct a complete test suite and to select a minimal subset of this test suite having the same fault detection power. On the other hand, the proposed method can be used to reduce a test suite given by human experts while preserving its fault detection power.

6 REFERENCES

[Chow 78] *T. S. Chow, Test software design modelled by finite state machines,* IEEE Trans. SE-4, No. 3, pp. 178-187, 1978.

[Fujiwara 91] *S. Fujiwara, G. v. Bochmann, F. Khendek, M. Amalou, A. Ghedamsi , Test selection based on finite state models,* IEEE Trans., SE-17, No. 6, 1991, pp. 591-603.

[Hopkroft 79] *J. E. Hopkroft, J. D. Ulman, Introduction to automata theory, languages and computation,* 1979, Addison-Wesley, NY.

[Huang 96] *S. Huang, D. Lee, M. Stauskas, Validation-based test sequence generation for networks of extended finite state machines,* Proc. of the 1st Joint Intern. Conf. on Formal Description Techniques for Distributed Systems and Communication Protocols., and Protocol Specification, Testing, and Verification. 1996, pp. 403-418.

[IS-9646 91] *Information Technology, Open Systems Interaction, Conformance Testing Methodology and Framework,* International Standard IS-9646. ISO, 1991.

[Kim 72] *J. Kim, M. Newborn, The simplification of sequential machines with input restrictions,* IEEE Trans. on Computers, C-20, pp. 1440-1443, 1972.

[Kim 95] *M. C. Kim, S. T. Chanson, S. Yoo, Design for testability of protocols based on formal specifications,* Proc. of the 8th Intern Workshop on Protocol Test Systems, 1995.

[Lee 96] *D. Lee, K. K. Sabnani, D. M. Kristol, S. Paul, Conformance testing of protocols specified as communicating finite state machines - a guided random walk based approach,* IEEE Trans. on Communication, 1996, vol. 44, 5, pp. 631-640.

[Lima 97] *L. P. Lima, A. R. Cavalli, A pragmatic approach to generating test sequences for embedded systems,* Proc. of the 10th Intern. Workshop on Protocol Test Systems, 1997.

[Lima 98] *L. P. Lima, A. R. Cavalli, Application of embedded testing methods to service validation,* submitted to 2nd IEEE International Conference on Formal Engineering Methods, Brisbane, Australia, 1998.

[Merlin 83] *P. Merlin et al. On the construction of submodule specifications and communication protocols,* ACM Trans. on Programming Languages and Systems. Vol. 5, No. 1, 83.

[Parrow 89] *J. Parrow, Submodule construction as equation solving in CCS,* Theoretical Computer Science, 68, 1989, pp. 175-202.

[Petrenko 97] *A. Petrenko, N. Yevtushenko, Testing faults in embedded components,* Proc. of the 10th Intern. Workshop on Protocol Test Systems, 1997.

[Petrenko 96-1] *A. Petrenko, N. Yevtushenko, G. v. Bochmann, Testing deterministic implementations from nondeterministic FSM specifications,* Proc. of the 9th Intern Workshop on Protocol Test Systems, 1996.

[Petrenko 96-2] *A. Petrenko, N. Yevtushenko, G. v. Bochmann, Fault models for testing in context,* Proc. of the 1st Joint Intern. Conf. on Formal Description Techniques for Distributed Systems and Communication Protocols and Protocol Specification, Testing, and Verification. 1996, pp. 125-140.

[Petrenko 92] *A. Petrenko, N. Yevtushenko, Test suite generation for a given type of implementation errors,* Proc. of the Intern. Conf. on Protocol Specification, Testing, and Verification. 1992.

[Phalippou 94] *M. Phalippou, "Relations d'implantation et hypothèses de test sur des automates à entrées et sorties",* Ph.D. Thesis, Université de Bordeaux I, 1994.

[PT 96] *Component testing for mobile and broadband telecommunications,* COIMBRA, COPERNICUS project Proposal, 1996.

[Tretmans 92] *J. Tretmans, "A Formal Approach to Conformance Testing",* Ph.D. Thesis, 1992.

[Vasilevsky 73] *M. P. Vasilevsky, Failure diagnosis of automata. Cybernetics,* Plenum Publishing Corporation, NY, No. 4, 1973, pp. 653-665.

[Vuong 89] *S. T. Vuong, W. W. L. Chan, M. R. Ito, The UIOv-method for protocol test sequence generation,* Proc. of IFIP TC6 Second Inter. Workshop on Protocol test Systems. 1989, pp. 161-175.

[Watanabe 93] *Y. Watanabe, R. K. Brayton, The maximal set of permissible behaviours for FSM networks.* Proc. of the IEEE/ACM Intern. Conf. on Computer-Aided Design, 1993, pp. 316-320.

[Yannakakis 95] *M. Yannakakis, D. Lee, Testing finite state machines: fault detection,* Journal of Computer and System Sciences, 1995, 50, pp. 209-227.

7 BIOGRAPHY

Nina Yevtushenko received the Diploma degree in radio-physics in 1971 and Ph.D. in computer science in 1983, both from the Tomsk State University, Russia. She is currently a Professor at that University. Her research interests include the automata and FSM theory and testing problems.

Luiz Paula Lima Jr. is currently a PhD student at INT (*Institut National des Télécommunications*), Evry, France and he received his MSC degree in 1994 at UNICAMP (State University of Campinas), Brazil. His current research interests include object-oriented distributed systems and platforms (ODP/CORBA) and testing methods for these architectures.

Ana Rosa Cavalli received the Doctorat d'État es Mathematics and Computer Science in 1984 from the University of Paris VII, Paris, France. From 1985 to 1990, she was a staff research member at the CNET (*Centre National d'Etudes des Télécommunications*), where she worked on software engineering and formal description techniques. Since 1990, she joined the INT (*Institut National des Télécommunications*) as professor. Her research interests include formal description techniques, validation of protocols and services, computing methodology and testing methods for distributed architectures.

PART FIVE

Test Architectures and Concurrent Systems
Invited Talk

Test Architectures for Distributed Systems - State of the Art and Beyond

T. Walter
Computer Engineering and Networks Laboratory (TIK)
Swiss Federal Institute of Technology Zurich
Gloriastrasse 35, CH-8092 Zürich, Switzerland
Phone: (+41 1) 632 7007, Fax: (+41 1) 632 1035
http://www.tik.ee.ethz.ch

I. Schieferdecker
GMD FOKUS
Kaiserin-Augusta-Allee 31, D-10589 Berlin, Germany
Phone: (+49 30) 3463 7241, Fax: (+49 30) 3463 8241
http://www.fokus.gmd.de/tip

J. Grabowski
Institute for Telematics (ITM)
Medical University of Lübeck
Ratzeburger Allee, D-23538 Lübeck, Germany
Phone: (+49 451) 500 3723, Fax: (+49 451) 500 3722
http://www.itm.mu-luebeck.de

Abstract

A generic test architecture for conformance, interoperability and performance testing of distributed systems is presented. The generic test architecture extends current test architectures with respect to the types of systems that can be tested. Whereas in the conformance testing methodology and framework the focus is on testing protocol implementations, the generic architecture focuses on testing real distributed systems whose communication functions are implemented on different real systems and whose correctness can only be assessed when tested as a whole. In support of the latter requirement, a test system itself is regarded as being a distributed system whose behaviour is determined by the behaviour of components and their interaction using a flexible and dynamic communication structure.

Keywords

CTMF, TTCN, test architecture, types of testing, advanced distributed systems

1 INTRODUCTION

A distributed processing system is a system which can exploit a physical architecture consisting of multiple, autonomous processing elements that do not share memory but cooperate by sending messages over a communications network (Blair et al, 1998). Distributed processing is information processing in which discrete components may be located in different places, or where communication between components may suffer delay or may fail.

Distributed systems offer several advantages in comparison to centralized systems such as the ability to share resources, to be dynamically extended with new ones and to potentially increase availability and performance. Typically, distributed systems are heterogeneous in terms of interconnected networks, operating systems and middleware platforms they are based on, as well as in terms of programming languages used to develop individual components.

The goal of open distributed systems is to enable access to components of a distributed system from anywhere in the distributed environment without concern of its heterogeneity. Openness includes openness to various levels of the distributed system architecture: communication network, middleware platform and application level. Openness requires the definition of interfaces to the components on the different levels of distributed systems and a notion of compliance to these interfaces in order to ensure compatibility, interoperability and portability.

The rapid growth of distributed processing systems has led to a need for coordination and for standardized frameworks:

- The ISO/ITU-T Reference Model for Open Distributed Systems (RM-ODP) (ISO International Standard 10746, 1991), (Blair et al, 1998) provides a framework by describing a generic object-oriented architecture that is based on five different viewpoints. The viewpoints enable the description of distributed systems from various perspectives: enterprise, information, computation, engineering, and technology viewpoint. The framework also defines functions required to support ODP systems and transparency prescriptions showing how to use the ODP functions to achieve distribution transparency. The ODP framework is provided in terms of architectural concepts and terminology and is a rather general and generic framework to enable concrete standards for open distributed systems to emerge.

- The OMG Common Object Request Broker Architecture (CORBA) Object Management Group, 1995) provides an object-oriented framework for distributed computing and mechanisms that support the transparent interaction of objects in a distributed environment. The basis of the architecture is an object request broker (ORB) which provides mechanisms by which objects transparently invoke operations and receive results.
- The Telecommunication Information Networking Architecture (TINA) (TINA-C, 1998) aims at providing a framework for future telecommunication networks. The kernel of the architecture is a distributed processing environment (DPE) that adopts the concepts of the RM-ODP computational and engineering model. The basis for the TINA DPE is a CORBA ORB.
- The Open Group's Distributed Computing Environment (DCE) (Digital Equipment Corporation, 1992), (Open Software Foundation, 1992), (Lockhart, 1994), (Schill, 1996) provides a communication network and middleware platform independent platform for distributed processing systems. It is based on a client-server architecture and does not support object-oriented technologies.

The main concept which should ensure compatibility, interoperability and portability of the various components in a distributed processing system is conformance. Conformance testing is the main tool to check if components of a distributed processing system are able to interwork. ODP, CORBA, TINA and DCE have different definitions of conformance:

In ODP conformance is a relationship between specifications and implementations. The relationship holds when specific requirements in the specification are met by the implementation. ODP does not define conformance for specific requirements but rather identifies conformance points at which an implementation should be tested. Conformance testing in ODP means to test all requirements that are given by the viewpoint models of a distributed system.

Conformance in CORBA is defined in terms of compliance points, which refer to the interfaces of the CORBA architecture. Emphasis is given to compliance points from the ORB Interoperability Architecture (with GIOP, IIOP and ESIOP) in order to improve the interoperability between ORBs.

Currently, in the TINA framework there are no explicit definitions for conformance. However, work is in progress to use reference points, e.g., the retailer reference point, as the basis for defining conformance.

The DCE approach towards conformance is based on system specifications and associated conformance test suites. An implementation will only be claimed DCE conforming if it passes the conformance tests.

As a summary it can be said that the problem of conformance testing for ODP, CORBA, TINA and DCE based applications is not resolved. Research, industry and standardization bodies are working on solutions for providing unique and comparable test specifications.

However, the most successful conformance testing standard is the Conformance Testing Methodology and Framework (CTMF) (ISO International Standard 9646-1, 1995), (Linn, 1989), (Sarikaya, 1989), (Baumgarten et al, 1994) defined by ISO/IEC for the test of communication protocols and services that are in accordance with the OSI (Open Systems Interconnection) basic reference model (ISO International Standard 7498, 1984). Beside others, CTMF defines several test architectures, called abstract test methods, and the Tree and Tabular Combined Notation (TTCN) (ISO International Standard 9646, 1996), (Probert et al, 1992) for the specification of test cases. In this paper, the terms test architectures and abstract test methods are used as synonyms.

The CTMF test architectures and TTCN have been used successfully also outside the OSI conformance testing area, e.g., for testing ATM protocols. But, it has been recognized that the CTMF definitions, in general, cannot cope with other types of testing, e.g., real-time and performance testing, and with applications based on new architectures and frameworks (ODP, CORBA, TINA and DCE). For opening the scope of TTCN for real-time and performance testing several language extensions have been proposed (Schieferdecker et al, 1997), (Walter et al, 1997).

This paper intends to open the discussion on test architectures. A definition of a generic test architecture is proposed which can be adapted to different types of testing and to testing of applications based on new architectures and frameworks.

The paper is organized as follows: Section 2 presents the state of the art in test architectures. In Section 3 the requirements on test architectures from advanced distributed systems are identified. A new generic test architecture for distributed systems is proposed in Section 4. The application of the new architecture model is shown in Section 5. The paper concludes with a summary.

2 STATE OF THE ART IN TEST ARCHITECTURES

For a couple of years, the international standard 9646 (Conformance Testing Methodology and Framework, CTMF) (ISO International Standard 9646-1, 1995), (ISO International Standard 9646-2, 1995) has been the reference for test architectures. In its first and second part, it defines a number of abstract test methods. Abstract test methods describe how an implementation under test (IUT) is to be tested, i.e., what outputs from the IUT are observed and what inputs to the IUT can be controlled.

2.1 Abstract test methods

Taking the possible variety of real open systems configuration into account, a number of abstract test methods have been defined. Abstract test methods resemble a number of concepts that have been introduced in the Open Systems Interconnection Basic Reference Model (OSI BRM) (ISO International Standard 7498, 1984). In particular, a system under test (SUT) is generally a system that uses standardized OSI protocols in all layers, from the physical to the application layer. However, CTMF is also applicable to partial open systems, i.e., systems that use OSI protocols only in some layers to provide an OSI service, and CTMF is applicable to relay systems at network and application level. Within the SUT the IUT is subject to conformance testing and may comprise several adjacent protocol layers. The IUT is assumed to rest on an underlying service provider connecting SUT and test system.

Given the situation of an IUT which is a single layer protocol implementation. The interfaces between IUT and its adjacent layers within the SUT are referred to as points of control and observation (PCO). As the name suggests, at these points the behaviour of the IUT in performing communication tasks can be controlled and observed. As stated above control and observation are in terms of ASPs (which map to OSI service primitives) and PDUs embedded in ASPs. Similar to the PCOs located in the SUT, corresponding PCOs can be identified in the test system and the underlying service provider, i.e., PCOs may be allocated somewhere below and above the IUT. Even if the PCO through which the IUT is accessed is on the remote test system, it is said that control and observation of the IUT takes place from below.

Although PCOs can be located somewhere in the SUT below and above the IUT, not all conceivable arrangement of PCOs must be supported by the SUT. A few specific abstract test methods have been defined which clearly identify the requirements on the SUT and IUT concerning access to interfaces and location.

These abstract test methods use one or two PCOs. In case of two PCOs, one is below and one is above the IUT, otherwise only the PCO below the IUT is used for controlling and observing the IUT. Even if the PCO above the IUT is available, this does not imply the control and observation of this interface to the IUT is done from within the SUT. In the local test method (Figure 1) the upper PCO is managed from within the test system. In the distributed test method (Figure 1) this is done from within the SUT.

The active components in a test system which perform control and observation of the IUT are named lower and upper tester (LT & UT). Coordination of these components is defined in a test coordination procedure (TCP), which similar to a protocol specification, defines the rules for the cooperation of LT and UT. In addition, the underlying service provider is regarded as an active component which is assumed to be sufficiently reliable so that control and observation of an IUT can be done remotely.

The coordinated test method (Figure 2) explicitly uses a test management protocol (TMP) as TCP. Although an UT is sitting on top of the IUT, no PCO is used in this test methods. The UT is assumed to implement parts of the functionality of the test management protocol.

The remote test methods puts the least requirements on IUT, SUT and availability of PCOs (Figure 2). Only the PCO below the LT is available. No UT is used in the remote test method. However, for the purpose of test case specification, some UT functions which may be present, can be used if necessary. No assumptions are made

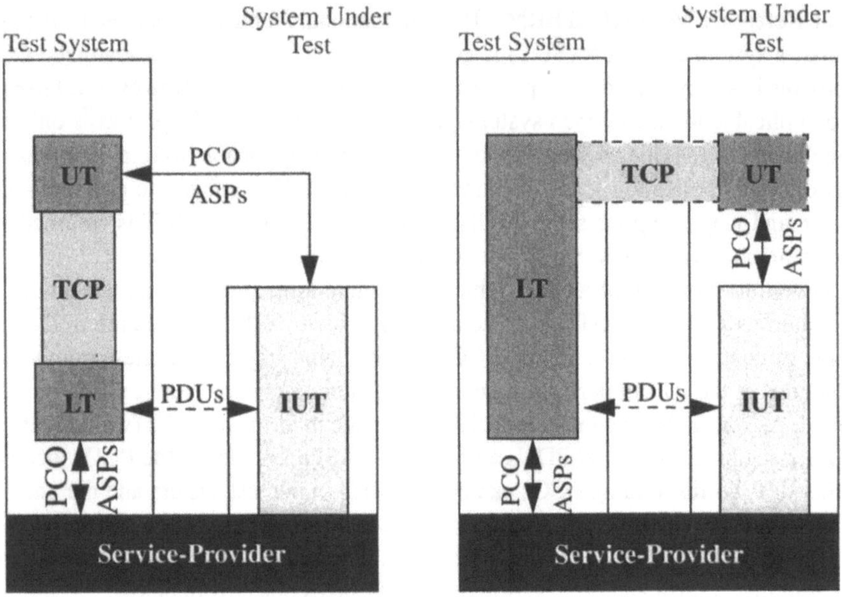

Figure 1 Local and distributed abstract test methods.

on the feasibility or implementability of these function in a real system.

In (Zeng et al, 1989) the Ferry Clip approach is presented. It provides implementation support for abstract test methods. Its purpose is to provide access to the lower and upper boundaries of an IUT in the SUT. The test system is remote from the SUT, but a specific component, called passive ferry clip, is installed on the SUT. It controls and observes the respective IUT interfaces. The passive ferry clip communication with its counterpart in the test system, called active ferry clip, uses a ferry control protocol and ferry transfer service. UT and LT run on the test system and communicate with the active ferry clip. The test case logic is still in the test components, but the ferry clips perform as mediators between test system and SUT and IUT, respectively. This approach makes the implementation of abstract test methods rather simple because the complexity of implementing a passive ferry clip is lower than implementing a complete UT.

2.2 Multi-party testing context

A generalization of the above discussed abstract test methods, also known as single-party testing context, is given by the multi-party testing context (Figure 3). In this setting, an IUT is supposed to communicate simultaneously with several real open systems; a networt of application relays, for instance, maintains communication links with several peers at the same time. In the multi-party testing context more than

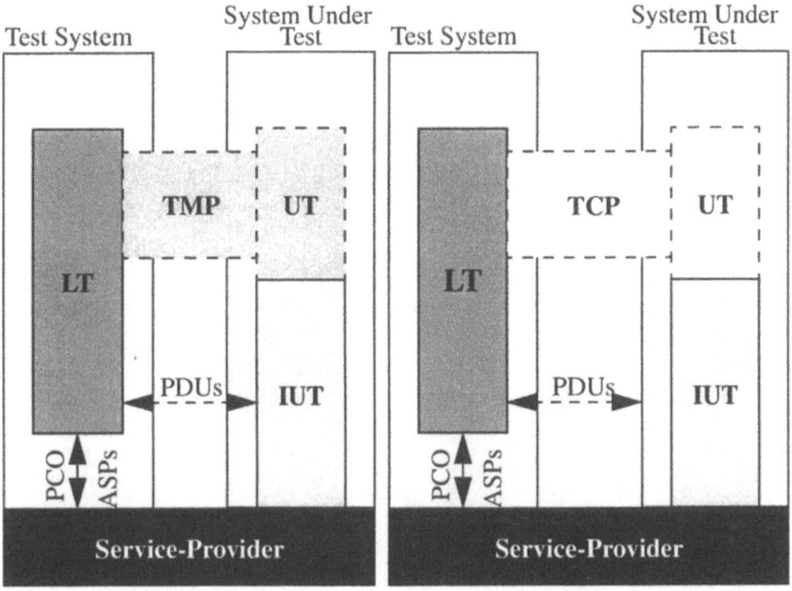

Figure 2 Coordinated and remote test methods.

one LT and zero, one or more UTs are active and control and observe the IUT. The coordination of LTs is performed by an entity referred to as lower tester control function (LTCF). In particular, the LTCF is the entity that determines the final test verdict after test execution. All LTs are required to return a preliminary test result to the LTCF after they have stopped test case execution. The LTCF is also responsible for starting all LTs. Coordination of LTCF, LTs and UTs is defined in a TCP. Communication between LTCF, LTs and UTs is supported by coordination points (CP) which, like PCOs, are modelled as two FIFO queues for inputs and outputs.

LTs communicate with the IUT and possibly with an UT as in the single-party context, the same rules for identifying points of control and observation apply.

2.3 Test Architecture beyond Conformance

In (ISO International Standard 9646-1, 1995), the following statement can be found: "The primary purpose of conformance testing is to increase the probability that different implementations are able to interwork.... Even if two implementations conform to the same protocol specification, they may fail to interwork fully. Trial interworking is therefore recommended." According to (ISO International Standard 9646-1, 1995) and (Gadre et al, 1990), conforming implementations may fail to interwork for the following reasons:

- protocol specifications contain options and implementations may differ in the options being supported

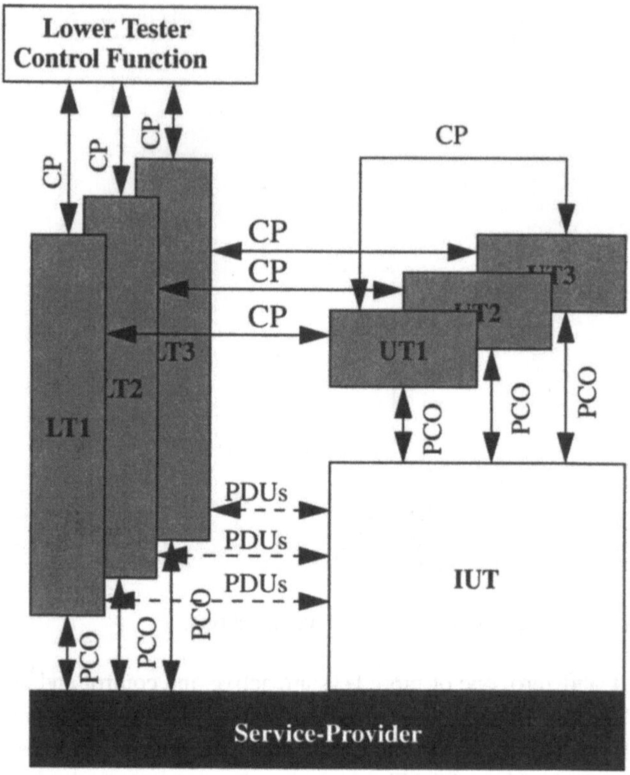

Figure 3 Multi-party testing contex.

- factors outside the scope of conformance testing and OSI, e.g., performance requirements, may impact the behaviour of real systems which is not foreseen by the specification.

In the following subsections, approaches to testing interoperability, performance and quality-of-service are discussed. Only since the evolution of multimedia applications has started some years ago, the latter type of testing has become an issue.

2.3.1 Interoperability Testing

Interoperability testing evaluates the degree of interoperability of implementations with one another. It involves testing both, the capabilities and the behaviour of an implementation in an interconnected environment and checking whether an implementation can communicate with another implementation of the same or of a different type. Interoperability testing is not standardized. However, a couple of interoperability testing proposals and guidelines exist (Buehler, 1994), (Myungchul et al, 1996), (EWOS ETG 028, 1993), (Hopkinson, 1996), (Gadre et al, 1990) (Hogrefe, 1990).

The two main approaches to interoperability testing in the context of OSI are passive and active testing (Gadre et al, 1990) (Hogrefe, 1990). The difference is that active testing allows controlled error generation and a more detailed observation of the communication, whereas passive testing on the other side involves testing valid behaviour only. OSI interoperability testing should be done by using a reference implementation (RI) of the protocol entity to be tested. Within the SUT RI and IUT are combined and the tester functions play the role of service users. The behaviour of the RI is correct by definition. In case that IUT and RI do not interwork this is due to an error in the IUT.

Figure 4 shows a generic passive interoperability test system architecture where two implementations are interconnected and control and observation are performed by two UTs.

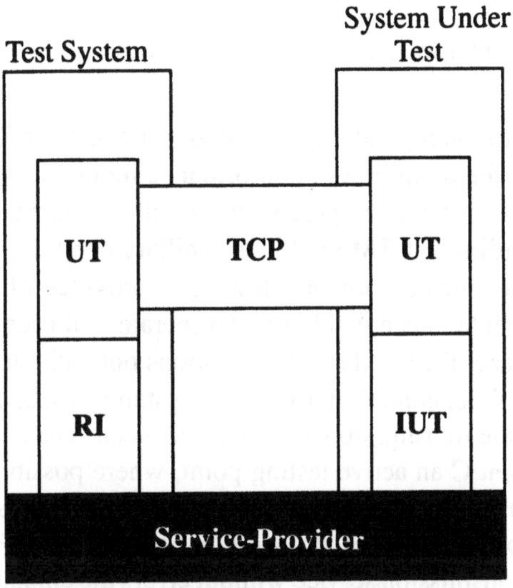

Figure 4 Passive Interoperability Testing Architecture.

Practice has shown that in most cases no RI is available. As a consequence, two IUTs are used and the communication between the IUTs is monitored or emulated. A test configuration following this approach and proposed by the ATM Forum in (Buehler, 1994) is shown in Figure 5.

However, the test configuration most often used for interoperability testing is simply the interconnection of the network components which have to interoperate. The message exchange between the network components is stimulated at the edges by a test components or by any suitable application. This complies with Figure 5 without monitor point C. The observed reactions on the edges were used to assign a verdict about the interaction between the SUTs. This form of testing has been termed pure interoperability testing (Myungchul et al, 1996). Set-up of such an pure interoperability test session is possible with limited hardware and software requirements, and thus it is very popular within the test labs. The main disadvantage of this test configuration is the missing monitor point C between the SUTs. It does not allow a statement about the protocol between the two systems and makes locating errors very difficult in the case of failure.

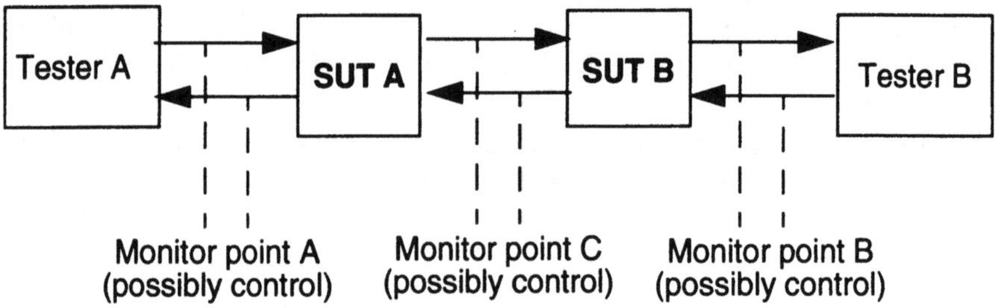

Figure 5 ATM Forum interoperability test architecture

Other approaches to interoperability testing use the test configuration shown in Figure 5, but differ in the definition of conformance requirements. Interoperability testing with strong conformance requirements on all monitor points realizes conformance testing according to CTMF with the ancillary condition of two IUTs. This includes static and dynamic conformance testing. It provides a high level of confidence that the implementations are able to interoperate, but causes problems if implementations to not meet the standard specifications but still interoperate.

In order to have a direct control on the tested systems, a specific test component is used for emulating the transmission service in between. The emulator component makes the monitor point C an active testing point, where possibly impairments are generated in accordance with the properties of the transmission service, i.e., errors such as loss, corruption, disordering, or insertion are generated. Please note, that in the case of a reliable transmission service without errors the emulator test component is not needed. An extended interoperability architecture is shown in Figure 11.

2.3.2 Performance Testing

The main objective of performance testing (Schieferdecker et al, 1997) is to test the performance of a network component under normal and overload situations. Performance testing identifies performance levels of the network component for different ranges of parameter settings and assesses the measured performance of the component. A performance test suite describes precisely the performance characteristics that have to be measured and procedures how to execute the measurements. In addition, the performance test configuration including the configuration of the network component, the configuration of the underlying networt, and the network load characteristics are described.

Depending on the characteristics of the network component under test, different types of performance test configurations are defined: end-user telecommunication application, end-to-end telecommunication service and communication protocol (Figure 6). Foreground test components (FT) implement control and observation of the network under test. Background test components (BT) generate continuous streams of data to load the network component under test. Monitor components are used to monitor the real network load during the performance test. BTs do not control or observe directly the network under test but implicitly influences the network under test by putting the network into normal and overload situations.

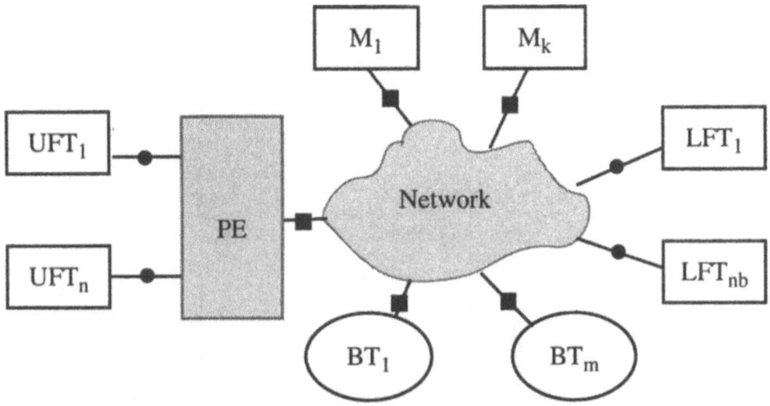

Performance Test Components ● Points of Control and Observation

SUT Components ■ Measurement Points

UFT: Foreground Tester for Emulated Protocol User
LFT: Foreground Tester for Emulated Peer-to-Peer Protocol Entity
BT: Background Tester
M: Monitors of Real Network Load
PE: Tested Protocol Entity

Figure 6 Performance test configuration for a communication protocol.

2.3.3 Quality-of-Service Testing

Quality-of-service (QoS) (ITU-T Recommendation X.200, 1989), (Danthine et al, 1992), (Danthine et al, 1993) refers to a set of parameters that characterize a connection between communication entities across a network. QoS parameters are performance and reliability oriented such as throughput, delay, jitter or error rates and failure probabilities, respectively. QoS parameters are negotiated by service users and service provider at connection set-up and should be maintained during the life-time of the connection. A QoS semantics defines the procedures how QoS parameters are negotiated and how negotiated QoS parameters are to be handled. In particular, a QoS semantics defines how QoS parameter violations are to be processed. It is mainly the service provider who is in charge of maintaining negotiated QoS parameters.

QoS testing refers to assessing the behaviour of protocol implementations performing QoS maintenance. However, it is not necessary to control and to observe the behaviour of the implementation directly. It suffices if the tester can eventually observe the specified behaviour according to the agreed QoS semantics.

Figure 7 presents a testing architecture for QoS testing (Grabowski et al, 1995). As can be seen, the testing architecture is very similar to the passive interoperability testing architecture (Figure 4). The obvious difference is that in QoS testing the IUT is distributed. Furthermore, some QoS parameter tests, e.g., error rate tests and delay tests, require the active involvement of the network. For this, the network has to be configurable. The testing architecture, therefore, provides a communication link between testers and network for the exchange of configuration information.

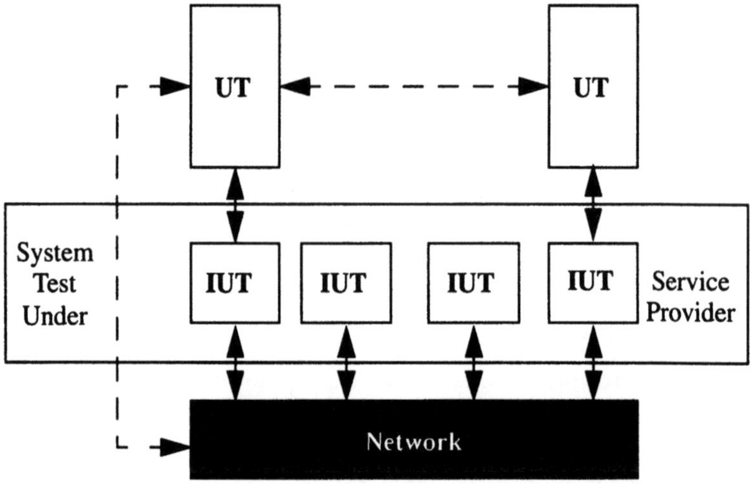

Figure 7 QoS testing architecture.

2.3.4 An Evaluation of Test Architectures

As the previous discussion of the different test architecture has shown, interoperability, performance and QoS test architecture are straightforward extensions of the CTMF abstract test methods. This is particularly true for the interoperability test architecture where the lower tester functions are provided by a reference implementation, which in turn is driven by a test component on top. Similarly, the QoS test architecture transforms the interoperability test architecture into a distributed test architecture, where the functions of the IUT are distributed over several real systems. Control and observation of the distributed IUT are done by UTs. Additionally, UTs can control the behaviour of the underlying network. This, however, makes the QoS test architecture different. Whereas in the CTMF test methods and interoperability test architectures the underlying network is regarded as a black-box, the QoS test architecture weakens this assumption because certain behaviour of the SUT is observable only if network behaviour changes in a controlled manner. As such, the performance test architecture does something similar. In order to force the IUT into specific states, additional network load is generated by specifically designed test components. Thus, the network is being controlled by the test system as well.

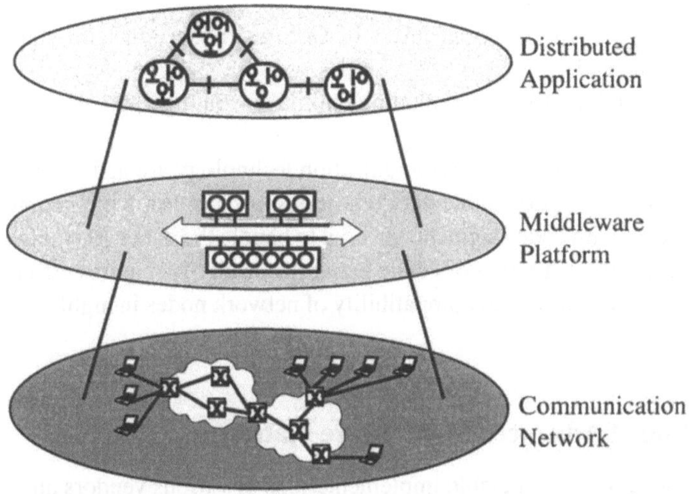

Figure 8 Distributed Systems Architecture.

3 REQUIREMENTS FROM ADVANCED DISTRIBUTED SYSTEMS

A simplified view of distributed system architectures is given in Figure 8. A distributed system consists of application level objects that interact through well-defined interfaces. The middleware platform is a distributed computing environment that supports the implementation of the distributed application by offering various distribution transparencies such as access and location transparency. Example middleware platforms are OMG CORBA or OSF DCE. The communication network with various network nodes and end-systems offers transmission services to the middleware platform that are used to support the communication between the components of the distributed system.

Testing one of the three layers imposes different requirements on testing. Subsequently, testing the communication network, the distributed processing environment, and the distributed applications level is discussed.

3.1 Testing of Advanced Communication Networks

The IETF activities on defining communication protocols, services and mechanisms in internetworks are the driving forces in defining advanced network technologies. The following aspects of the current Internet (in particular of the Integrated and Differentiated Service Architecture (IETF Differentiated Service Working Group, 1998), (IETF Integrated Services Working Group, 1998)) impose new requirements on testing network nodes as well as end-to-end services:

- a variety of group communication scenarios such as multicast with dynamic join and leave in multicast groups,
- stream support with different levels of QoS guarantees and soft-state resource reservations, and
- variety of routing protocols including multilayer routing approaches.

Another direction in advanced communication technology are active networks. Active networks use programmable network nodes or capsules, which combine communication data with code fragments on how to handle the data in the network nodes (Tennenhouse, 1997). Testing of active networks primarily requires efficient means to test the interoperability and compatibility of network nodes in highly dynamic environments.

3.2 Testing of Advanced Middleware Platforms

Due to the need for interoperable implementations of various vendors and due to the complex nature of middleware platforms such as CORBA, there is a need for a methodology that middleware platforms can be analysed with respect to their compliance to the respective specifications. This is done in order to increase the likelihood that they can interoperate.

In essence, a middleware platform is used by a distributed application like a black box with various service access points. However, in order to evaluate for example a CORBA ORB as such (what includes testing the ORB Core, the Interoperability Reference Points, CORBA Services, and CORBA Facilities), a grey-box testing approach has to be taken, which supports the test access to ORB internal interfaces (Rao et al, 1997).

3.3 Testing of Advanced Distributed Applications

With the development of ODP and TINA -- being an instantiation of ODP for telecommunication systems -- and the provision of new and complex services such as telecommunication, management and information services which may be deployed in the context of various distributed object computing environments, the need to validate and test large and heterogeneous object systems is becoming increasingly important. Testing may be used to check

- components of the applications individually,
- conformance to reference points, and
- to check individual service components working together in a multi-service environment.

3.4 Requirements of Testing Advanced Distributed Systems

As a result of the above analysis, the following requirements of testing advanced distributed systems can be identified:

- development of distributed testing architectures with means for synchronizing distributed test components;
- support of dynamically configurable and scalable test architectures;
- ability to express test configurations for different communication scenarios;
- possibility to use grey-box testing with access to internal components and interfaces;
- support of real-time, performance and QoS testing for distributed systems to test time-related aspects of distributed systems;
- support of interoperability testing to focus on essential interoperability aspects;
- development of a test methodology coherent with object-oriented technologies which are used for the development of distributed systems;
- test architectures that make use of management, monitoring and measurement systems;
- methods that support testing in the pre-deployment, deployment and usage phase of distributed systems;
- efficient testing methods in order to deal with the complexity of distributed systems.

4 GENERIC TEST ARCHITECTURE FOR DISTRIBUTED SYSTEMS

The requirements of advanced distributed systems with respect to testing architectures are not met by CTMF. There exists today no unified approach for a flexible and adaptable testing architecture. Since the grade of distribution defines the complexity of testing and determines applicable test architectures, a taxonomy of testing is given in Table 1. In this section a definition of a generic test system architecture is presented which equally well fits the different testing types in Table 1.

Table 1: Testing taxonomy (with respect to the grade of distribution)

System under Test	Test System	Selected Approaches
centralized	centralized	CTMF peer-to-peer
centralized	distributed	CTMF multi-party testing context

Table 1: Testing taxonomy (with respect to the grade of distribution)

System under Test	Test System	Selected Approaches
distributed	centralized	Test execution of telecommunications services (Lima et al, 1997)
distributed	distributed	General design rules for distributed tester

4.1 Generic Test Architecture Model

The idea for a generic test architecture is a tool box of elements, which can be combined generically to a test architecture suitable for a specific application or system to be tested. A test architecture comprises several instances of different types of components. The types are:

- implementation under test (IUT), i.e., the implementation or a part of the distributed application to be tested;
- interface component (IC), i.e., a component which is needed for interfacing IUTs, e.g., an underlying service or an application in which an IUT is embedded;
- test component (TC), i.e., a component which contributes to the test verdict by coordinating other TCs or controlling and observing IUTs.
- controlled component (CC), i.e., a component which does not contribute to the test verdict but provides SUT specific data to TCs or the SUT, e.g., a load generator, an emulator or a simulator;
- communication point (CoP), i.e., a point at which communication takes place and at which communication can be observed, controlled or monitored;
- communication link (CL), i.e., a means for describing possible communication flows between TCs, IUTs, NCCs and CCs;
- system under test (SUT), i.e., a combination of ICs and IUTs.

For describing a test architecture in an intuitive and understandable manner a graphical representation might be preferable. The different types of components are explained by using the test architecture shown in Figure 9.

4.1.1 Implementation Under Test (IUT)

An IUT is meant to be a piece of software or hardware to be tested. The entire application to be tested may comprise several IUTs. The IUTs may have different functionality or be different instances of the same type, i.e., symmetrical peer entities of a protocol. An IUT is a black box which can be observed and controlled either di-

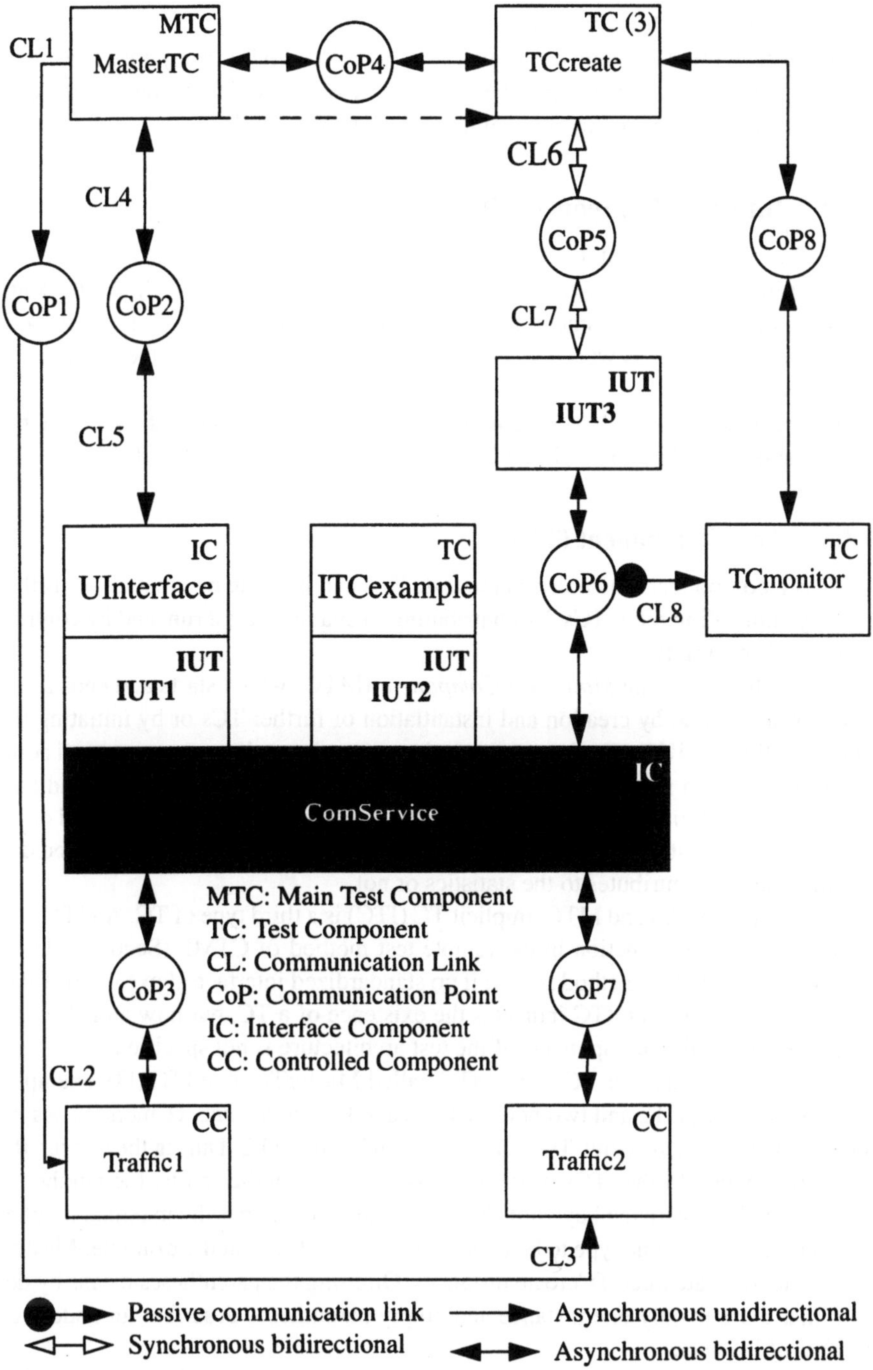

Figure 9 Generic test architecture.

rectly via CLs, or indirectly via ICs.

The test architecture shown in Figure 9 includes three IUTs. They are represented by means of boxes which are inscribed with the keyword IUT in the upper right corner and an IUT name, i.e., in our example IUT1, IUT2 and IUT3.

4.1.2 Interface Component (IC)

An IUT may be embedded in other applications or may be only interfaced via underlying services. The components are termed Interface Components (IC). An IC is not controlled by the test equipment, it is only used to interface the IUT. For testing, it is assumed that an IC is working correct. Within a test architecture there may exist several ICs.

Figure 9 includes two ICs. IC ComService provides a communication service between the three IUTs and IC UInterface describes an interface to the IUT IUT1.

4.1.3 Test Component (TC)

A TC is a component which drives the test. This can be done by creation of further TCs, by controlling other TCs, by contributing to evaluate a test run, and by controlling and observing IUTs.

There should be one *Main Test Component* (MTC) which starts and ends a test run. A start is done by creation and instantiation of further TCs or by initiating the first stimuli to the IUTs. Ending a test run does not necessarily mean to stop all communication and to stop all TCs. It means that the MTC should be able to indicate when the run is finished. When a test run is finished a final verdict is assigned or, in case that for statistical evaluation several runs are required, it should be decided whether the run contributes to the statistics or not.

In addition to TCs and MTC, implicit TC (ITC) is a third type of TC. An ITC corresponds to the UT function in the remote test method of CTMF (Section 2.1). In CTMF the remote test method is used if no standardized interface above the IUT can be used. This means, an ITC indicates the existence of a TC, but how an ITC communicates with other components of the test architecture is not specified.

Figure 9 presents four TCs, one MTC called MasterTC, one ITC (ITCexample) used to interface IUT2 and two normal TCs called TCcreate und TCmonitor. As indicated by the dashed arrow, TCcreate is created by the MTC. During the test run the MTC may create further TCs of the same type. This is indicated by the number in parentheses following the keyword TC. This number describes the maximal number of instances of the same type to be created during a test run. In the example, MasterTC is able to create three TCcreate instances. Omitting the parentheses means by default that there is only one instance and empty parentheses describes an undefined number of instances.

Use and specification of TCs needs special support in the used test specification language. In TTCN, for example, the handling of the UT function within the remote test method is supported by means of implicit send signals. Other points to be handled in the used test specification language are the creation of initially existing TCs when a test run starts, the stopping of TCs when a test run finishes, the evaluation of test runs by TCs, and the communication among TCs, e.g., addressing in case of dynamic TC creation.

4.1.4 Controlled Component (CC)

A CC is a component which is used to set-up the test case specific environment or is used by TCs to control test execution. Examples of CC types are load generators for providing background load, emulators which may be used instead of ICs, or simulators which can be used for comparing the reactions of an IUT with the output of a simulator.

From an abstract point of view a CC may be seen as a TC. Our intention for introducing the CC type is to distinguish explicitly between control and environment of a test. In performance testing (ATM Forum, 1994) a similar distinction is made by using the terms foreground and background load.

Figure 9 includes the CCs Traffic1 and Traffic2. Both generate and manage background load for the IC ComService.

4.1.5 Communication Point (CoP)

A CoP in the generic test architecture corresponds to the PCOs in CTMF. It denotes a point in the test architecture where communication can be observed, controlled, and, in addition to CTMF, be monitored. It is allowed to access several communication flows at the same CoP. From this point of view, a CoP bundles communication flows. For simplicity, no special semantics are assigned to CoPs; note that in CTMF PCOs have a FIFO queue semantics. The semantics of the communication is given to CLs.

In Figure 9 eight CoPs, named CoP1, CoP2, ..., CoP8, are used. They are used to describe the CoPs relevant for testing between TCs, IUTs, ICs, and CCs. Points of communication which cannot be accessed, e.g., between IUT1 and UInterface, or from which the test specification may have to abstract, e.g., between IUT2 and ITCexample, are not described.

4.1.6 Communication Link (CL)

IUTs, ICs, TCs and CCs are connected with CoPs by using CLs. A CL describes a possible communication and the kind of communication which may take place. Active communication can be classified by the kind, i.e., either synchronous or asynchronous, and the direction, i.e., either unidirectional or bidirectional. In addition, a passive CL allows to monitor communication, i.e., to listen at a CoP.

The test architecture shown in Figure 9 includes 18 CLs (only the CLs used in the discussion below are annotated with labels) describing several types of communication. For example, asynchronous unidirectional communication from MasterTC to

Traffic1 and Traffic2 is realized by using CL1, CL2 and CL3 via CoP1. This communication will only be used to start and stop the CCs (for instance, in form of broadcast messages). Bidirectional asynchronous communication between MasterTC and UInterface is described by using CL4 and CL5. Bidirectional synchronous communication between IUT3 and TCcreate is defined by CL6 and CL7 (via CoP5). CL8 describes a passive CL. TCmonitor monitors the communication between IUT3 and ComService.

For the dynamic creation of TCs it is assumed that CLs are also created dynamically and that the CoPs used for communication are all known before test execution, i.e., CoPs cannot to be created dynamically. For the example in Figure 9 this means that during a test run there may exist up to three instances of CL6.

For the different testing requirements, this basic components are combined so that a testing architecture results which fits the specific needs of an application. It has to be noted that for some specific cases, for instance, real-time testing, additional test specific information has to be coded in the dynamic behaviour of a test case. Not all information relevant for testing can be mapped to testing architecture only.

4.1.7 System Under Test (SUT)

In CTMF an IUT together with ICs is called system under test (SUT). The term SUT is used, but in contrast to CTMF, an SUT includes several IUTs. An example of the use of SUTs within a test architecture for interoperability testing is shown in Figure 10.

4.2 Evaluation of Requirements from Advanced Distributed Systems

In the following an assessment is given whether and how the defined requirements of testing advanced distributed system are met by the generic test architecture:

- Development of distributed testing architectures with means for synchronizing distributed test components: With the generic test architecture a support for distributed test architectures is given. Test system and SUT may be arbitrarily distributed over real systems. Support for synchronizing distributed test components is given. Communication of synchronization information along communication links is supported.
- Support of dynamically configurable and scalable test architectures: Firstly, test components can be created on demand by the main test component. Secondly, more than one instance of a specific test component may exist in the system. Together, this gives a test case specifier some initial degree for a dynamic configuration of test architectures.
- Ability to express test configurations for different communication scenarios (unicast, multicast, or broadcast): Communication links support unicast communication, either one-way or two-way. By adding the concept of communication points, multicast communication scenarios can be defined. Attaching a test component to a communication point gives this component access to all data going

through the communication point.

- Possibility to use grey-box testing with access to internal components and interfaces: By introducing passive communication links a means for monitoring even inside the SUT is given. However, an CoP has to be defined within the SUT.
- Support of real-time, performance and QoS testing for distributed systems to test time-related aspects of distributed systems: The generic test architecture is an essential component which is required in the mentioned types of testing. In performance testing, for instance, background test components are used for overloading the SUT. These test components map to controlled components. However, the challenges in real-time, performance and QoS testing are more on how to define test cases; which notation to be used and how test cases are to be implemented.
- Support of interoperability testing to focus on essential interoperability aspects: See Section 5.
- Development of a test methodology that is coherent with object-oriented technologies which are used for the development of distributed systems: In this paper the focus is on test architectures, and not on a test methodology which is more than just a test architecture, e.g., does also consider test case specification and generation, test implementation, test execution, test result analysis etc.
- Test architectures that make use of management and/or monitoring and measurement systems: These specific components can be integrated in a specific instantiation of the generic test architecture. The basic support is given, for instance in terms of passive communication links to monitor and to measure systems.
- Methods that support testing in the pre-deployment, deployment and usage phase of distributed systems: This again is a question of methodology. Testing as understood by CTMF is pre-deployment testing.
- Efficient testing methods in order to deal with the complexity of distributed systems: The generic test architecture has been invented with this requirement in mind. In the following section it is shown how problem specific test architectures are built from the basic components of the generic test architecture.

5 TEST ARCHITECTURES FOR DISTRIBUTED SYSTEMS

For proving the generality of our approach the different test architectures described in Section 2 have to be mapped to the generic model. The mapping of the conformance testing architectures according to CTMF is straightforward because the development of the generic model starts with the CTMF concepts and there is a one-to-one mapping for most of the CTMF concepts onto the generic model. Therefore, in the following an extended interoperability testing architecture and a performance testing architecture are described by using our generic model.

5.1 Interoperability Testing Architecture

Figure 10 maps the interoperability testing architecture proposed by the ATM Forum (Section 2.4.1 Figure 5) to the generic model. The CoPs A, B, C specify explicitly the points where communication is monitored. For the task of the monitoring TCs MA, MB and MC have been introduced.

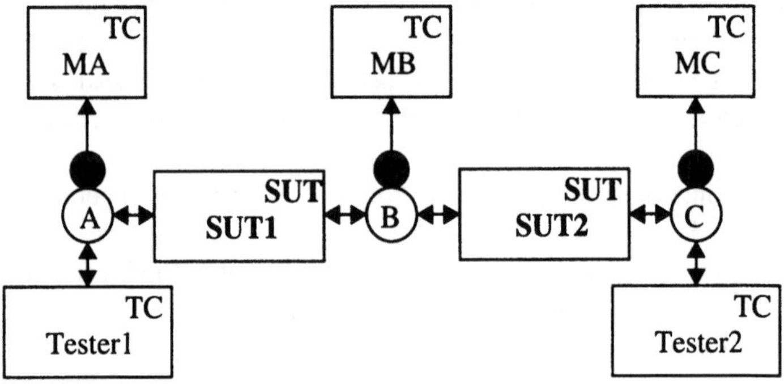

Figure 10 ATM Forum interoperability architecture.

In Section 2.4.1 an extended test architecture for interoperability testing is proposed. Figure 11 shows a transformation of the interoperability test architecture to the generic model. In order to have control over the communication of the IUTs which have to interoperate, the used communication service is emulated by the CC Emulator. The component Emulator communicates with TC Monitor, thus reporting all communication between the IUTs and to receive commands in case the emulator should actively influence the communication, e.g., decrease the performance or corrupt data packets.

5.2 Performance Testing Architectures

Figure 12 shows an instance of the performance testing architecture sketched in Figure 6 (Section 2.4.2) by using the generic model. Background Tester (BT1, BT2) are meant to be load generators and therefore are mapped to CCs. The CC Monitor measures of the real network load during a test run. The Foreground Tester (UFT1, UFT2, LFT1, LFT2) are mapped onto TCs. PCOs and measurement points are mapped onto CLs, their physical interface is described by using CoPs. Additionally, an MTC communicates with the other components in order to start and stop a test run.

6 CONCLUSIONS

In this paper a generic test architecture for conformance, interoperability, perform-

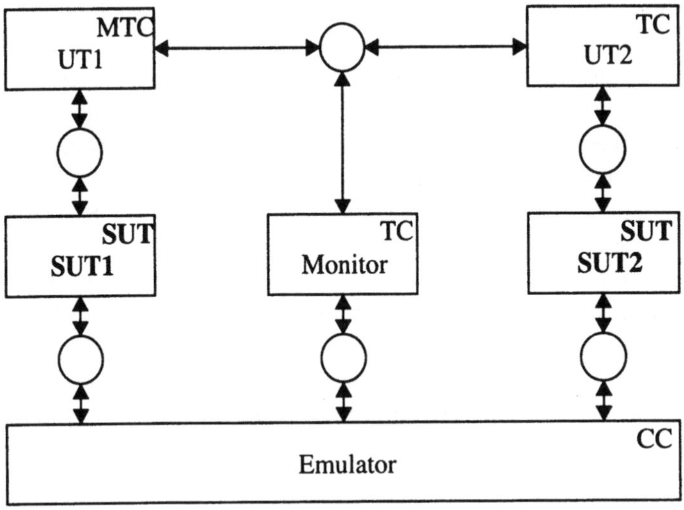

Figure 11 Interoperability Test Architecture.

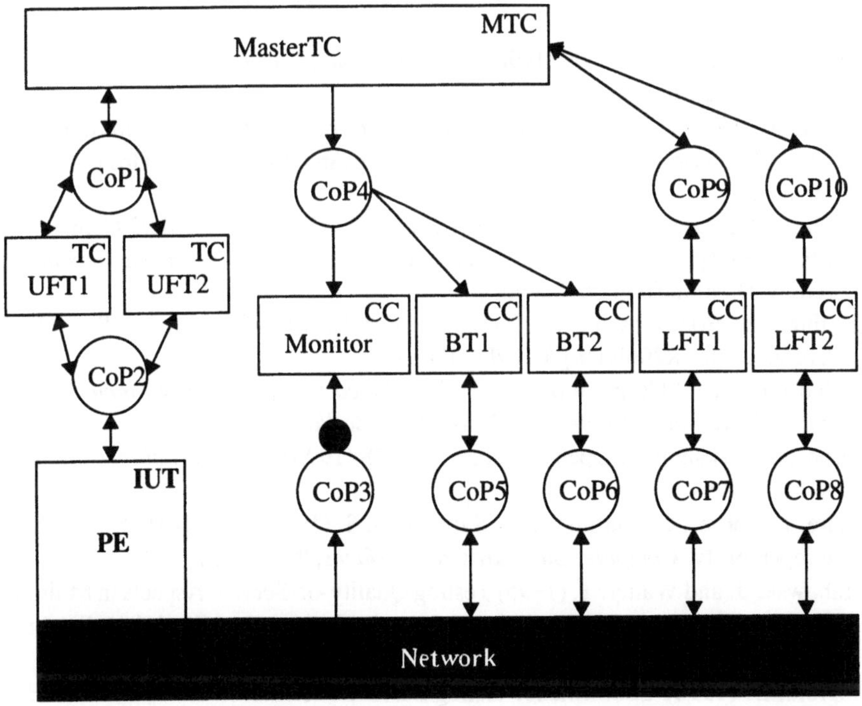

Figure 12 Performance test architecture.

ance and real-time testing has been proposed. The development of the generic test architecture has been motivated by the observations that (1) the abstract test methods defined in the CTMF context are too restrictive with respect to the types of systems that can be tested; and (2) the up-coming new kind of advanced distributed (object, real-time, safety-critical) systems require a flexible and adaptable test architecture. The proposed generic test architecture has been designed as a toolbox whose components can be configured as needed and, thus, provide the required flexibility to set up any required test system configuration.

The discussion of the state-of-the-art in test architectures has been done looking at a test system as a distributed system that consists of active components, named test components, service provider or IUT in CTMF, and passive components PCOs or CPs. The latter define a static communication structure over the active components. In the generic test architecture this point of view has been developed further. Its applicability to different test scenarios has been demonstrated.
For the future a language support for the new test architecture will be investigated.

7 REFERENCES

ATM Forum (1994) *Introduction to ATM Forums Test Specifications.* af-test.0022.pdf, http://www.atmforum.com/atmforum/specs/approved.html.

Baumgarten, B. and Giessler, A. (1994) *OSI Conformance Testing Methodology and TTCN*, Elsevier.

Blair, G.S. and Stefani, J.-B (1998) *Open distributed processing and multimedia.* Addison Wesley Longman Ltd.

Buehler W. (Ed.) (1994) *Introduction to ATM Forum Test Specifications, Version 1.0.* ATM Forum Technical Committee, Testing Subworking Group, af-test-0022.000.

Danthine, A., Baguette, Y., Leduc, G., Léonard, L. (1992) The OSI 95 Connection-Mode Transport Service - The Enhanced QoS, in *High Performance Networking* (ed. A. Danthine and O. Spaniol), IFIP.

Danthine, A., Bonaventure, O. (1993) From Best Effort to Enhanced QoS. CIO Deliverable, No. R2060/ULg/CIO/DS/P/004.

Digital Equipment Corporation (1992) *Distributed Computing Environment Application Development Reference.* Maynard, Maryland, U.S.A.

European Workshop for Open Systems ETG 028 (1993) *Interoperability Classification.* Brussels.

Gadre, J., Rohrer, C., Summers, C., Symington, S. (1990) A COS Study of OIS Interoperability. *Computer Standards & Interfaces*, **9**, 217-237.

Grabowski, J. and Walter, T. (1995) Testing Quality-of-Service Aspects in Multimedia Applications, in *Protocols for Multimedia Systems 2nd Workshop*, Salzburg, Austria.

Hogrefe, D. (1990) Conformance testing based on formal methods, in *FORTE 90 Formal Description Techniques* (ed. J. Quemada, A. Fernandez).

Hopkinson T. (Ed.) (1996) *Scoping Further Activity for Interoperability Testing.* European Workshop for Open Systems EWOS/TA PT N034, Final Report Version 1.0 EWOS EGCT/96/130 R1, Brussels.

IETF Differentiated Services Working Group (1998) *An Architecture for Differentiated Services*. Internet-Draft, draft-ietf-diffserv-arch-00.txt, http://www.ietf.org/html.charters/diffserv-charter.html.

IETF Integrated Services Working Group (1998) *Integrated Services* http://www.ietf.org/html.charters/intserv-charter.html.

ISO International Standard 7498 (1984) *Information processing systems - Open Systems - Basic Reference Model.*

ISO International Standard 9646-1 (1995) *Information technology - Open Systems Interconnection - Conformance Testing Methodology and Framework Parts 1: General concepts.*

ISO International Standard 9646-2 (1995) *Information technology - Open Systems Interconnection - Conformance Testing Methodology and Framework Parts 2: Abstract Test Suite specification.*

ISO International Standard 9646-3 (1996) *Information technology - Open Systems Interconnection - Conformance Testing Methodology and Framework Parts 3: The Tree and Tabular Combined Notation (TTCN).*

ISO International Standard 10746 (1991) *Information processing systems – Open Systems Interconnection: Reference Model for Open Distributed Processing, Part 2: Descriptive Model.*

ITU-T Recommendation X.200 (1989) *Data Communication Networks Open Systems Interconnection (OSI) Service Definitions Recommendations X.200 - X.219.*

Lima Jr., L.P. and Cavali, A. R. (1997) Test Execution of telecommunications services, in *Proc. of IFIP FMOODS '97*, Canterbury, UK.

Linn, R. (1989) Conformance Evaluation Methodology and Protocol Testing. *IEEE Journal on Selected Areas in Communications*, **7**, 1143 – 1158.

Lockhart, H.W. (1994) *OSF DCE – Guide to Developing Distributed Applications.* McGraw-Hill, New York, U.S.A.

Myungchul K., Gyuhyeong K., Yoon D. C. (1996). *Interoperability Testing Methodology and Guidelines.* Digital Audio-Visual Council, System Integration TC, DAVIC/TC/SYS/96/06/006.

Object Management Group (1995) *The Common Object Request Broker: Architecture and Specification; Revision 2.0.* Framingham, Massachusetts, U.S.A.

Open Software Foundation (1998). *http://www.opengroup.org.*

Probert, R. and Monkevic, O. (1992), TTCN: The International Notation for Specifying Tests of Communications Systems. *Computer Networks and ISDN Systems*, **23**, 111 – 126.

Rao, S., BeHanna, Ch., Sun, M., Forys, F. (1997) *CORBA Service Test Environment.* NEC Systems Laboratory, Inc.

Sarikaya, B. (1989) Conformance Testing: Architectures and Test Sequences. *Computer Networks and ISDN Systems*, **17**.

Schieferdecker, I., Stepien, B., Rennoch, A. (1997) PerfTTCN, a TTCN language extension for performing testing, in *Testing of Communication Systems Volume 10* (ed. Myungchul Kim, Sungwon Kang, Keesoo Hong), Chapman & Hall, 21-36.

Schill, A. (1996) *DCE - das OSF Distributed Computing Environment, Einführung und Grundlagen*. Springer Verlag.

Tennenhouse, D., Smith, D., Sincoskie, D., Wetherall, D., Minden, G. (1997) A survey of active networks research. *IEEE Communications Magazine*, **35**, 80-86.

TINA-C (1998) *http://www.tinac.com.*

Walter, T. and Grabowski J. (1997) Real-time TTCN for testing real-time and multimedia systems, in *Testing of Communication Systems Volume 10* (ed. Myungchul Kim, Sungwon Kang, Keesoo Hong), Chapman & Hall, 37-54.

Zeng, H.X., Chanson, S.T., Smith, B.R. (1989) On Ferry Clip Approaches in Protocol Testing. *Computer Networks and ISDN Systems*, **17**, 77-88.

Schütz, W. *On the Testability of Distributed Real-Time Systems*. Report of the ESPRIT Basic Research Project 3092 *Predictably Dependable Computer Systems*.

8 BIOGRAPHY

Thomas Walter received his Diploma in Informatics from University of Hamburg and his Doctorate degree in electrical engineering from the Swiss Federal Institute of Technology in Zurich (ETHZ) in 1987 and 1993, respectively. Since 1986 he is with the Computer Engineering and Networks Laboratory (TIK) of ETHZ, and in 1994 he became lecturer at ETHZ. His current research interests include formal methods for specification and validation of real-time systems. Besides this he is also active in setting up an infrastructure for teleteaching at ETHZ.

Ina Schieferdecker studied mathematical computer science at the Humboldt University in Berlin and received her Ph.D. from the Technical University in Berlin in 1994. She attended the postgraduate course on open communication systems at the Technical University in Berlin. Since 1993, she is a researcher at GMD Fokus and a lecturer at Technical University Berlin since 1995. She is working on testing methods for network components, and carries out research on formal methods, performance-enhanced specifications and performance analysis.

Jens Grabowski studied computer science and chemistry at the University of Hamburg, Germany, where he graduated with a diploma degree. From 1990 to October 1995 he was research scientist at the University of Berne, Switzerland, where he received his Ph.D. degree in 1994. Since October 1995 Jens Grabowski is researcher and lecturer at the Institute for Telematics in Lübeck, Germany.

Using partial-orders for detecting faults in concurrent systems

A. Petrenko[a], A. Ulrich[b], and V. Chapenko[c]

a *Centre de Recherche Informatique de Montreal (CRIM), 1801 Av. McGill College, Suite 800, Montreal, H3A 2N4, Canada, petrenko@crim.ca*

b *Department of Computer Science, University of Magdeburg, PF 4120, 39016 Magdeburg, Germany, ulrich@cs.uni-magdeburg.de*

c *Institute of Electronics and Computer Science, University of Latvia, 14 Dzerbenes street, Riga, LV-1006, Latvia, chapenko@edzi.lza.lv*

Abstract

The paper suggests test derivation approaches to obtain test suites for concurrent systems based on the concept of fault coverage criteria in opposition to structural test coverage criteria. Using a partial-order model, called *Mazurkiewicz Trace Machine* (MTM), for test derivation, the state explosion problem can be alleviated. The derived test suites are characterized by their small size compared to test suites from traditional test derivation approaches and exhibit a defined degree of fault coverage according to certain fault models. The fault models of concurrent systems considered in the paper are based on the most common faults, acceptance, refusal, and transfer faults. A scenario of test execution in concurrent systems, including a suitable test architecture, is discussed that explains the application of a test suite derived from an MTM in a test run.

Keywords

Concurrent systems, specification-based testing, test derivation, fault coverage.

1 INTRODUCTION

1.1 Motivation

Testing is an important means in the development cycle of software. It comprises the derivation of a *test suite* from a suitable specification of the software system and the application of the test suite in a *test run* to test whether an implementation of a system conforms to the specification. If the behaviors of the implementation and of the test suite differ from each other, an error probably exists in the implementation. In order to derive test suites from a formal specification of a system, two different approaches are known: (1) *structural testing* based on coverage criteria in the specification and (2) *fault testing* based on an assumed fault model of the implementation [3]. The first approach selects test suites according to a coverage criterion that is defined over the syntactical structure of the formal specification. The coverage criteria are based on empirical knowledge that have been proven some usefulness in practice. Examples of test coverage criteria are the statement coverage, edge coverage or path coverage criterion [9] or other coverage criteria defined in particular for concurrent systems [29].

On the other hand, fault testing uses a designated *fault model* of the implementation of a system that is decoupled from the syntactical structure of the system. Such fault models in the realm of labeled transition systems (LTSs) are, for instance, acceptance and refusal fault models or next-state fault models [3]. The advantage of fault testing over structural testing is that after a successful test run, one can conclude that a certain class of faults definitely will not appear in an implementation.

The exact knowledge about faults that may occur in a faulty implementation allows an effective reduction of test data to a minimum amount necessary to detect these faults. Test derivation approaches for fault testing have been introduced before mainly for testing telecommunication systems [2], [4], [7], [23], [25], [34], [35]. Results of this research are test derivation methods, like the transition tour method, W-method, or HSI-method.

In this paper we propose to apply fault testing also in the general context of testing concurrent systems consisting of a collection of sequential modules. A prerequisite of fault testing of concurrent systems is the existence of a suitable description model for them. The traditional model is the reachability graph that is obtained when the sequential modules of the concurrent system are combined together step-by-step in an interleaving framework. Latest developments in verification techniques for concurrent systems avoid the construction of a reachability graph in favor of a partial-order model though. One partial-oder model is the so-called *Mazurkiewicz Trace Machine* (MTM), first introduced in [11] and [12]. The MTM is a reduced reachability graph that still preserves the safety properties of the concurrent system.

So far the model of an MTM has been exploited only in the verification process of concurrent systems. Their application to test derivation, however, was underestimated. To end this situation, this paper exploits the MTM model for the purpose of test derivation. It shows that this partial-order model is also a favorable model in testing since test suites derived from an MTM are much smaller than ones from the traditional reachability graph under the assumption of an equal degree of fault coverage.

1.2 Related work

First research in the area of concurrent systems was done by developing debugging methods for them. Debugging focuses on the process of isolating an already known error in the concurrent system. This is different to testing where the emphasis lies in detecting errors first, which can be debugged later. The paper [21] addressed the problem of unpredictable system runs due to concurrency. In [28], an approach based on Ada was developed to provide a method that allows a deterministic replay of a system run. The approach can be adopted to support a suitable test architecture for concurrent systems (see Section 2.2).

Test derivation methods for concurrent systems that systematically cover the behavior of a concurrent system and provide test suites of a defined fault coverage are still quite new. In [16], an hierarchy-based finite state machine (FSM) construction approach is used to perform a structural test ofa system consisting of several concurrent FSMs. This technique was refined in [17]. It describes an incremental, bottom-up-oriented approach and assumes that each subsystem considered can be tested separately. Test derivation is based on degrees of *test coverage* (opposite to *fault coverage* as considered in our paper). The selection of test suites, however, is left open by referring to the work in [29]. A further approach for structural test derivation presented in [6] is based on a specification-based selection of test data. The paper gives hints on the selection of suitable test data to provide a useful degree of test coverage. The selection process must be assisted by a test expert though.

Early test derivation approaches that systematically derive test suites according to a certain fault coverage and try to avoid state explosion during the generation of test suites are given in [15], [32], and [33]. A general drawback of these approaches is that they exploit a complicated concurrency model as the basis for test derivation that cannot, with exception of [33], be efficiently computed in all cases.

At this point we extend the work previously done in testing concurrent systems by suggesting test derivation methods that generate test suites with fault coverage guarantee. Based on the partial-order model of an MTM, our paper discusses fault models that supports acceptance, refusal, and transfer faults. As a result, test suites are generated which are quite short in many cases, but still able to detect all faults of the associated fault model. The application of an MTM in testing requires specific measures for test architectures of concurrent systems. This issue is a further discussion point in this paper.

The paper is organized as follows. First, Section 2 introduces necessary assumptions and definitions of a concurrent system and discusses aspects of a test architecture that can support test execution based on partial orders. Section 3 introduces a framework for fault detection used in this paper. Section 4 deals with test derivation algorithms that generate test suites according to a particular fault model. Finally, Section 5 concludes the paper.

2 PRELIMINARIES

2.1 Interleaved-based models vs. partial-order models of a concurrent system

A *concurrent system* $\mathfrak{I}$ is defined as a parallel composition $\mathfrak{I} = M_1 \parallel \ldots \parallel M_n$ of n finite labeled transition systems (LTSs) communicating synchronously. The same message can be exchanged between two or more component LTSs at a single synchronization (multi-rendezvous). Transmitting messages and their receipt through interaction points are referred to *actions* in an LTS.

Definition 1. A *labeled transition system* (LTS or *machine* for short) M is defined by a quadruple $(S, A, \rightarrow, s_0)$, where S is a finite set of states; A is a finite set of actions (the alphabet); $\rightarrow \subseteq S \times A \times S$ is a transition relation; and $s_0 \in S$ is the initial state.

A transition $(s_1, a, s_2) \in \rightarrow$ is also written as $s_1 \text{–} a \rightarrow s_2$. An LTS is *deterministic* if there are no two transitions $s \text{–} a \rightarrow s_1$ and $s \text{–} a \rightarrow s_2$ for any start state s and action a

with $s_1 \neq s_2$. With no loss of generality, we assume that each component LTS is initially connected. Furthermore, we consider concurrent systems to be composed of deterministic LTSs only.

As usual, $s{=}\alpha{\Rightarrow}s'$ denotes a *trace* starting from state s and ending in state s'. The end state s' is reached by the sequence of actions a_i in the trace $\alpha \in A^*$ of M. The trace α traverses a set of intermediate states: $s{-}a_1{\rightarrow}s_1{-}a_2{\rightarrow}s_2{-}\ldots{\rightarrow}s'$. The set of traces of state s is defined as the set $\{\alpha \in A^* \mid \exists s' \in S{:}\ s{=}\alpha{\Rightarrow}s'\} = Tr(s)$. The set of traces of M is $L(M) = Tr(s_0)$, i.e. the language accepted by LTS M starting from its initial state [14].

Definition 2. State s_1 of LTS M_1 and state s_2 of LTS M_2 are *trace-equivalent*, $s_1 \approx_{tr} s_2$, if $Tr(s_1) = Tr(s_2)$; states s_1, s_2 are *distinguishable*, $s_1 \neq s_2$, if $Tr(s_1) \neq Tr(s_2)$. Trace equivalence (distinguishability) of LTSs is defined as trace equivalence (distinguishability) of their initial states s_{01} and s_{02}, i.e., $M_1 \approx_{tr} M_2$, if $Tr(s_{01}) = Tr(s_{02})$ and $M_1 \neq M_2$, if $Tr(s_{01}) \neq Tr(s_{02})$.

Henceforth, we assume that each component of a concurrent system is minimal, i.e., every two of its states are distinguishable.

The joint behavior of a concurrent system $\Im = M_1 \parallel \ldots \parallel M_n$ can be described by means of a *composite machine* defined over $A_\Im \subseteq A_1 \cup \ldots \cup A_n$, the (global) alphabet of system $\Im$. Components execute *shared* actions that require rendezvous of several component LTSs along with *local* actions that are executed by a component and its environment only. We assume that different rendezvous on the same event have distinct names in the concurrent system, i.e., names of local and shared actions in $A_\Im$ are globally unique. Let $ID = \{1, 2, \ldots, n\}$ be the set of indexes of the LTSs in $\Im$. For each $a \in A_\Im$, $\boldsymbol{id}(a)$ denotes the *occurrence set* $\{i \in ID \mid a \in A_i\}$, i.e. the set of components involved in a rendezvous over a.

Definition 3. A *composite machine* of a given concurrent system $\Im$ of n LTSs $M_i = (S_i, A_i, \rightarrow_i, s_{0i})$ is the quadruple $(S_\Im, A_\Im, \rightarrow_\Im, s_\Im)$, where $S_\Im$ is a global state space, $S_\Im \subseteq S_1 \times \ldots \times S_n$; $A_\Im \subseteq A_1 \cup \ldots \cup A_n$ is the set of actions (the global alphabet), $s_\Im = (s_{01}, \ldots, s_{0n})$ is the initial global state, and the transition relation $\rightarrow_\Im$ is defined as follows: Let $a \in A_\Im$ and $(s_1, \ldots, s_n) \in S_\Im$. $((s_1, \ldots, s_n), a, (s_1', \ldots, s_n')) \in \rightarrow_\Im$, where $s_j' = s_j$ for all $j \notin \boldsymbol{id}(a)$, if there exists $s_i' \in S_i$ such that $(s_i, a, s_i') \in \rightarrow_i$ for all $i \in \boldsymbol{id}(a)$.

We further assume that the global state space comprises only states reachable from the initial state. We use $C_\Im$ to denote the composite machine $(S_\Im, A_\Im, \rightarrow_\Im, s_\Im)$ of a given concurrent system $\Im$, and $s(i)$ to denote a local state of component M_i in a global state s of $C_\Im$. Let $L(C_\Im)$ denote the set of all traces executed by the composite machine from its initial state, i.e., $L(C_\Im)$ is the language of the concurrent system $\Im$. A way to represent the composite machine is by means of the reachability graph. It represents the behavior of a concurrent system $\Im$ in the interleaved-based semantics. In partial-order semantics, the same behavior may have a more compact representation. We define such a representation based on the trace theory of Mazurkiewicz [20], [24].

Definition 4. The *independence relation* over the global alphabet $A_\Im$ is the relation $I = \{(a, b) \in A_\Im \times A_\Im \mid \boldsymbol{id}(a) \cap \boldsymbol{id}(b) = \emptyset\}$.

$(A_\Im, I) = \Lambda$ is the *concurrent alphabet* of $\Im$, as defined in the trace theory of Mazurkiewicz. Let "." denote the concatenation of words over $A_\Im$. We define the *equivalence of traces* over Λ as the least congruence relation $\equiv_\Lambda \subseteq A_\mathfrak{S}^* \times A_\mathfrak{S}^*$ w.r.t. the concatenation operator and including $\forall\, a, b \in A_\mathfrak{S}$: $(a, b) \in I \Rightarrow a.b \equiv_\Lambda b.a$.

Definition 5. A *Mazurkiewicz trace* (*M-trace*) over $\Lambda = (A_\mathfrak{S}, I)$ is an equivalence class of $\equiv_\Lambda$ of traces over $A_\mathfrak{S}^*$.

An M-trace is fully characterized by one of its traces α and the concurrent alphabet $(A_\mathfrak{S}, I)$ and is denoted by $[\alpha]$, where α is a representative trace of the class. By successively permuting adjacent independent actions in α, one can obtain all other traces in $[\alpha]$. An M-trace $[\alpha]$ is a partial order over $A_\mathfrak{S}^*$, traces in $[\alpha]$ are called *linearizations*. Since an M-trace is sufficiently represented by a single linearization over a concurrent alphabet, the behavior of a concurrent system can be represented by one linearization for each M-trace only, instead of giving all possible traces the system can perform in the composite machine. In fact, certain submachines of the composite machine contain the required linearizations.

Definition 6. Given two machines $M_1 = (S, A, \rightarrow_1, s_{01})$ and $M_2 = (S', A', \rightarrow_2, s_{02})$. M_2 is said to be a *submachine* of M_1 if $S' \subseteq S$, $s_{02} = s_{01}$, $A' \subseteq A$, and $\rightarrow_2 \subseteq \rightarrow_1$.

Clearly, $L(M_2) \subseteq L(M_1)$ for all submachines M_2 of M_1. We immediately extend this definition for machines with a state isomorphism. Namely, if a machine M_3 is isomorphic to M_2 that, in turn, is a submachine of M_1, then we say that M_3 is a submachine of M_1.

Definition 7. Let $C_\mathfrak{S} = (S_\mathfrak{S}, A_\mathfrak{S}, \rightarrow_\mathfrak{S}, s_\mathfrak{S})$ be the composite machine of concurrent system $\mathfrak{S}$. A submachine $M_\mathfrak{S} = (S_M, A_\mathfrak{S}, \rightarrow_M, s_\mathfrak{S})$ of $C_\mathfrak{S}$ is a *Mazurkiewicz trace machine* (MTM) of $\mathfrak{S}$, if for all traces $\alpha \in L(C_\mathfrak{S})$, there exists a trace $\alpha' \in L(M_\mathfrak{S})$ such that α' is a linearization of an M-trace defined by an extension of α, i.e., $\alpha \in$ *Pref*$([\alpha'])$.

It follows from the definition that an MTM completely characterizes the traces of the concurrent system, i.e.

$$L(C_\mathfrak{S}) = \bigcup_{\alpha \in L(M_\mathfrak{S})} Pref([\alpha]),$$

where *Pref*(Γ) denotes the set $\{\alpha \in A_\mathfrak{S}^* \mid \exists \beta \in A_\mathfrak{S}^*: \alpha\beta \in \Gamma\}$ for a given $\Gamma \subseteq A_\mathfrak{S}^*$. The definition of an MTM is inspired by [12]. The main advantage of an MTM over the model of a composite machine is that the number of states in the MTM is usually much less than in the composite machine. Results e.g. from [8], [13], and [36] show that the saving rate in the number of states can be huge for many examples (27 to 90 per cent of reduction), although the state space size is still exponential in the worst case since the overall complexity of state space exploration remains PSPACE-complete. Also, the computational complexity of the MTM construction from a set of concurrent modules is mostly smaller compared to the construction of the reachability graph [13].

The construction of an MTM is similar to that of a composite machine, except that in each global state, we select among all independent actions only one action for inclusion into the MTM, see [12]. Since such a selection is an arbitrary process, several MTMs, probably with a different number of states, can be obtained. An MTM possesses a number of attractive properties, which are used to improve verification techniques and, as we shall demonstrate in this paper, can also be exploited for test derivation purposes.

Proposition 1. Let $C_{\mathfrak{I}}$ be the composite machine of concurrent system $\mathfrak{I}$, and let $M_{\mathfrak{I}}$ be an MTM for this system. For all components M_i, for all states $s_{ij} \in S_i$, where S_i is the set of states in M_i, s_{ij} is reachable from the initial state in $C_{\mathfrak{I}}$ iff s_{ij} is reachable from the initial state in $M_{\mathfrak{I}}$.

In short, all local states s_{ij} reachable in $C_{\mathfrak{I}}$ are also reachable in $M_{\mathfrak{I}}$ and vice versa. The proof is given in [12]. We can also easily prove the following.

Proposition 2. Let $C_{\mathfrak{I}} = (S_{\mathfrak{I}}, A_{\mathfrak{I}}, \rightarrow_{\mathfrak{I}}, s_{\mathfrak{I}})$ be the composite machine of concurrent system $\mathfrak{I}$, and let $M_{\mathfrak{I}} = (S_M, A_{\mathfrak{I}}, \rightarrow_M, s_{\mathfrak{I}})$ be an MTM for this system. For all actions $a \in A_{\mathfrak{I}}$, and all transitions $(s, a, s') \in \rightarrow_{\mathfrak{I}}$, there exists a transition $(p, a, p') \in \rightarrow_M$ such that $p(i) = s(i)$ for all $i \in \boldsymbol{id}(a)$.

This proposition indicates that all unexecutable transitions in a concurrent system can be already detected when an MTM is constructed. For the purpose of test derivation, we may well assume that these transitions have been deleted from the given system $\mathfrak{I}$, henceforth, $A_{\mathfrak{I}} = A_1 \cup \ldots \cup A_n$ is taken for granted.

As an example, we consider the concurrent system $\mathfrak{I} = A \parallel B$, which consists of the two component machines A and B (Figure 1). Actions a and c are shared, the remaining actions are local. Figure 1 also shows the composite machine $C_{\mathfrak{I}}$. Note that, in this example, the number of states of the composite machine reaches its maximum, and all global states are reachable from the initial state. The independence relation of this system is $I = \{(b, e), (b, e), (d, e), (e, d)\}$. Given the trace *abed*, for example, one can, by successively permuting adjacent independent actions in *abed*, obtain all other sequences in [*abed*], i.e. [*abed*] = {*abed*, *abde*, *aebd*}. Two MTMs of the example system are shown in Figure 2, they use different numbers of states to represent the same M-traces.

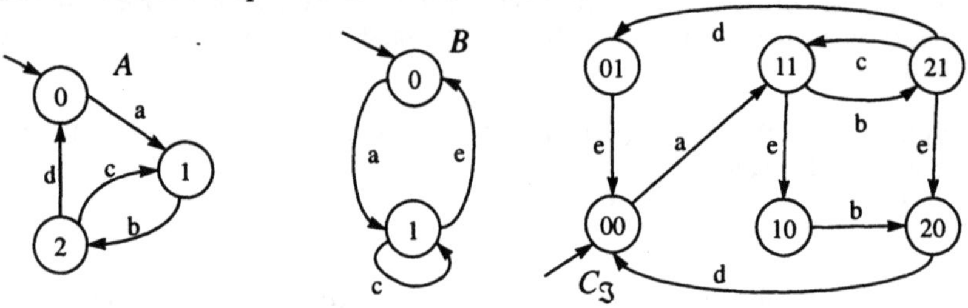

Figure 1. The concurrent system $\mathfrak{I} = A \parallel B$ and its composite machine $C_{\mathfrak{I}}$.

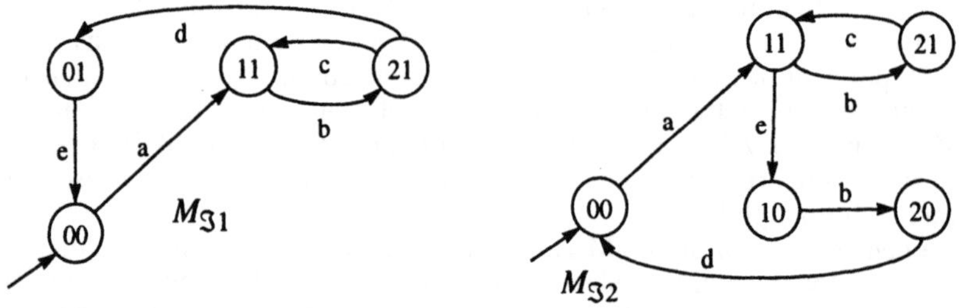

Figure 2. Two MTMs from the composite machine $C_{\mathfrak{I}}$.

2.2 Assumptions on a test architecture

A *test architecture* is defined as a parallel, synchronous composition of the *implementation under test* (*IUT*) of the concurrent system $\mathfrak{S}$ with its tester. If a test suite comprises several test cases, a separate tester is built for each test case. It is required that the number of LTSs in the IUT of the concurrent system is constant when the test run takes place (static system). Furthermore, the structure of the system $\mathfrak{S}$, i.e. its composition of n LTSs, is preserved in the implementation. From the testing perspective, it should be stated what actions can be observed and controlled. To avoid nondeterministic execution of the IUT due to nonobservability, we assume that *all* local and shared actions are observable during testing (grey-box testing approach).

In testing concurrent systems the information about action names observed during a test run is not sufficient to assess conformance between specification and IUT. Due to the existence of multi-rendezvous among component LTSs of the IUT, the tester must also know what components participate in a specific multi-rendezvous. Furthermore, the issue of true concurrency among actions of the IUT requires that the tester has the power not only to observe an action of the IUT, but must also control each occurrence of a multi-rendezvous. Thus, the crucial point in testing concurrent systems is to perform a *deterministic test run*.

The problem can be solved by applying *instant replay techniques* used for debugging concurrent systems [28]. The proposal in [28] assumes a global controller that is asked for permission by the components of the IUT before they are allowed to interact with each other. To achieve this, control code, called *probes*, is added into the source code of the components before each interaction invocation. Only after the tester received all requests to execute a certain interaction from participating components, it grants this interaction. If not or if a wrong component asks for permission, an error in the IUT has occurred.

Adopting this technique to our purposes, it means that the tester collects the action names together with their independence information from the IUT, i.e., it controls and observes the global alphabet and the occurrence of the multi-rendezvous of the system. Only if the action name *and* its occurrence set observed from the IUT equal to the corresponding ones in the specification, a correct multi-rendezvous happened. Otherwise, the IUT does not conform to the specification. In Section 3.2 we give a model of such a tester in the context of a conformance relation between the IUT and its corresponding specification.

3 A FRAMEWORK FOR FAULT DETECTION

3.1 General fault model

Fault models are usually required to guarantee the detection of certain types of faults by means of a finite tester [4]. As usual in specification-based testing, a tester is derived from a given specification of the system. In addition, the tester requires a certain criterion to assess a test run over the IUT. This criterion is customary defined as *conformance relation* between the specification and an implementation of a certain domain [5], [30], [31]. The conformance relation classifies all possible implementations into a class of implementations conforming to the specification, i.e., a test run of an implementation with a test case derived according to the conformance relation yields the verdict *pass*, and into a class of non-conforming implementations, i.e., a test run yields the verdict *fail*. The implementation domain is usually a finite set of implementations that can be derived from the specification by performing a number of mutations representing faults of a certain type, i.e., the implementation domain is defined implicitly by properties common to all implementations. Similar to the realm of testing sequential systems, we assume as a minimal prerequisite for the fault domain that all actions an IUT submits during testing are known in advance, i.e. that the global alphabet of an IUT is a subset of the given global alphabet. This assumption appears fundamentally in the context of fault-driven testing of sequential systems [10]. We believe that the above concepts are pertinent to a parallel composition of sequential systems as well, and following [25] we define a fault model for a concurrent system as the triple

<specification, conformance relation, implementation domain>.

Based on this concept, we are ready to state what constitutes a sound and complete test suite: A test suite is a finite set of finite test cases consisting of actions accessible for testing. A test suite is *sound* w.r.t. a fault model if any conforming implementation passes the test suite. A sound test suite is *complete* w.r.t. a fault model if any non-conforming implementation from the implementation domain fails it.

A general fault model and a complete test suite for a concurrent system corresponding to it might be derived by the following approach originally developed for sequential systems. In particular, given a concurrent system $\mathfrak{I} = M_1 \parallel \ldots \parallel M_n$, we may treat a composite machine $C_{\mathfrak{I}}$ as the specification and the trace-equivalence of composite machines as a conformance relation. The implementation domain could take several forms according to the existing test derivation methods for LTS specifications with guaranteed fault coverage [23]. As an example, the implementation domain is often defined as the universe $C(A_{\mathfrak{I}}, N_{max})$ of all LTSs defined over some actions of the given alphabet $A_{\mathfrak{I}}$ with at most N_{max} states. The fault model $<C_{\mathfrak{I}}, \approx_{tr}, C(A_{\mathfrak{I}}, N_{max})>$, where $N_{max} \geq |S_{\mathfrak{I}}|$, is a classical fault model most frequently used for fault detection in state machines [4], [7], [23], [34], [35]. It corresponds to the most general type of faults that may occur in a sequential system. Following this approach, a test suite for the concurrent system $\mathfrak{I}$ that is complete with respect to this fault model can be obtained by applying, for example, the method presented in [30].

Trace-equivalence of composite machines is, however, not sufficiently strong enough for testing concurrent systems, even for deterministic ones as considered here. It does not discriminate between a sequential system M and the concurrent system $M \parallel M$ composed of two instances of M, since they have isomorphic composite machines. However, any tester with control over multi-rendezvous, as outlined in Section 2.2, does. Speaking more generally, we wish to distinguish two instances of the same action with different occurrence sets. Apparently, the problem can be easily fixed by decorating each action in composite machines with the names of component LTSs executing it.

Given the composite machine $C_{\mathfrak{I}} = (S_{\mathfrak{I}}, A_{\mathfrak{I}}, \rightarrow_{\mathfrak{I}}, s_{\mathfrak{I}})$ of concurrent system $\mathfrak{I} = M_1 \parallel \ldots \parallel M_n$, where $A_{\mathfrak{I}} = A_1 \cup \ldots \cup A_n$, and the set of names of components ID, we use $\hat{a}$ to denote the pair <*a*, ***id***(*a*)>. Replacing the global alphabet $A_{\mathfrak{I}}$ by the set $\hat{A}_{\mathfrak{I}} = \{\hat{a} \mid a \in A_{\mathfrak{I}}\}$ in the composite machine, we obtain an isomorphic LTS $\hat{C}_{\mathfrak{I}} = (S_{\mathfrak{I}}, \hat{A}_{\mathfrak{I}}, \rightarrow_{\mathfrak{I}}, s_{\mathfrak{I}})$, called the *augmented composite machine* of $\mathfrak{I}$. Now, the trace equivalence can immediately be employed to define the conformance relation for (deterministic) concurrent systems.

Definition 8. Given two concurrent systems $\mathfrak{I}$ and $\mathfrak{R}$, let $\hat{C}_{\mathfrak{I}}$ and $\hat{C}_{\mathfrak{R}}$ be the augmented composite machines of $\mathfrak{I}$ and $\mathfrak{R}$, respectively. $\mathfrak{I}$ and $\mathfrak{R}$ are *trace-equivalent*, denoted $\mathfrak{I} \approx \mathfrak{R}$, if $\hat{C}_{\mathfrak{I}} \approx_{tr} \hat{C}_{\mathfrak{R}}$. $\mathfrak{I}$ and $\mathfrak{R}$ are *distinguishable*, $\mathfrak{I} \neq \mathfrak{R}$, if $L(\hat{C}_{\mathfrak{I}}) \neq L(\hat{C}_{\mathfrak{R}})$.

In other words, two systems $\mathfrak{I}$ and $\mathfrak{R}$ are trace-equivalent if their composite machines are trace-equivalent, and for all actions $a \in A_{\mathfrak{I}}$ in all traces of $L(C_{\mathfrak{I}})$, the names of components involved in the execution of a are identical. Henceforth, we will use the notions of a composite machine and an augmented composite machine interchangeably whenever no confusion arises. Similarly, we define an augmented MTM $\hat{M}_{\mathfrak{I}}$ for system $\mathfrak{I}$.

It remains now to show how an implementation domain for concurrent systems can be defined. We assume that an implementation $\mathfrak{R}$ is modeled as parallel composition of a number of component LTSs. The number of components is exactly the same as in the specification $\mathfrak{I}$, since the tester establishes a communication link to every component to perform test runs (Section 2.2). This implies that the underlying communication subsystem is assumed to be perfect and its possible faults are modeled by faults inside the components of the IUT.

As mentioned earlier, we assume that $A_{\mathfrak{R}} \subseteq A_{\mathfrak{I}}$ for all possible implementations $\mathfrak{R}$. It implies that in the general case, the local alphabet of any component LTS is also a subset of $A_{\mathfrak{I}}$, but does not necessarily coincide with the local alphabet in the specification. Some components might be assumed to be fault-free, and their alphabets remain intact.

3.2 The model of a tester

Next, we present a model of testers used to verify whether or not a system $\mathfrak{R}$ is trace-equivalent to the system $\mathfrak{I}$. A test run should stop as soon as deadlock occurs. Thus, test cases for $\mathfrak{I}$ can be defined as valid traces $\hat{\sigma} \in L(\hat{C}_{\mathfrak{I}})$ and invalid traces $\hat{\gamma}.\hat{a}$, where $\hat{\gamma} \in L(\hat{C}_{\mathfrak{I}}) \cup \{\hat{\varepsilon}\}$, $\hat{\varepsilon} = \langle\varepsilon, \varnothing\rangle$, an empty symbol; $\gamma.a \notin L(C_{\mathfrak{I}})$, $\hat{a} = \langle a, id\rangle$, $a \in A_{\mathfrak{I}}$, and id is a non-empty subset of the set ID. Intuitively, the valid traces are used to verify whether or not an IUT possesses traces of $\mathfrak{I}$ (valid behavior), whereas the invalid traces check the IUT for invalid behavior.

Given a valid trace $\hat{\sigma}$, we define the corresponding tester $T(\hat{\sigma})$ as an acyclic LTS $T(\hat{\sigma}) = (S_T, \hat{A}_{\mathfrak{I}}, \rightarrow_T, s_T)$ such that its language is the set $Pref(\hat{\sigma}) \subseteq \hat{A}_{\mathfrak{I}}$ and the set of its states S_T contains $|\hat{\sigma}| + 1$ states. A test run of $\hat{\sigma}$ against the IUT $\mathfrak{R}$ can be modeled as a parallel composition $T(\hat{\sigma}) \parallel \hat{C}_{\mathfrak{R}}$. The system $T(\hat{\sigma}) \parallel \hat{C}_{\mathfrak{R}}$ is used to assign the verdict of a test run, *pass* or *fail*, to states of the tester according to a labeling function ***ver***: $S_T \rightarrow \{pass, fail\}$. Note that we are dealing with deterministic composite machines, for which there is no need to use the verdict *inconclusive*. The system $T(\hat{\sigma}) \parallel \hat{C}_{\mathfrak{R}}$ executing trace $\hat{\sigma}$ comes to a deadlock since the tester $T(\hat{\sigma})$ reaches its final state; this state is the only one labeled with *pass*, the remaining states are labeled with *fail*. The IUT passes test case $\hat{\sigma}$ if the system $T(\hat{\sigma}) \parallel \hat{C}_{\mathfrak{R}}$ deadlocks only when the state *pass* of the tester is reached.

Similarly, we define the tester $T(\hat{\gamma}.\hat{a})$ corresponding to an invalid trace $\hat{\gamma}.\hat{a} \notin L(\hat{C}_{\mathfrak{I}})$. The tester is modeled with $|\hat{\gamma}.\hat{a}| + 1$ states. Since the system $T(\hat{\gamma}.\hat{a}) \parallel \hat{C}_{\mathfrak{I}}$ must deadlock after $\hat{\gamma}$ is executed, the state of the tester reached after $\hat{\gamma}$ is labeled with verdict *pass*, whereas all others are labeled with *fail*. In both cases, the test verdict of a test run is obtained from the label of the last state reached by the tester.

4 TEST DERIVATION

4.1 Detecting missing transitions

We assume here that certain transitions possibly not implemented at some states of component LTSs are the only implementation faults. That means, an action caused by a transition in the specification of a component may be refused in a faulty implementation. At the same time, we assume that an accepted action causes a transition into a correct state. Thus, whatever an implementation of each component does it conforms to the specification; it may reduce the specification though.

Formally, we define the acceptance fault model as $<\mathfrak{I}, \approx, \mathfrak{R}_{acc}>$, where $\mathfrak{R}_{acc}$ is the set of all implementation systems $\mathfrak{R}$ consisting of n initialized LTSs such that they are submachines of the corresponding machines in specification $\mathfrak{I}$. The *acceptance set* [18] of any local state in $\mathfrak{R}$ is a subset of that set of the corresponding local state in $\mathfrak{I}$. The action set $A_{\mathfrak{R}}$ of any concurrent system $\mathfrak{R} \in \mathfrak{R}_{acc}$ is a subset of the action set $A_{\mathfrak{I}}$ of $\mathfrak{I}$. Intuitively, the detection of acceptance faults can be achieved by using a transition cover of a machine [22].

Definition 9. Given an initially connected LTS $M = (S, A, \rightarrow, s_0)$, a set of traces $TC(M)$ is said to be a *transition cover* of M if for all transitions $(s, a, s') \in \rightarrow$, there exists $\beta a \in Pref(TC(M))$ such that $s_0 = \beta \Rightarrow s$.

By definition, the global states of composite machine $C_{\mathfrak{I}}$ contain all local states of all component machines of $\mathfrak{I}$. Moreover, all executable transitions between local states are present in the composite machine (it has no unexecutable transition by our assumption). To check whether or not all transitions at a particular local state are preserved in the IUT, we should try to execute all actions accepted at a corresponding global state. If an action a is executed in a global state and if the observed occurrence set coincide with the one in the augmented composite machine $\hat{C}_{\mathfrak{I}}$, then the action causes transitions at all local states defined by ***id***(a). Consequently, a transition cover of $\hat{C}_{\mathfrak{I}}$ is already a complete test suite w.r.t. the fault model $<\mathfrak{I}, \approx, \mathfrak{R}_{acc}>$. We demonstrate that this statement holds even when the augmented composite machine $\hat{C}_{\mathfrak{I}}$ is replaced by an augmented MTM $\hat{M}_{\mathfrak{I}}$. A shorter test suite detecting acceptance faults is then obtained, provided that we can find a transition cover of $\hat{M}_{\mathfrak{I}}$ shorter than that of $\hat{C}_{\mathfrak{I}}$.

Proposition 3. Given a concurrent system $\mathfrak{I}$, an augmented MTM $\hat{M}_{\mathfrak{I}}$, and the fault model $<\mathfrak{I}, \approx, \mathfrak{R}_{acc}>$, let $TC(\hat{M}_{\mathfrak{I}})$ be a transition cover of $\hat{M}_{\mathfrak{I}}$. Then $TC(\hat{M}_{\mathfrak{I}})$ is a test suite complete w.r.t. $<\mathfrak{I}, \approx, \mathfrak{R}_{acc}>$.

Proof. Consider a transition cover $TC(\hat{M}_{\mathfrak{I}})$ of the augmented MTM $\hat{M}_{\mathfrak{I}}$. By assumption that the system $\mathfrak{I}$ has no unexecutable transitions and $A_{\mathfrak{I}} = A_1 \cup \ldots \cup A_n$, and by virtue of Proposition 2, the corresponding transition cover $TC(M_{\mathfrak{I}})$ of MTM $M_{\mathfrak{I}}$ executes all transitions in each component LTS in $\mathfrak{I}$. In other words, the projection of all sequences of $TC(\hat{M}_{\mathfrak{I}})$ into a component M_i of $\mathfrak{I}$ is a transition cover $TC(M_i)$ of this component. The system $\mathfrak{R} \in \mathfrak{R}_{acc}$ consists of exactly n components LTSs, as the system $\mathfrak{I}$ does. If $\mathfrak{R}$ passes the test suite $TC(\hat{M}_{\mathfrak{I}})$, it means that for all components in $\mathfrak{R}$, the traces of $TC(M_i)$ are valid traces of the i-th component in $\mathfrak{R}$. Moreover by definition of $\mathfrak{R}_{acc}$, each component in $\mathfrak{R}$ is a submachine of the corresponding component machine in $\mathfrak{I}$. It means that each component machine of $\mathfrak{R}$ is isomorphic to the corresponding machine of $\mathfrak{I}$. As result, $\mathfrak{R} \approx \mathfrak{I}$.

◊

Thus, the problem of deriving a minimal test suite for a given concurrent system complete w.r.t. acceptance faults can be reduced to that of finding a minimal transition cover of an MTM of the given system. Consider our example (Figure 1). Each MTM (Figure 2) has a minimal transition tour of 6 actions although they have different number of states. Take one of them, e.g. *a.b.c.b.d.e*. Decorating actions with occurrence sets, we obtain the test sequence $<a, \{1, 2\}>.<b, \{1\}>.<c, \{1, 2\}>.<b,$

$\{1\}>.<d, \{1\}>.<e, \{2\}>$. The tester has seven states, the seventh state is labeled with *pass*, the other ones with *fail* (see Figure 3).

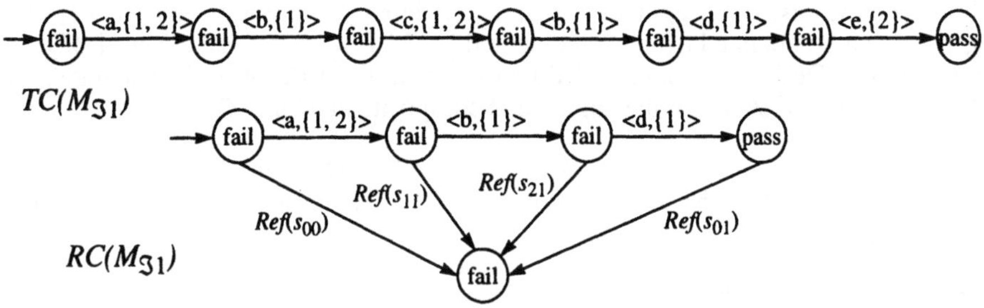

Figure 3. Test suites $TC(\hat{M}_{\mathfrak{I}1})$ and $RC(\hat{M}_{\mathfrak{I}1})$ for acceptance and refusal testing of the system $\mathfrak{I}$.

4.2 Detecting superfluous transitions

Assume now that any faulty implementation $\mathfrak{R}$ of a given specification $\mathfrak{I}$ extends the specification. In this case, we assume that superfluous transitions implemented at some local states of component LTSs are the only faults and that the specified transitions are correctly implemented in each component of the IUT.

We define the refusal fault model as $<\mathfrak{I}, \approx, \mathfrak{R}_{ref}>$, where $\mathfrak{R}_{ref}$ is defined as the set of all implementations consisting of n initialized LTSs such that the components of specification $\mathfrak{I}$ are submachines of the corresponding components in implementation $\mathfrak{R}$. The action set of any component of $\mathfrak{R} \in \mathfrak{R}_{ref}$ is assumed to be a subset of the global alphabet $A_{\mathfrak{I}}$. Note that the number of states in a component may exceed that in the specification, however, we do not bound it here.

Similar to [18], we define the *refusal set* of a global state s in the augmented composite machine $\hat{C}_{\mathfrak{I}}$ as the set $Ref_{\mathfrak{I}}(s) = \{<a, id> \mid s{\not}{-}a{\rightarrow}_{\mathfrak{I}} \vee id \neq \mathbf{id}(a)\}$, where $s{\not}{-}a{\rightarrow}_{\mathfrak{I}}$ denotes the fact that action $a \in A_{\mathfrak{I}}$ is refused at state s and $id \subseteq ID$, $id \neq \emptyset$. For any $<a, id> \in Ref_{\mathfrak{I}}(s)$, the action a accepted at a corresponding global state in an IUT means that each component in the occurrence set id observed by the tester has a superfluous transition labeled with a. Intuitively, to detect implementation faults defined by the fault model $<\mathfrak{I}, \approx, \mathfrak{R}_{ref}>$, it is sufficient to check whether or not any action from $Ref_{\mathfrak{I}}(s)$ causes an invalid transition at a corresponding state of the IUT. This must be done for each global state s of the augmented composite machine $\hat{C}_{\mathfrak{I}}$. For this purpose, we use invalid traces as test cases. Alternatively, to save a number of test runs, we can define tests of a set of refusal traces if we assume that the tester is able to proceed after a deadlock (as in [18]).

Minimal invalid traces are constructed based on a minimal state cover. Given an initially connected LTS $M = (S, A, \rightarrow, s_0)$, we define a *state cover*, denoted $SC(M)$, of the LTS as a prefix-closed set of traces such that for all $s \in S$, there exists a transfer sequence $\alpha \in Pref(SC)$ such that $s_0{=}\alpha{\Rightarrow}s$. For a strongly connected machine, a state cover can be constructed as a state tour of the machine.

Definition 10. Given the augmented composite machine $\hat{C}_{\mathfrak{I}} = (S_{\mathfrak{I}}, \hat{A}_{\mathfrak{I}}, \hat{\rightarrow}_{\mathfrak{I}}, s_{\mathfrak{I}})$ and its state cover $SC(\hat{C}_{\mathfrak{I}})$, we define a *refusal cover* of $\hat{C}_{\mathfrak{I}}$, denoted $RC(\hat{C}_{\mathfrak{I}})$, as

the set
$\{\hat{\gamma}.\hat{a} \mid \hat{\gamma} \in Pref(SC(\hat{C}_{\mathfrak{I}})) \wedge s_{\mathfrak{I}}=\hat{\gamma}\Rightarrow s \wedge \hat{a} \in Ref_{\mathfrak{I}}(s)\}$.

Similar to the case of acceptance faults, a complete test suite w.r.t. fault model <$\mathfrak{I}$, $\approx$, $\mathfrak{R}_{ref}$> can be obtained from an augmented MTM $\hat{M}_{\mathfrak{I}}$. A refusal cover of $\hat{M}_{\mathfrak{I}}$ is defined as in Definition 10, provided that the refusal set $Ref_M(s)$ of each state $\hat{s}$ of $\hat{M}_{\mathfrak{I}}$ is the one of state s in the augmented composite machine $\hat{C}_{\mathfrak{I}}$.

Proposition 4. Given concurrent system $\mathfrak{I}$, an MTM $\hat{M}_{\mathfrak{I}}$, and the fault model <$\mathfrak{I}$, $\approx$, $\mathfrak{R}_{ref}$>, let $RC(\hat{M}_{\mathfrak{I}})$ be a refusal cover of $\hat{M}_{\mathfrak{I}}$. Then the set $RC(\hat{M}_{\mathfrak{I}})$ is a test suite complete w.r.t. <$\mathfrak{I}$, $\approx$, $\mathfrak{R}_{ref}$>.

Proof. Consider an arbitrary concurrent system $\mathfrak{R} \in \mathfrak{R}_{ref}$. Similar to Proposition 3, we claim that each component of $\mathfrak{R}$ is isomorphic to the corresponding one of $\mathfrak{I}$. We prove Proposition 4 by contradiction.

Assume that $\mathfrak{R}$ is not a conforming implementation of $\mathfrak{I}$, i.e. $L(\hat{C}_{\mathfrak{I}}) \neq L(\hat{C}_{\mathfrak{R}})$, but $\mathfrak{R}$ passes $RC(\hat{M}_{\mathfrak{I}})$. In this case, the composite machine $\hat{C}_{\mathfrak{R}}$ has at least one trace $\hat{\sigma}.\hat{a}$ that is not a valid trace of $C_{\mathfrak{I}}$, i.e. $\hat{\sigma} \in L(\hat{C}_{\mathfrak{R}}) \cap L(\hat{C}_{\mathfrak{I}})$ and $\hat{\sigma}.\hat{a} \in L(\hat{C}_{\mathfrak{R}}) \setminus L(\hat{C}_{\mathfrak{I}})$. Let s be a global state of $\hat{C}_{\mathfrak{I}}$ such that $s_0=\hat{\sigma}\Rightarrow s$, $\hat{a} \in Ref_{\mathfrak{I}}(s)$ and $\hat{a}$ = <a, $\hat{id}$>. It means that in $\mathfrak{R}$ for all $i \in \hat{id}$, the action a causes a supplementary transition from a local state that corresponds to $s(i)$ in $\mathfrak{I}$. By virtue of Proposition 1, if state $s(i)$ is reachable in LTS $C_{\mathfrak{I}}$, then it is also reachable in LTS $M_{\mathfrak{I}}$. Specifically, let s' be a global state of $\hat{M}_{\mathfrak{I}}$ such that $s'(i) = s(i)$. As a result, we obtain $\hat{a} \in Ref_M(s')$. The MTM $\hat{M}_{\mathfrak{I}}$ is initially connected, hence there exists a trace $\beta \in SC(\hat{M}_{\mathfrak{I}})$ such that $s_0=\beta\Rightarrow s'$. By construction, the test suite $RC(\hat{M}_{\mathfrak{I}})$ has an invalid trace $\beta.\hat{a}$, where $\hat{a} \in Ref_M(s')$. Thus if $\mathfrak{R} \in \mathfrak{R}_{ref}$ and $L(C_{\mathfrak{R}}) \neq L(C_{\mathfrak{I}})$, then $\mathfrak{R}$ cannot pass the test case $\beta.\hat{a} \in RC(\hat{M}_{\mathfrak{I}})$.

◊

We illustrate the process of test derivation for refusal faults using our example again (Figure 1 and Figure 2). The sequence <a, {1, 2}>.<b, {1}>.<d, {1}> can be used as a state cover of the augmented MTM $\hat{M}_{\mathfrak{I}_1}$. The corresponding refusal cover is $Ref(s_{00})$.<a, {1, 2}>.$Ref(s_{11})$.<b, {1}>.$Ref(s_{21})$.<d, {1}>.$Ref(s_{01})$, where for example the refusal set of state s_{21} is $Ref(s_{21})$ = {<a, {1}>, <a, {2}>, <a, {1, 2}>, <b, {1}>, <b, {2}>, <b, {1, 2}>, <c, {1}>, <c, {2}>, <d, {1}>, <d, {2}>, <d, {1, 2}>, <e, {1}>, <e, {1, 2}>} (see Figure 3).

4.3 Detecting both missing and superfluous transitions

In a more general case, it is likely that both faults, missing and superfluous transitions, may occur in an IUT of a concurrent system simultaneously. The two types of faults are independent from each other. Assume that $Acc(s)$ and $Ref(s)$ are the acceptance and refusal sets of state s of an LTS, respectively. Clearly, $Acc(s) \cap Ref(s) = \emptyset$. Thus, a test suite can be derived to cover both faults by combining test suites for accepting and refusal faults.

We define the acceptance/refusal fault model as <$\mathfrak{I}$, $\approx$, $\mathfrak{R}_{mix}$>, where $\mathfrak{R}_{mix}$ is the set of all implementation systems $\mathfrak{R}$ consisting of n initialized LTSs according to specification $\mathfrak{I}$ such that the components of an implementation $\mathfrak{R}$ contain zero or more missing transitions as well as zero or more superfluous ones. The set of local actions of a component in $\mathfrak{R}$ is a subset of the global alphabet $A_{\mathfrak{I}}$.

Due to the independence property between acceptance and refusal faults, a test suite that is complete w.r.t. the fault model can now be obtained by combining a transition cover and a refusal cover of the augmented composite machine, $TC(\hat{C}_{\mathfrak{I}})$

$\cup$ $RC(\hat{C}_{\Im})$. As carried out above, a shorter test suite for the combined fault model can be obtained if the augmented MTM $\hat{M}_{\Im}$ is used instead, $TC(\hat{M}_{\Im}) \cup RC(\hat{M}_{\Im})$.

Both test suites $TC(\hat{M}_{\Im})$ and $RC(\hat{M}_{\Im})$ can be merged into a single test suite $TRC(\hat{M}_{\Im})$ under the assumption that the transition cover used in $TC(\hat{M}_{\Im})$ is also a state cover in $RC(\hat{M}_{\Im})$. Since the composite machine is initially connected, this property is always fulfilled. Taking our example of Figure 1 and Figure 2, a test suite complete w.r.t. $<\Im, \approx, \Re_{mix}>$ can be given, for example, as a single test sequence $TRC(\hat{M}_{\Im_1}) = Ref(s_{00}).<a,\{1, 2\}>. Ref(s_{11}).<b,\{1\}>.Ref(s_{21}).<c, \{1, 2\}>.<b,\{1\}>.<d, \{1\}>.Ref(s_{01}).<e, \{2\}>$, where $Ref(s_x)$ is the refusal set of the global state s_x in $M_{\Im_1}$.

4.4 Detecting transfer faults

Faults considered in the previous sections do not exhaust the variety of potential faults which may occur in implementations of a given concurrent system. We demonstrate that an MTM can also be useful to derive tests detecting other types of faults, in particular, transfer faults in component machines.

An implementation $\mathcal{M}_i$ of a component machine M_i of the concurrent system $\Im = M_1 \parallel \ldots \parallel M_n$ is said to have *transfer faults* if $\mathcal{M}_i$ can be obtained from M_i by changing only the tail state of some transitions. We denote with $\aleph(M_i)$ the set of all possible implementations with transfer faults. The set $\aleph(M_i)$ is finite since any implementation $\mathcal{M}_i$ has at most $|S_i|$ states. In testing compound systems, one often assumes that faults are located in a single component only. These assumptions lead us to the fault model $<\Im, \approx, \Re(M_i)>$, where $\Re(M_i)$ is the set of all concurrent systems $\Re = M_1 \parallel \ldots \parallel \mathcal{M}_i \parallel \ldots \parallel M_n$, with $\mathcal{M}_i \in \aleph(M_i)$, including the system $\Im$ itself.

Transfer faults that are considered when a component is tested in isolation require a state identification facility. In the realm of I/O-FSMs, a characterization set, harmonized state identifiers, a distinguishing sequence, or UIO-sequences serve as examples of this facility [4], [23]. These notions were also redefined for the LTS model assuming trace semantics [30]. In particular, a *characterization set* for machine $M = (S, A, \rightarrow, s_0)$ is the set $W(M) \subseteq A^*$ such that for all $s_k, s_l \in S$, $k \neq l$, there exists a sequence $\alpha \in Pref(W(M)) \cap (Tr(s_k) \oplus Tr(s_l))$. The idea in constructing a test suite complete w.r.t. transfer faults of M in isolation with the remaining machines in $\Im$ is to concatenate a sequence of actions covering a given transition with every sequence of the characterization set $W(M)$ and repeat this process for each transition of M (this is the W-method). The characterization set $W(M)$ allows to identify the tail state of each transition in the IUT.

Adopting this approach to our needs of testing a concurrent system, we define a set of transfer sequences of system $\Im$ that cover all transitions in the component M_i as follows. Given an augmented MTM $\hat{M}_{\Im}$, let (p, a, p') be a local transition of M_i and $(s, \hat{a}, s')$ be the corresponding global transition, i.e. $s(i) = p$, $s'(i) = p'$, and $\hat{a} = <a, \boldsymbol{id}(a)>$. To cover the local transition (p, a, p'), a sequence $\hat{\gamma}.\hat{a}$ must be used, where $\hat{\gamma} \in Pref(SC(\hat{M}_{\Im}))$ such that $s_{\Im} = \hat{\gamma} \Rightarrow s$. The union of such sequences over all local transitions of M_i constitutes a *cover of local transitions* of M_i in $\hat{M}_{\Im}$, denoted $TC_i(\hat{M}_{\Im})$. Thanks to Proposition 1 and Proposition 2, we can always determine such a cover for a concurrent system without unexecutable local transitions directly from an MTM, avoiding thus the exploitation of the composite machine.

Proposition 5. Given a concurrent system $\Im$, an augmented MTM $\hat{M}_{\Im}$, and the fault model $<\Im, \approx, \Re(M_i)>$. Let $TC_i(\hat{M}_{\Im})$ be a cover of local transitions of M_i in

$\hat{M}_{\Im}$. If a characterization set $W(M_i)$ comprises only local actions of M_i, then the set $\{\hat{\gamma}.\hat{a}.\hat{\sigma} \mid \hat{\gamma}.\hat{a} \in TC_i(\hat{M}_{\Im}) \wedge \sigma \in W(M_i)\}$ is a test suite complete w.r.t. $\langle \Im, \approx, \Re(M_i) \rangle$.

The case when a characterization set $W(M_i)$ uses shared actions is a bit more involved and we will report on it in an extended version of this paper. In our example, we use the augmented MTM $\hat{M}_{\Im 1}$ to derive a test suite to detect transfer faults in module *B*. There are three global transitions that represent all local transitions of *B* labeled with *a*, *c*, and *e* in Figure 2. In the MTM $\hat{M}_{\Im 1}$, we use the three transfer sequences, namely *a*, *abc*, and *abde*. The two states of *B* are distinguished by the local action *e*, i.e. *W*(*B*) = {*e*}. The set of sequences {<*a*, {1, 2}>.<*e*, {2}>, <*a*, {1, 2}>.<*b*, {1}>.<*c*, {1, 2}>.<*e*, {2}>, <*a*, {1, 2}>.<*b*, {1}>.<*d*, {1}>.<*e*, {2}>.<*e*, {2}>} yields a test suite that detects any transfer fault in implementations of module *B*.

5 CONCLUSIONS

The paper proposed the application of a partial-order model MTM for test derivation according to a chosen fault model. It was shown that test suites derived from an MTM exhibit the same degree of fault coverage as test suites from the composite machine of a concurrent system assuming the fault models of acceptance, refusal, and transfer faults. The advantage of an MTM over the composite machine is due to the fact that an MTM has a largely reduced state space in many cases resulting in much smaller test suites. However, testers are required that possess a higher degree of controllability to perform a deterministic execution of a specific test case. It was shown how such testers can be constructed using the concept of occurrence sets of actions.

We considered a tester that exercises global control over all actions in a concurrent system, shared and local ones. It would be interesting to consider somewhat more restrained testers which have no control over certain actions. It seems that the results of this paper can easily be applied to the case when certain local actions cannot be observed. We could simply replace these actions by a non-observable action and determinize the LTSs obtained. The situation with non-observable shared actions seems a bit more complicated. Once certain actions become invisible, a tester cannot directly observe a fault, which in turn might later be tolerated by other components. The problem seems similar to testing in context considered in the realm of I/O-FSMs [25], [26], but more research is required in this direction.

Acknowledgment. This work was in part supported by the NSERC grant OGP0194381.

6 REFERENCES

[1] A. V. Aho, A. T. Dahbura, D. Lee, M. Ü. Uyar: *An optimization technique for protocol conformance test generation based on UIO sequences and rural Chinese postman tours*; IEEE Transactions on Communications, vol. 39, no. 11, 1991; pp. 1604-1615.

[2] B. S. Bosik, M. Ü. Uyar: *Finite state machine based formal methods in protocol conformance testing: from theory to implementation*; Computer Networks and ISDN Systems 22 (1); 1991; pp. 7-33.

[3] G. v. Bochmann, A. Petrenko: *Protocol testing: Review of methods and relevance for software testing*; Invited Paper, ACM International Symposium on

Software Testing and Analysis (ISSTA'94), USA, 1994.

[4] G. v. Bochmann, A. Petrenko, M. Yao: *Fault coverage of tests based on finite state machines*; 7th IWPTS; Tokyo, Japan; 1994; pp. 55-74.

[5] E. Brinksma: *A theory for the derivation of tests*; 8th IFIP Int'l Conference on Protocol Specification, Testing, and Verification (PSTV'88); 1988, pp. 63-74.

[6] R. H. Carver, K. C. Tai: *Test sequence generation from formal specifications of distributed programs*; 15th ICDCS; Vancouver, Canada; 1995; pp. 360-367.

[7] T. S. Chow: *Test design modeled by finite-state machines*; IEEE Transactions on Software Engineering, SE-4, No.3, 1978, pp. 178-187.

[8] C. Corbett: *An empirical evaluation of three methods for deadlock analysis of Ada tasking programs*; International Symposium on Software Testing and Analysis (ISSTA'94); Seattle, USA; 1994.

[9] C. Ghezzi, M. Jazayeri, D. Mandrioli: *Fundamentals of software Engineering*; Prentice-Hall, 1991.

[10] A. Gill: *Introduction to the theory of finite state machines*; McGraw-Hill, NY; 1962.

[11] P. Godefroid, P. Wolper: *Using partial orders for the efficient verification of deadlock freedom and safety properties*; Formal Methods in System Design, vol. 2, no. 2 (April 1993); pp. 149-164.

[12] P. Godefroid: *Partial-order methods for the verification of concurrent systems*; LNCS 1032; Springer, 1996.

[13] P. Godefroid, D. Peled, M Staskauskas: *Using partial-order methods in the formal validation of industrial concurrent programs*; International Symposium on Software Testing and Analysis (ISSTA'96); San Diego, USA; 1996; pp. 261–269.

[14] J. E. Hopcroft, J. D. Ullman: *Introduction to automata theory, languages, and computation*; Addison-Wesley; 1979.

[15] M. C. Kim, S. T. Chanson, S. W. Kang, J. W. Shin: *An approach for testing asynchronous communicating systems*; 9th IWTCS 1996; Germany; Sep. 1996.

[16] P. V. Koppol, K. C. Tai: *Conformance testing of protocols specified as labeled transition systems*; 8th International Workshop on Protocol Test Systems (IWPTS'95); Paris, France; 1995; pp. 143-158.

[17] P. V. Koppol, K. C. Tai: *An incremental approach to structural testing*; International Symposium on Software Testing and Analysis (ISSTA'96); San Diego, California, USA; 1996; pp. 14-23.

[18] R. Langerak: *A testing theory for LOTOS using deadlock detection*; 9th PSTV 1989; Enschede, The Netherlands; North Holland, 1990.

[19] D. Lee, K. K. Sabnani, D. M. Kristol, S. Paul: *Conformance testing of protocols specified as communicating FSMs*; IEEE INFOCOM'93; USA; 1993.

[20] A. Mazurkiewicz: *Trace theory*; Advances in Petri Nets (part II); LNCS 255; Springer, 1986; pp. 279-324.

[21] Ch. E. McDowell, D. P. Helmbold: *Debugging concurrent programs*; ACM Computing Surveys, vol. 21, no. 4 (Dec. 1989); pp. 593-622.

[22] S. Naito, M. Tsunoyama, Fault detection for sequential machines by transition tours; Fault Tolerant Computing Systems, 1981, pp. 238-243.

[23] A. Petrenko, G. v. Bochmann and M. Yao: *On fault coverage of tests for finite state specifications*; Special issue on Protocol Testing, Computer Networks and ISDN Systems, vol. 29, 1996, pp. 81-106.

[24] D. Peled, W. Penczek: *Using asynchronous Büchi automata for efficient automatic verification of concurrent systems*; 15th PSTV 1995; Warsaw, Poland; Chapman & Hall, 1995; pp. 305-321.

[25] A. Petrenko, N. Yevtushenko, G. v. Bochmann: *Fault models for testing in*

context; 9th FORTE 1996; Chapman & Hall, 1996; pp. 163-178.

[26] A. Petrenko, N. Yevtushenko, G. v. Bochmann and R. Dssouli: *Testing in context: framework and test derivation*; Computer Communications, 19, 1996, pp. 1236-1249.

[27] K. C. Tai, R. H. Carver: *Testing of distributed programs*; in A. Zomaya (ed.): *Handbook of Parallel and Distributed Computing*; McGraw Hill; 1995; pp. 956-979.

[28] K. C. Tai, R. H. Carver, E. E. Obaid: *Debugging concurrent Ada programs by deterministic execution*; IEEE Transactions on Software Engineering, vol. 17, no. 1 (Jan. 1991); pp. 45-63.

[29] R. N. Taylor, D. L. Levine, Ch. D. Kelly: *Structural testing of concurrent programs*; IEEE Transactions on Software Engineering, vol. 18, no. 3; 1992; pp. 206-215.

[30] Q. M. Tan, A. Petrenko, G. v. Bochmann: *Checking experiments with labeled transition systems for trace equivalence*; 10th IFIP Workshop on Testing Communicating Systems (IWTCS'97), Korea, 1997.

[31] J. Tretmans: *Conformance testing with labelled transition systems: Implementation relations and test generation*; Special issue on Protocol Testing, Computer Networks and ISDN Systems, Vol.29, 1996.

[32] A. Ulrich, S. T. Chanson: *An approach to testing distributed software systems*; 15th PSTV 1995; Warsaw, Poland; pp. 107-122; 1995.

[33] A. Ulrich, H. König: *Specification-based testing of concurrent systems*; IFIP Joint Int'l Conference on Formal Description Techniques, and Protocol Specification, Testing, and Verification (FORTE/PSTV'97); Osaka, Japan; 1997.

[34] M. P. Vasilevski: *Failure diagnosis of automata*; Cybernetics, Plenum, New York; No. 4, 1973, pp. 653-665.

[35] M. Yannakakis and D. Lee: *Testing finite state machines*; in Proceedings of the 23d Annual ACM Symposium on Theory of Computing, 1991, pp. 476-485.

[36] S. Duri, U. Buy, R. Devarapalli, and S. M. Shatz: *Application and experimental evaluation of state space reduction methods for deadlock analysis in Ada;* ACM Trans. Software Engineering and Methodology, Vol.3, No. 4, Oct. 1994, pp.340-380.

PART SIX

Industrial Applications

Conformance Testing of a Multimedia System Using PHACT

L.M.G. Feijs.
Philips Research Laboratories Eindhoven,
Eindhoven University of Technology.
EUT, P.O. Box 513, 5600 MB Eindhoven, The Netherlands.
Tel. +31 402742953, Email: feijs@natlab.research.philips.com

F.A.C. Meijs, J.R. Moonen.
Philips Research Laboratories Eindhoven.
PRLE, Prof. Holstlaan 4, 5656 AA Eindhoven, The Netherlands.
Tel. +31 402743932, Email: {meijs, jmoonen}@natlab.research.philips.com

J.J. van Wamel.
Centre for Mathematics and Computer Science.
CWI, P.O. Box 94079, 1090 GB Amsterdam, The Netherlands.
Tel. +31 205924099, Email: Jos.van.Wamel@cwi.nl

Abstract

In this paper we report on the applicability of a number of ISO-9646 techniques, concepts and automatic test-generation tools in the area of multimedia processing (MPEG2). Several concepts, such as PICS and PCO, and EFSM-based TTCN generation carried over to this new domain very well. The PHACT tool environment, originally developed at Philips for hardware (VHDL) protocol testing, was adapted and used for test generation and test execution in the multimedia processing domain. The paper also highlights a number of interesting issues related to the usage of (E)FSMs, such as dealing with *reactive* behavior at a software API PCO and dealing with stream-PCOs. Multimedia streams were modeled as FSMs. This turned out to be a new application area for test generation from composite or product FSMs, a technique originating from the embedded testing method. The process followed for test preparation and test execution consists of seven phases, which are surveyed in the paper.

Keywords

Test Generation, PICS, EFSM, TTCN, MSC, Conformance Testing, Multimedia Testing, MPEG, DVB, Digital Television

1 INTRODUCTION

It is widely recognised that more and more CE (consumer electronics) products are becoming part of distributed, open systems. From the area of telecommunication systems, we known that for interoperability of the constituting parts of a system, it is essential that these parts are tested on functional conformance with respect to internationally agreed standards. This has led to the ISO9646 methodology [12], the language TTCN [13], and a wealth of techniques for automatic test generation, notably (E)FSM-based test-generation [1, 10, 17, 20]. Thus it seemed promising to apply these concepts and techniques to the CE domain, which used to be based on analog techniques, but which is progressing into digital coding and transmission techniques. Furthermore, we consider functional conformance testing to be as important for hardware as it is for software.

At Philips a one-chip DVB (Digital Video Broadcast) source decoder is being developed. DVB systems are digital television systems based on MPEG2-compressed audio and video, which have inherent technical advantages over conventional systems [2]. The core of the Philips DVB decoder system is the DIVA5 IC, which integrates all functionality required for receiving and decoding MPEG2 *transport streams*, including descrambling, demultiplexing, audio and video decompression, overlay graphics provisions, and analog television encoding. It also contains an embedded microprocessor and several peripheral interfaces. The DIVA5 is therefore capable of performing all controller tasks in digital television applications such as *set-top boxes*.

Conformance testing usually concerns system characteristics which are of interest for interconnectivity with other systems. In the DIVA5 project, this means that we are interested in whether the system correctly processes MPEG2-compliant bit streams, as specified by the international standards for MPEG2 [11] and DVB [7, 8, 9].

We started from a test tool environment for protocol conformance testing of hardware designs in VHDL called PHACT (PHilips Automated Conformance Tester), which was developed during 1995 and 1996, as reported in IWTCS'97 [18]. PHACT was applied to the development of both the DIVA5 hardware and software. During the project the test environment was adapted to obtain an environment specifically suited for DIVA5 compliance testing, while the generic overall structure of the tool environment was maintained.

The main experiences we would like to report in the present paper are the following. Firstly, we found several ISO-9646 concepts such as PICS and PCO applicable in the multimedia processing area. Secondly, we found EFSM-based test generation very useful, but it had to be complemented with other techniques, notably for data-aspects (stream construction). Next to that, the paper highlights a number of interesting issues related to the usage of (E)FSMs, such as dealing with *reactive* behavior at a software API PCO (Point of Control and Observation), and dealing with a stream-PCO. Modeling multimedia streams as FSMs also turned out to be a new application area for test generation from *composite* or *product* EFSMs, a technique originating from the embedded testing method.

Section 2 surveys the followed test preparation and execution process. Section 3 discusses test preparation in more detail. In Section 4 the various types of tests needed for MPEG conformance testing are treated. Section 5 is about the test tools (PHACT) and the way they were used. In Section 6 the application of composite EFSMs is discussed. Section 7 contains a general discussion on test coverage. Finally, Section 8 contains some concluding remarks.

2 SURVEY OF THE TEST PROCESS

Although there are systematic test sets for *video* decoders (e.g., the *Sarnoff stress tests* [19], as sold by Doctor Design Inc.), we were, and are, not aware of any commercially available, systematic test sets that cover the MPEG2 *systems* layer. Furthermore, our job was not only to test the MPEG2 systems stream processing in isolation, but also the various programmable control registers and software of the DIVA5 chip. This meant that we had to start from scratch. The approach is schematically shown in Figure 1.

In the conformance testing of communication protocols, Protocol Implementation Conformance Statement (PICS) proformas are used, in order to make statements about which capabilities and options of a protocol have been implemented by a product supplier [14, 16]. A completed PICS proforma can therefore be used as a basis for the definition of tests for a product. In an analogous way, we formulated a PICS proforma for DVB source decoders such as the DIVA5 system. In that document, system requirements were listed which were derived from the international standards for MPEG2 and DVB.

The following step was to describe a large number of *test purposes*, which are informal descriptions of tests, in such a way that the PICS items are adequately covered. In the formulation of test purposes we also used the DIVA5 systems specification, in order to provide sufficient detail for the implementation of the actual tests. Implementation details often turned out to generate specific test purposes.

The next step was the definition of the actual tests that were run on the implementation by filling out so called *test templates*. Each completed test template aims to cover a number of individual test purposes, in order to keep the total number of executed tests manageable.

A test template may contain an Extended Finite State Machine (EFSM) for modeling interactive aspects of the DIVA5 system, as it takes place at the software Application Programmers Interface (API) or hardware/software interface. These EFSMs were used as a basis for *automatic test generation* in the TTCN (Tree and Tabular Combined Notation) formalism [13]. In the test construction phase, the EFSMs were implemented, TTCN tests were generated and subsequently connected to the DIVA5 by additional software, which we will subsequently call *mapping software*. TTCN tests were *executed* by our test environment PHACT.

The results of EFSM-based tests are visualised using Message Sequence Charts (MSCs) [4]. For any test, the result is summarised by a *verdict*, which is either *pass*

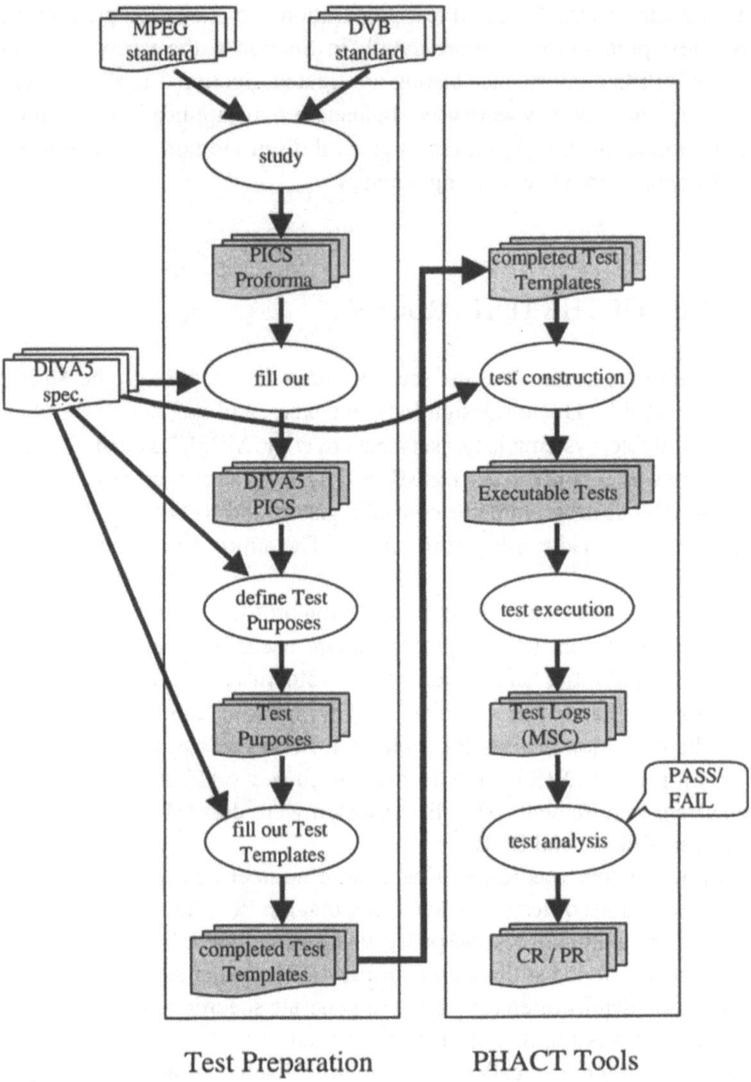

Figure 1 Survey of the Test Process.

or *fail*. 'Fails' generally lead to *Change Requests* (CRs) or *Problem Reports* (PRs) for the implementation under test.

3 TEST PREPARATION

In this section we describe the necessary steps that precede the actual testing. The test preparation process consists of three phases:

1. design and filling out of a PICS proforma;
2. identification of test purposes;
3. filling out of test templates.

We discuss each phase now, starting with the PICS proforma, which we had to design for MPEG2/DVB decoders. The proforma contains requirements from three categories:

1. MPEG2 decoder characteristics (hardware claims),
2. DVB specific features (application software claims),
3. DIVA5 API functionality (system software claims),

based on the ISO 'Systems' and 'Compliance' MPEG-2 standards [11], the ETSI DVB standards, and the Philips DIVA5 API specification, respectively. In Table 1, some PICS items are given as an example.

ITEM	REQUIREMENT	STAT.	REF.	SUPP.
A.1.13.a	For error conditions that do not result in unacceptable decoding artefacts, does the system maintain the system time base in the event of errored or missing packets that carry PCRs?	M	[11] : 8.1.8.2	Y / N
A.5.2	`transport_error_indicator`: If this bit is set, does the system invoke a suitable concealment or error recovery mechanism?	O	[7] : 5.4	Y / N
A.10.4	Does the system operate over the full range of the system clock frequency specified in [11] : 2.4.2.1 (27 MHz. ± 810 Hz.)?	M	[7] : 5.3	Y / N

Table 1 Examples of PICS items

Next we discuss the test purposes, which follow the PICS as a kind of step-wise refinement. A PICS may contain a large number of hardware and software claims, but it states in no way how such claims can be verified. In order to fill this gap, the following step was to informally describe a set of tests, such that all relevant PICS items were adequately covered. Since it now came to relate PICS items to hardware and implementation issues, the DIVA5 specification had to be scrutinised.

The informal test definitions were called *test purposes*. Test purposes state what DIVA5 functions have to be tested, and possibly how. The test purposes are related to specific system components and control registers of the DIVA5, to the relevant parts of the software API, and to specific PICS items, or conformance requirements.

The test purposes are grouped into so-called *test groups*. Each test purpose was described as a record, using keywords, as in the following example taken from Test Suite 11: Handling of duplicate, errored, and missing transport stream packets.

Test purpose 11.5	: Section filter response to duplicate packets.
Test method	: Inspection of buffer contents.
Streams	: DPS with pairs of duplicate (P)SI packets.
DIVA5 component	: PID filter (CC), section filter.
Pics items	: A.5.5.a.
Project stages	: VHDL.

A DPS is a Dedicated Packet Sequence, a short stream that is constructed for testing some specific functional behaviour of the system, with no real-time audio or video contents. Please note that EFSM-based testing is only one of the test methods, next to auditory/visual inspection and inspection of control registers or memory contents. Other keywords were used as well, including *Interrupts, R, W, R/W-Register(s)* (Read, Write, Read/Write control registers). Next to the VHDL stage, there are test purposes best covered in the QuickTurn phase (QT), where a hardware emulator, the QuickTurn, is used, or in the IC phase where the real silicon is used.

Finally we discuss the last step of the test preparation phase, the completion of so called *test templates*. These are documentation forms, used to describe a single test. For practical reasons we limited the total number of test templates and test streams by allowing a single test template to cover a number of individual test purposes. The test template form reflects the fact that a typical test is a pair $\langle s, t \rangle$, where s is a stream, for example a DPS, and where t is a test script, for example an EFSM together with additional mapping software to translate concrete events and data-structures into the abstract events of the EFSM, and vice versa. For cross-referencing purposes, the templates have an entry where the test purposes that are covered by the test are included.

4 TYPES OF TESTS

Before presenting the tools proper, we must discuss the distinct types of tests. To execute a test on the DIVA5, both an MPEG2 *test stream* and a *test script* are necessary. The stream carries MPEG audio, video and data, and is played in the test execution environment by a bit-stream player (BSP). During testing, the BSP plays the role of an MPEG/DVB broadcaster. The test script controls the settings of various parameters of the DIVA5 system, specific for a test. It plays the role of a set-top box application on top of the software API of DIVA5.

A distinction can be made between tests where the focus is on the 'stream' part, and tests where the focus is on the 'script' part. In general, a test script consists of a prescribed interaction pattern: a list of (stimulus, reaction) tuples. Execution of such a script basically boils down to the application of the stimuli to the system and observation of the system's reactions. When the interaction pattern is as prescribed, the test verdict will be pass, otherwise it will be fail. In our case, test scripts take the form of a TTCN test suite generated from an EFSM specification of the system.

Tests where the focus is on the *script* part typically test the system's programmable control registers and interrupts. While a stream is playing, the control registers are reprogrammed and the system's reaction is observed either via interrupts or via status registers. The reprogramming as well as the monitoring is controlled by the test script.

Since we aim at automatic test execution and analysis, the system's reaction should be detectable by our test software. For some types of tests, however, this is not a feasible or the most convenient method. For example, in order to determine the correct presentation of audio and video in the presence of *time-base discontinuities*, we also constructed tests that have to be verified by simply observing the audio and video as the system presents it. For such tests, the test script degenerates to an initial configuration of the system, and test verdicts will have to be issued by the human observer. These tests are called *stream tests*, and EFSM-based test script generation will not be used here.

Another type of stream tests are tests where there is no direct, observable reaction to a stimulus supplied to the system. For example, for testing whether some filter extracts the right data from a stream, it is more convenient to simply program the filter, play the stream, route the output of the filter to memory and inspect the memory contents afterwards.

For script tests, we could often abstract from parts of the contents of the stream; the packet *payload*. For example, for a test of the 'MPEG-Systems' buffer management only the size of the buffered data matters, not the specific contents. This does not imply that the stream may be constructed completely independent of the test script. For example, when the control registers of the descrambler module are tested, the stream must contain scrambled data at appropriate positions in the stream. But, in general, streams for tests focussed on scripts tend to be less complex than for stream tests.

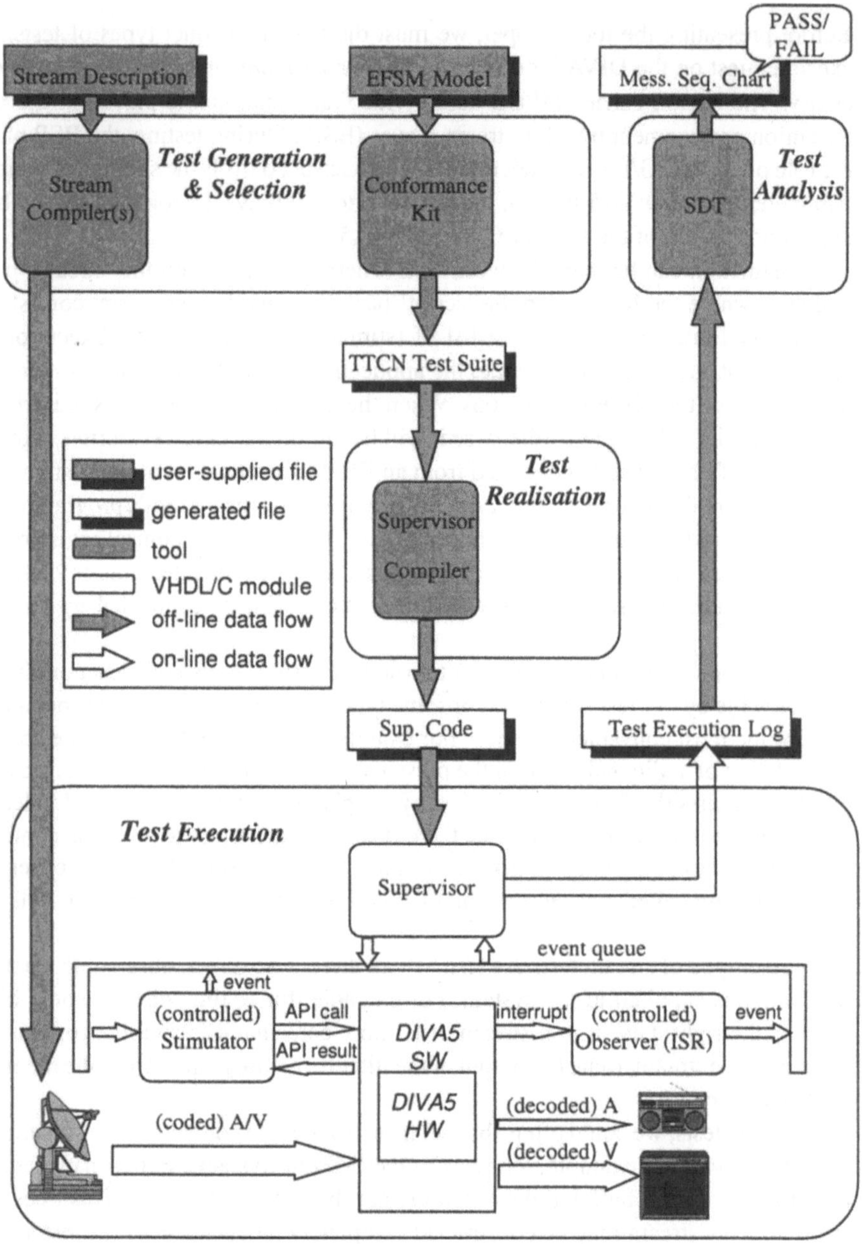

Figure 2 The PHACT test tool environment

5 TOOLS

Since *both* a stream and a test script are necessary to execute at test, the tools used in test development therefore belong to two different categories: (1) stream generation tools, (2) test script generation and execution tools. The first category is needed for the construction and validation of MPEG2-compliant bit streams, the second category serves for control of the test execution.

PHACT, as it has evolved now, is shown in Figure 2. The design of PHACT has been reported on in [18], hence we will not explain all details here. We only mention that the Conformance Kit [3] is used to generate a set of tests in TTCN format from an EFSM specification. An example TTCN test case is given below.

Test Case Dynamic Behaviour

Test Case Name:	test_2
Test Group:	buffers3/pt/filling_odd/
Purpose:	test input "allocate_even_buffer" in state "filling_odd"
Defaults Reference:	general_default

Nr	Behaviour Description	Constr. Ref.	Verdict
1	+ ts_filling_odd		
2	System ! allocate_even_buffer		
3	System ? odd_buffer_full		
4	System ! allocate_odd_buffer		
5	System ? even_buffer_full		PASS

Detailed Comments:

line 1: *transfer to state "filling_odd"*
line 2: *check input "allocate_even_buffer"*
line 4: *SIOS on "filling_even"*

This TTCN is subsequently compiled and combined with mapping information (as explained below) in order to produce an executable test. The test execution environment consists of a stimulator and observer, associated with the software API PCO, a stream-input PCO (coded audio, video and data) and a stream-output PCO (decoded audio and video). The stream-input PCO poses some limitations: it can only be offered an off-line prepared stream. In practice, bit-stream players have no capability to dynamically choose between packets, construct packets on-the-fly, etc. Concerning the stream-output PCO, it should be noted that it is often not feasible to check the output automatically and the decoded stream has to be checked audio-visually (i.e., by a human observer).

We are aware that, due to the Conformance Kit tool, standard formal specifications like SDL are not usable without a translation into the input language of this tool. However, since standard formal specification languages are not commonly used in the CE domain, we did not consider this to be a serious disadvantage.

Tests have been executed on three subsequent platforms: modeled or simulated hardware only, emulated hardware with software and the final hardware product with software.

We also like to mention that the output of the test execution phase is a test log in a format that can be visualised as a Message Sequence Chart (MSC) by the (commercial) tool SDT [21], see Figure 3. Next to the TTCN events flowing between the tester process and the IUT (Implementation Under Test) process, all interrupts that occur during a test are also shown as arrows from the IUT to the ISR (Interrupt Service Routine) process, even when these interrupts are not directly related to a TTCN input. An ISR is a function which gets executed when a specific interrupt occurs. The time at which the events occurred are also shown. TTCN verdicts are shown as MSC conditions. Using MSCs for test logging, as opposed to the more common use of MSCs for test goal specification, turned to out to be very useful, especially for long test runs. Tools used for MPEG stream construction are outside the scope of the paper.

The rest of this section is devoted to the issue of mapping software. In EFSM specifications the input and output events may be described at a very abstract level. For instance, a complicated filter initialisation may be covered by a single input event `setup`. Such abstraction is often useful to obtain a manageable set of meaningful tests. But when one wants to use the TTCN test suite defined in terms of these abstract events, extra information has to be added for linking the abstract events to concrete software API calls (or hardware register access), and interrupts of the system under test. We call the software which defines this relation mapping software, and mapping software has to be supplied in VHDL for the VHDL-based execution environment, and in C for the C-based execution environments (for QT & IC).

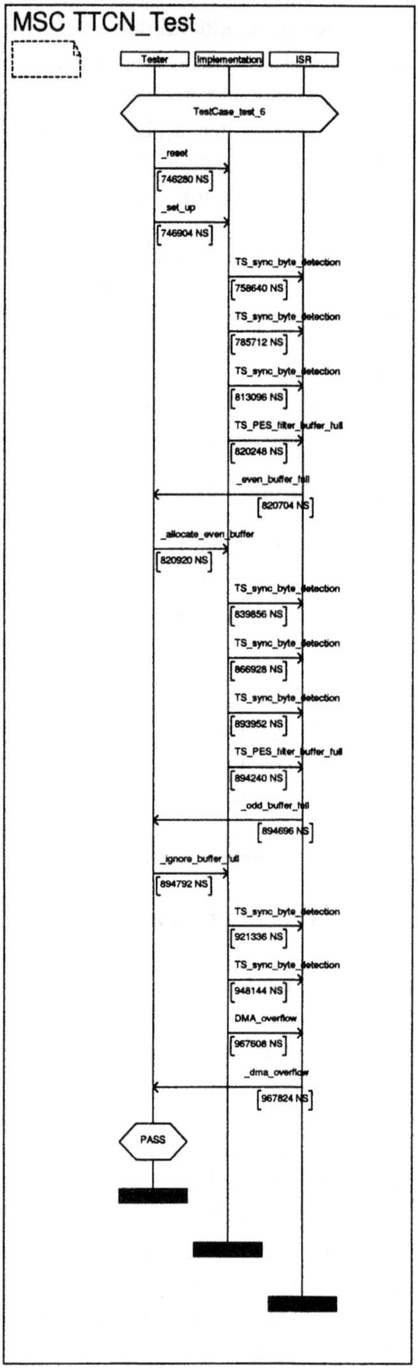

Figure 3 Example Message Sequence Chart

Basically, mapping software for stimulus events takes the form of case statements which map abstract events to sequences of (lower-level) software API calls. Sometimes, the returned results of these software API calls are the (lower-level) representation of TTCN input events. In such a case, the input event is communicated to the supervisor component (see Figure 2) via the *event queue* (in Figure 2: the upward pointing arrow leaving the stimulator). For example:

```
void stimulate(int ttcn_event)
{
    switch(ttcn_event)
    {
      case set_up:
        pid_program(100);
        pid_link(100, section_filter);
        pid_enable(100);
        break;
      case check_scrambling_info:
        BOOL scrambled = read_scrambling_info();
        if (scrambled)
          observe(SCRAMBLED_EVENT):
        else
          observe(NOT_SCRAMBLED_EVENT);
        break;
      /* ... */
    }
}
```

In the above example of stimulus mapping software in a C environment, both types of stimulus mappings are shown. The first type of stimulus is the most common; a stimulus event is simply mapped to a number of software API calls. The second type of stimulus is actually a read action, where the result of the read is the low level representation of a TTCN input event. The function `observe` indicates the occurrence of the TTCN input event.

Mapping software for events generated by the system (system output, TTCN input) can also assume the form of a set of interrupt service routines (ISRs), which contain code in their function bodies that indicates the occurrence of a TTCN input event. For example:

```
void buffer_full_isr(void)
{
    EVENODD p = read_parity();
    if (p == ODD)
      observe(ODD_BUFFER_FULL_EVENT);
    else
      observe(EVEN_BUFFER_FULL_EVENT);
}
```

In the above example of observer mapping software in a C environment, there are two abstract events (`ODD/EVEN_BUFFER_FULL_EVENT`) related to one concrete interrupt (`buffer_full`). The function `read_parity` is used to obtain extra information needed to determine which of the two possible abstract events occurred. This piece of code is installed as an interrupt service routine for the `buffer_full` interrupt before the test execution starts. Most ISRs tend to be one-liners which directly map a physical interrupt to an abstract event, but this does not always have to be the case, as the (real-life) example demonstrates. Figure 4 relates the control and data flow in the above mapping software examples to the test-execution architecture of PHACT.

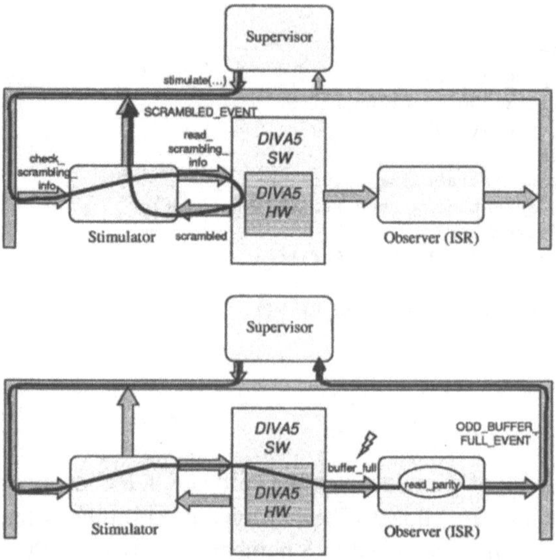

Figure 4 Mapping software control and data flow.

6 ON THE USE OF COMPOSITE EFSMS

Usually, there is a close connection between the EFSM used for test script generation and the test stream that is offered to the system during test execution. In such cases, the EFSM becomes a description of some very specific system behaviour, and not general system behaviour description. As a result, many of such specific EFSMs may be needed to achieve a 'good' test coverage (see Section 7). Such EFSMs are also harder to interpret, since they depend on detailed stream information.

In order to make more generic and comprehensible EFSMs we have employed a technique called *composite EFSMs*. The idea is to specify *two* (E)FSMs: one EFSM which describes (to a certain extent) a system aspect independent of the stream: the *system EFSM*, and one FSM which describes a test stream: the *stream FSM*.

Now the product or composite EFSM is the EFSM which describes the behaviour of the system when a stream is decoded which conforms to the stream FSM. Symbolically: *Composite EFSM = system EFSM × stream FSM.* The Conformance Kit can now generate TTCN from this composite EFSM, where only transitions that originate from the system EFSM – so no transitions from the stream FSM – are taken as test goal* This technique originates from the embedded testing method, and is described in reference [15]. The stream FSM can be considered as test context of the system EFSM.

An increased coverage (compared to the single stream situation) of a specific system aspect can now be obtained by developing a single system EFSM and a number of pairs (stream FSM, stream implementation). TTCN is generated from the composition of the system EFSM and a stream FSM, and each generated TTCN test suite is executed on the system while the corresponding stream is played by a bit stream player in the test execution environment.

We obtain a stream FSM by considering a stream simply as a sequence of packets partitioned or categorised in sequences of a particular packet *type*. We can then describe a stream by a regular expression in terms of these packet-sequence types. For example, consider a stream that starts with a packet that contains a PCR sample (some kind of MPEG time stamp), which is followed by alternating *dummy* packets – packets with no relevant payload – and packets with *section* data. This stream can be described by a regular expression: *pcr;(dummy;section)**, and also by the lower FSM of Figure 5.

A system EFSM can be made generic with respect to a set of stream FSMs when no assumptions are made about the *order* of the various packet-sequences in the stream, but only on the *types* of packet-sequences in the stream. As a consequence, the system EFSM will have to specify, for all of its states, the reaction upon each 'packet-sequence event' induced by the stream FSM.

A tester now has the option to construct streams that consist of various packet orderings; he only has to update the stream FSMs accordingly. The system EFSM can be used for all streams, without modifications, for EFSM composition and TTCN generation.

For a system EFSM, input events are either connected to the stream FSM ('packet-sequence events'), or unconnected, in which case they correspond to software API calls given to the system. All EFSM output events are unconnected to the stream FSM, and correspond to interrupts or software API call results generated by the system.

In Figure 5 an example composite EFSM is shown. The emphasised events are the 'packet-sequence' events. Reserved event `spontaneous` is used to model 'output-only' transitions.

*Transitions originating from the stream FSM are internal, technical artefacts introduced for modeling purposes and are therefore not treated as test goals.

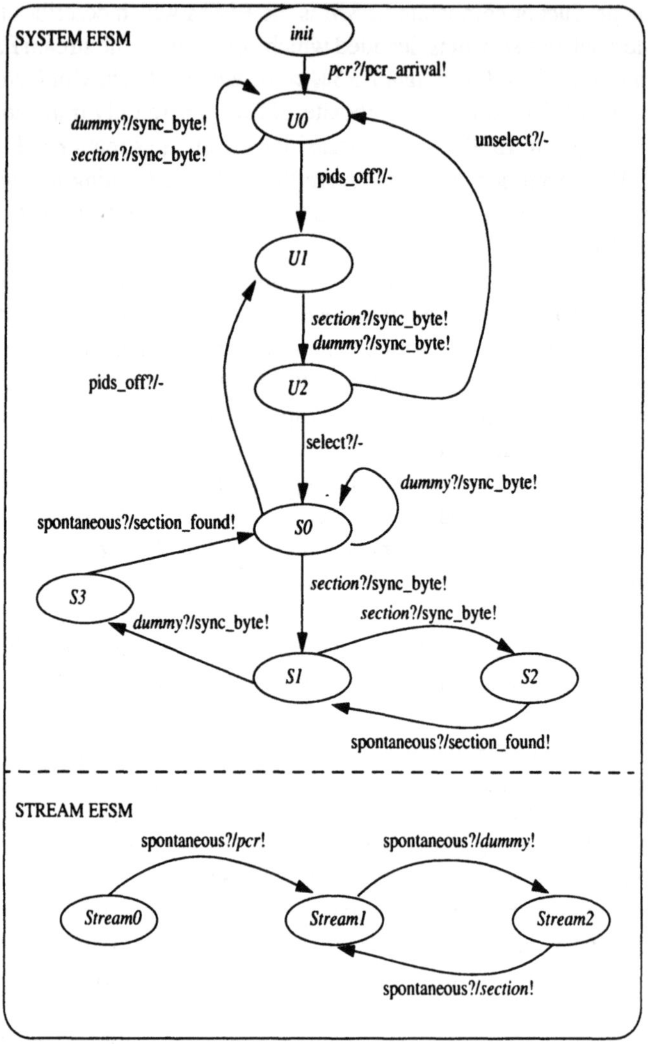

Figure 5 Composite EFSM example

7 COVERAGE

This section is devoted to a short discussion on coverage. Although the tests were derived by means of a systematic and step-wise approach, starting from the relevant standards, parts of the hardware specification, and the software API specification, the tests do not give us absolute certainty that the hardware and software are correct and fully compliant. Although some of the considerations may be well-known to the IWTCS audience, we consider it worthwhile to give them here.

- Firstly, there is the fundamental point put forward by Dijkstra [6], that testing can only reveal the presence of errors, never show their absense. This holds for computer programs, and thus it also holds for the combination of hardware and software.
- Secondly, the specifications of the IUT (implementation under test) are not given as formal models with mathematical rigour. Yes, we were able to cast certain aspects of the system into EFSM behaviour, but these were not all aspects. Moreover, the process of constructing mapping software is a non-exact process in itself.

The first point is much more fundamental than the second, although the present state-of-practice in specification techniques makes both equally relevant. In order to illustrate the first point, we present two arguments.

- Suppose that we would refrain from testing *all* behaviours of the IUT, but that we would be satisfied to find all things that could go wrong within one second. How many different streams exist which could be fed into the system in one second? With a rate of 40 Mbit/s, there are $2^{40,000,000}$ possible bitsequences that would have to be tried. Even considering only bitsequences conforming to the MPEG syntax, this is a number beyond comprehension.
- Even for those aspects which have been formalised as EFSMs and which have been tested by automatically generated TTCN, let us recall the basis for the completeness claims of this generation process. Even the best variants of the UIO (unique input output sequence)-based test-sequence generation algorithms (like UIOv [23]) only promise to find any defect under the assumption that the faulty implementation (E)FSM does not have more states than the specified (E)FSM. But in practice this assumption need not hold, for example a single integer variable declared and used by the implementor may cause a blow-up of the state space. However, examples where a faulty (E)FSM goes undetected by UIO-based test generation (the test-generated method using in the DIVA5 project) are somewhat academic.

It is also worthwhile to compare (black-box) conformance testing with traditional IC testing (see e.g. the IEEE ITC test conferences [5]). Traditional IC testing is concerned with finding physical defects, e.g. caused by dust particles or wafer anomalies, in ICs which are produced from correct (or assumed correct) designs. When the complexities of the ICs increased in the early 1970s, it was recognised that for finding physical defects, black-box testing was not the way ahead. In the so-called *structured testing* methods, the details of the design are taken into account. Fault-models are used, based on facts about connection paths which are or are not adjacent. In the DIVA5 project, however, we were testing for (functional) design errors, not for physical defects. Of course, measures have been taken to look into the implementation's source code and to try to find and eliminate errors, but that was not the main task of the test-project.

In summary, we believe that the obtained test coverage was increased, because of:

1. a *systematic test derivation process*, compared to traditional, ad-hoc testing, and
2. *automation by tool support where possible*, compared to purely manual testing.

8 CONCLUDING REMARKS

Although MPEG is strictly speaking not an interactive protocol but a data format, the idea of a PICS carried over very well to the multimedia domain. Considering the PICS, it is interesting to note that the MPEG/DVB standards rather define the compliance of streams than the compliance of decoders. Consequently we had to translate stream requirements to decoder requirements.

The intermediate step between PICS and test definition, the definition of test purposes, turned out to be indispensable, since a lot of implementation details are involved in decoding digital TV. We expect that the PICS will become a kind of reusable asset for other conformance testing activities in the area of digital TV.

Applying and adapting our original PHACT tool environment ([18]) to the DIVA5 gave valuable feedback and resulted in two new platforms for PHACT, while maintaining the overall tool structure. Currently PHACT runs on VHDL, pSOS/C (usable for both QT and IC) and UNIX/C environments.

Using MSCs for test logging, as opposed to the more common use of MSCs for test goal specification, turned to out to be very useful, especially for long test runs. Multimedia testing, in particular stream processing, gave rise to a new application area for composite EFSMs, a technique originating from the embedded testing method. Furthermore, one should realise multimedia tests, where the evaluation of test results is simply performed via the evaluation of image and sound quality, do not always allow for simple pass/fail verdicts. In some cases the comparison of various opinions may be necessary, in order to obtain a verdict.

The PICS proforma and the test purposes document led to a large set of tests. The tests themselves revealed various hardware defects and many software and specification/documentation defects. Besides defects directly concerning the test purposes covered by a test, many other defects were found during the process of "getting the test to run". For example, problems were found concerning development tools, bitstream players, peripherals and test-board issues. Finally, it is interesting to note that, although only 40% of our tests is EFSM-based (the rest was purely stream-based), all hardware errors which were found, we found with these EFSM-based tests.

Future Work. Interesting options for future work in this area are automatic test generation and verdict assignment for stream tests. More research should also go into guidelines for dividing a system into two composite EFSMs. Furthermore, there are other, more recent developments in hardware testing which maybe could be combined fruitfully with our EFSM based work, but we have not investigated this (yet). One such hardware testing technique is 'scan design', which provides that the flipflops in a circuit can be chained into a serial shift-register, called scan chain. Maybe this could be used to inspect an internal state (although this is not black-box, and also, reading the scan-chain destroys the system's state). Another hardware testing technique is testability analysis, where propagation of logical values and controllability of variables is analysed in order to select extra test points or partial-scan flip-flops. For a survey of these techniques we refer to [22]. Although we have no experience in ATM testing, we think that parts of our approach could be useful there as well. The MPEG-systems layer addressed in our project, can be compared best with certain variants of AAL, the ATM adaptation layer, notably AAL2.

Acknowledgements. We would like to thank Eric Gijsbers, Joris van den Hurk, Theo Kersjes and Remco Schutte, from Philips Semiconductors, and Ron Koymans, from Philips Research, for initiating and supporting the DIVA5 conformance testing project.

REFERENCES

[1] A.V. Aho, A.T. Dabhura, D. Lee, M. Umit Uyar. An optimisation technique for protocol conformance test generation based on UIO sequences and rural Chinese postman tours, In: S. Aggerwal, K. Sabnani (Eds.), PSTV VIII (1988).

[2] H. Benoit. Digital television, MPEG-1, MPEG-2 and principles of the DVB system. Arnold, 1997.

[3] S. van de Burgt, J. Kroon, E. Kwast, and H. Wilts. The RNL Conformance Kit, Proceedings 2nd *International Workshop on Protocol Test Systems*, pp.279-294, North-Holland, 1990.

[4] CCITT. Message Sequence Charts, CCITT Recommendation Z.120, ITU-T, 1996.

[5] R. Dekker, F. Beenker, L. Thijssen. Fault Modeling and Test Algorithm Development for Static Random Access Memories, Proceedings *Internal Test Conference*, IEEE Society Press, pp.343-352, 1988.

[6] E.W. Dijkstra, `http://alycia.dementia.org/~strong/public/quotes`.

[7] European Broadcasting Union / European Telecommunications Standards Institute. Digital broadcasting systems for television; Implementation guidelines for the use of MPEG2 systems; Video and audio in satellite and cable broadcasting applications. ETR 154 (draft), 1994.

[8] European Broadcasting Union / European Telecommunications Standards Institute. Digital Video Broadcasting (DVB). ETR 211 (final draft), 1996.

[9] European Broadcasting Union / European Telecommunications Standards Institute. Digital broadcasting systems for television, sound and data services; Specification for Service Information (SI) in Digital Video Broadcasting (DVB) systems. ETS 300 468, 1995.

[10] W. Hengeveld, J. Kroon. Using checking sequences for OSI layer conformance testing, In: H. Rudin, C.H. West (Eds.), PSTV VII (1987).

[11] International Organisation for Standardisation. Information technology – *Generic Coding of Moving Pictures and Associated Audio.* – Part 1:

Systems, Part 2: *Video*, Part 3: *Audio*, Part 4: *Compliance*. International Standard ISO/IEC 13818-{1,2,3,4}, 1995.

[12] International Organisation for Standardisation. Information technology – Open Systems Interconnection – *Conformance Testing Methodology and Framework* – Part 1: *General concepts.* ISO/IEC 9646-1, 1991.

[13] International Organisation for Standardisation. Information technology – Open Systems Interconnection – *Conformance Testing Methodology and Framework* – Part 3: *Tree and Tabular Combined Notation.* ISO/IEC 9646-3, 1991.

[14] International Organisation for Standardisation. Information technology – Open Systems Interconnection – *Conformance Testing Methodology and Framework* – Part 7: *Implementation Conformance Statements.* ISO/IEC 9646-7.

[15] H. Kloosterman, H. van Dam, E. Kwast. Test derivation for standardised test methods, In: J. Kroon, R.J. Heijink, E. Brinksma (Eds.), Protocol Test Systems IV (1992).

[16] K.G. Knightson. *OSI Protocol Conformance Testing.* McGraw-Hill Inc., 1993.

[17] D. Lee, D. Su. Modeling and testing of protocol systems, In M. Kim, S. Kang, and K. Hong, editors, *Testing of Communicating Systems*, volume 10, pp.339-364. Chapman and Hall, 1997.

[18] J.R. Moonen, J.M.T. Romijn, O. Sies, J.G. Springintveld, L.M.G. Feijs, and R.L.C. Koymans. A two-level approach to automated conformance testing of VHDL designs. In M. Kim, S. Kang, and K. Hong, editors, *Testing of Communicating Systems*, volume 10, pp.432-447. Chapman and Hall, 1997.

[19] D. Sarnoff Research Center. *Sarnoff Compliance Bitstreams for MPEG2 Video*. Version 1.1, user's manual, 1996.

[20] D.P. Sidhu, T.K. Leung. Formal methods in protocol testing, a detailed study, IEEE trans. on Softw. Eng., Vol. 15, No. 4, pp.413-426 (1989).

[21] Telelogic SDT3.1, Getting Started, Telelogic AB, nov. 1996.

[22] H.P.E. Vranken. Design for test and debug in harware/software systems, Ph.D. Thesis, Eindhoven University of Technology, ISBN 90-386-0440-8 (1998).

[23] S.T. Vuong, W.Y.L. Chan, M.R. Ito. The UIOv-Method for Protocol Test Sequence Generation, Proceedings 2nd *Internal Workshop on Protocol Test Systems*, pp.161-176, North-Holland, 1990.

An industrial experiment in automatic generation of executable test suites for a cache coherency protocol

Hakim Kahlouche, César Viho** & Massimo Zendri****

*IRISA - {*INRIA, **IFSIC-Université de Rennes I, Campus de Beaulieu, F35042 Rennes, France}, ***Bull Italia, Via ai Laboratori Olivetti, I-20010 Pregnana, Milanese (MI), Italy*
e-mail: {kahlouch,viho}@irisa.fr, massimo.zendri@bull.net

Abstract
In this paper, we present an end-to-end industrial case-study concerning the automatic generation of tests suites for the Cache Coherency Protocol of a Multiprocessor Architecture. It consists of the following stages : (1) formal specification of the architecture using Lotos language, (2) formal description of the test purposes, (3) automatic generation of abstract test suites using the prototype TGV, and (4) automatic generation and analysis of executable test suites. Through the description of each of the previous stages, this paper demonstrates that tools designed for protocol conformance testing can be efficiently used to generate executable tests for hardware concurrent systems.

Keywords
Conformance Testing, Test Generation, Lotos, Test Execution, Hardware Multi-Processor Architecture, Cache Coherency Protocol

1 INTRODUCTION

The aim of testing is to verify that the implementation of a system correctly realizes what is described in its specification. In this paper, we are particularly concerned with the so called *black box conformance testing*. In this testing approach, the behaviour of the implementation (otherwise called IUT for Implementation Under test) is known only by its interactions with the environment via its interfaces (called PCOs for Points of Control and Observation). Thus, testing consists in stimulating the IUT and observing its reactions on its PCOs. The prototype TGV has been [1] developed to generate test suites for communication protocols using the black box conformance testing approach. TGV is based on protocol verification algorithms and its main purpose is to fit as well as possible the industrial practice of test generation. It takes as entries the formal specification of the system to be tested and a formalization of a test purpose, and it generates an abstract test case. A test case is represented by a tree where each branch contains the interactions between the tester and the implementation. A verdict is associated to each branch. TGV has been experimented on the Drex military protocol [2]. The formal specification of this protocol is done in SDL formal description language. The comparison of the hand written test cases with those generated by TGV, has shown the interest and efficiency of TGV [3].

On another side, many tools have been developped in the area of Hardware testing, allowing simulation of the specifications, automatic synthesis of implementations and even tests generation. The hardware design is often based on hardware description languages such as VHDL. This is due to the ability of these languages to describe hardware-related details such as register-transfer, gate and switch level. These details may lead to over-specification and (most of all) they are not directly relevant to high-level functionalities specification, such as Cache Coherency Protocols, etc. Then it becomes fully justified to wonder whether the formal specification languages and associated tools designed in computer network area could be better appropriate for the description of these functionalities [4, 5].

The challenge for us, in the VASY action of the Dyade GIE Bull/Inria, is to demonstrate that TGV can also be used to generate tests for other systems than communication protocols: particularly, for the Cache Coherency Protocol of a Multiprocessor Architecture under construction at Bull Italia. In this experiment, we have had to deal with three main difficulties :

• the system to test is not a communication protocol but a Cache Coherency Protocol of a Hardware Multiprocessor Architecture.

• the formal description language used is not SDL but LOTOS.

• the habits and methodologies of test practicians in hardware architecture testing are not the same as in communication protocol testing.

First, we describe the fundamental aspects of TGV. We give some precisions on the Cache Coherency Protocol of the architecture. In this paper, we call this architecture the BULL'S CC_NUMA Architecture. Then, we indicate the appropriate abstractions made on its LOTOS formal specification, in order to make the test generation feasible. We show how we have used this formal specification to generate tests suites with TGV. We have also developped tools in order to make the abstract test cases generated by TGV executable in the real test environment of the BULL'S CC_NUMA Architecture. The last section is dedicated to the presentation of these tools. We end this paper by reporting results of the experiment which indicate how we have resolved the main difficulties enumerated above.

2 OVERVIEW OF TESTS GENERATION WITH TGV

The testing method with TGV consists in stimulating the IUT and observing its reactions on its PCOs. Depending on what is observed, a *verdict* is emitted indicating whether the IUT can be considered as a good implementation of the system or not. According to the importance of this verdicts, it is fundamental to give a precise specification of the system. It is also important to define the *conformance relation* between an implementation and the specification. Because a test activity of a complex system cannot be exhaustive, only particularly important aspects can be tested. This can be done by defining *test purposes* which help to choose the behaviours of the IUT to be tested.

In the testing methods where the test suites are hand written, all the objects enumerated before (specification, test purposes, conformance relation, verdicts) are described informally. This implies the problem of the correctness of these test suites, and therefore the problem of the confidence to put in the associated verdicts. The methods of automatic generation of test suites which are based on *formal description* of these different objects bring a solution to this problem. The prototype TGV we have developped in collaboration with the Spectre team of the Verimag laboratory [2, 1] is precisely situated in this context. TGV takes two main entries: the formal specification of the system and a the formal description of a test purpose (by an automaton) which represents an abstract form of the test case to be generated. From these objects, TGV gives as result a test case in form of a "decorated" DAG (Direct Acyclic Graph). The paths of this DAG (which can be unfolded into a tree) represent test sequences. Details on TGV algorithms can be found in [2, 1].

We present here the elements (described in Figure 1) which partake in the generation of a test case for the BULL'S CC_NUMA architecture.

The first main entry of TGV is the formal specification of the BULL'S CC_NUMA machine. The CAESAR.ADT compiler of the CADP toolbox [6, 7] is used to compile the *data part* of the specification. The CAESAR compiler produces the C file corresponding to the *control part*, including the functions (Init, Fireable, Compare,...) needed by TGV to manipulate "on-the-fly" the state graph of the system (without generating it) [8]. Then, the C compiler produces the corresponding object-file (CC_NUMA_spec.o in Figure 1).

Some observable interactions described in the LOTOS specification can be judged not important to the test activity point of view. Those interactions must be considered unobservable. This is done in TGV by a *hiding* mechanism (CC_NUMA_spec.hide on Figure 1) which contains all the interactions to be considered internal to the system.

The semantics of LOTOS (so do the CAESAR compiler) does not make distinction between input and output. In fact, interactions between processes are synchronization events. This puts in trouble TGV in which this distinction is needed to distinguish controllable events (from tester to implementation)

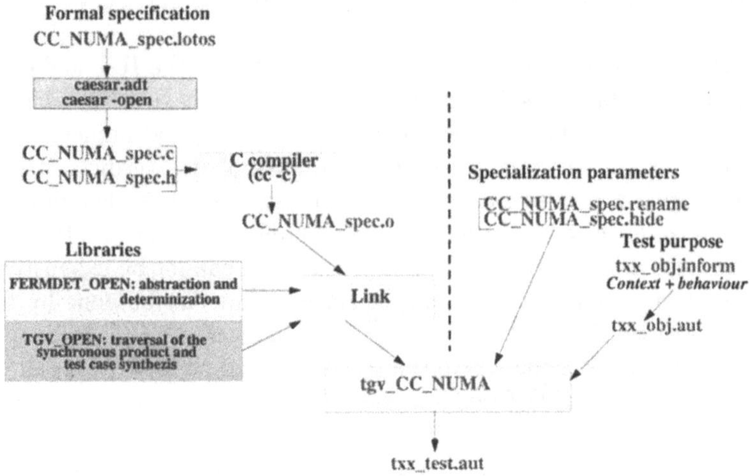

Figure 1 TGV General Architecture using LOTOS entry

from observable events (from implementation to tester) in the generated test cases. We introduce in TGV a *renaming* mechanism to resolve this problem.

The other main entry of TGV is the formal test purpose from which we have to generate a test case. It is formalized by an automaton in Aldebaran format (see an example in section 4.2).

The libraries FERMDET_OPEN and TGV_OPEN contain the functions which realize "on-the-fly" all the operations (abstraction, reduction, determinization and test case synthesizing) leading to the generation of the test case. This is a solution to the combinatory explosion problem which makes most of tools unable to generate test cases for complex system, as it is the case in the experiment we are describing in this paper.

Linking the object file together with the two libraries (FERMDET_OPEN and TGV_OPEN), produces an executable (tgv_CC_NUMA in Figure 1). Given a formal test purpose (txx_obj.aut) and the specialization files (described with two files CC_NUMA_spec.rename and CC_NUMA_spec.hide) as parameters of this executable, TGV generates the corresponding test case.

3 THE BULL'S CC_NUMA ARCHITECTURE: THE CACHE COHERENCY PROTOCOL

The BULL'S CC_NUMA architecture is a multiprocessor system based on a Cache-Coherent Non Uniform Memory Architecture (CC-NUMA), derived from Stanford's DASH multiprocessor machine [9]. It consists of a scalable interconnection of modules. The memory is distributed among the different modules. Each module contains a set of processors (see figure 2). The BULL'S CC_NUMA architecture key feature is its *distributed directory based cache*

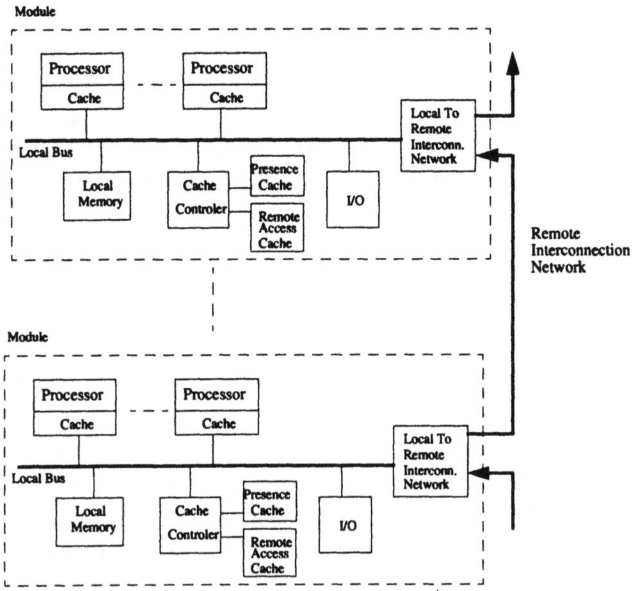

Figure 2 The BULL'S CC_NUMA General Architecture

coherency protocol using a Presence Cache and a Remote Cache in each module. The Presence Cache of a module is a cached directory that maps all the blocks cached outside the module. The global performance of the BULL'S CC_NUMA architecture is improved through the Remote Cache (RC) that locally stores the most recently used blocks retrieved from remote memories. Remote memory block can be in one of the following status: *uncached, shared, modified* which correspond to the possible RC status: *(INV)alid, (SH)ared, (MOD)ified.* So, the purpose of testing the Cache Coherency Protocol consists in ensuring that the status of the Presence Cache and Remote Cache are always correctly updated during the execution of any transaction in the BULL'S CC_NUMA architecture.

4 FORMALIZATION AND ABSTRACT TESTS GENERATION

4.1 Formal specification of the cache coherency protocol

The LOTOS language has been selected for the formal specification of the BULL'S CC_NUMA architecture because its underlying semantics model is based on the rendez-vous synchronization mechanism which is well suited for the specification of hardware entities [4, 5] such as processors, memory controllers, bus arbiter, etc. The communications between these components by sending electrical signals on conductors are better described by interactions between LOTOS processes, rather than infinite FIFO queues as in SDL.

The formal specification consists of about 2000 Lotos lines where 1000 lines describe the control part (13 processes) and the other half defines the ADT (Abstract Data Types) part. This specification is composed of two modules and has been debugged and verified with appropriate formal verification techniques, and is considered by TGV as the reference model of the system. In the following, we will call these modules M0 and M1. Each module contains one processor called P0. There are two block addresses in the system called A0 and A1, and two data D0 and D1. These blocks are physically located in module M0. Two main reasons bring us to make some abstractions in the formal specification:

• The first reason is due to the size and the complexity of the BULL'S CC_NUMA architecture, with as direct consequence the combinatory explosion problem even though TGV works "on-the-fly".

Thus, some causally dependent operations concerning the same transaction are collapsed. For example, from the testing point of view, the local response transaction always follows a local bus transaction in an atomic way (although if the real system can do something else between this two actions). These two transactions are collapsed in the Lotos specification. This reduces the complexity of the specification.

• The second reason is that in this work, we are interested in tests generation for the Cache Coherency Protocol. So, we make abstractions needed to hide all other operations which do not concern with this protocol.

4.2 Formalization of the test purposes

A test purpose in TGV is described by an automaton which represents an abstract view of the test case. So, in order to make TGV working, we have had to formalize each test purpose. The main test purposes to be applied to the BULL'S CC_NUMA architecture are informally described (in the shape of tables with comments) in the test plan. Seven *Test Groups* have been identified. In the experiment we are reporting in this paper, we are interested in two Test Groups concerning the test of the Cache Coherency protocol.

The Cache Coherency Test Groups: Some other definitions are needed to make what follows easier to understand. The *Requesting Processor* is the processor that initiates the transaction. The *Requestor* is the module that includes the Requesting Processor. The *Home Module* is the module which physically locates the requested block. The *Participant Modules* are modules which are requested by the Presence Cache to participate in the cache coherency protocol.

The test purposes described in the Test Group 3 are dedicated to Cache Coherency Testing (No Participants). This means that they aim to test interactions between two modules (the Requestor and the Home) which do not need interventions of other modules. For example, Table 1 describes an informal test purpose which means: "The block address is in M0. The CPU#0 of M1 executes a READ on this address. Verify that the Presence Cache (PC) status of Module#0 changes from Invalid to Shared." In this case, we can notice that the other modules are not concerned.

Cache Coherency Tests : set-PC-to-SH					Test group #3
Operation	Parameters	Source	Target	PC	Notes
READ	-	M1 CPU#0	M0	SH	PC Status of M0 changes from Inv. to Shared

Table 1 Presence Cache Status Setting to (SH)ared

Formal specification of test purposes: A test purpose is described with a labelled automaton in the Aldebaran syntax [6]. The format of a transition is: (from_state, label, to_state). A label is a LOTOS gate followed by a list of parameters. As said previously, TGV needs to distinguish between input and output actions of the system. This is achieved simply by the first occurrence of "?" (for input) or "!" (for output) in the label. The automaton corresponding to a test purpose describes a point of view of the system. As an example, we give hereafter the automaton which formalizes the test purpose described in Table 1:

```
des␣(0,␣11,␣4)
(0,"?BUS_TRANS␣!M1␣!READ␣!A0␣!PROCESSOR␣!FALSE",1)
(1,"?BUS_TRANS␣!M0␣!READ␣!A0␣!PROCESSOR␣!FALSE",2)
(1,"?BUS_TRANS␣!M0␣!READ␣!A1␣!PROCESSOR␣!FALSE",2)
(1,"?BUS_TRANS␣!M1␣!READ␣!A0␣!PROCESSOR␣!FALSE",2)
(1,"?BUS_TRANS␣!M1␣!READ␣!A1␣!PROCESSOR␣!FALSE",2)
(1,"?BUS_TRANS␣!M0␣!RWITM␣!A0␣!PROCESSOR␣!FALSE",2)
(1,"?BUS_TRANS␣!M0␣!RWITM␣!A1␣!PROCESSOR␣!FALSE",2)
(1,"?BUS_TRANS␣!M1␣!RWITM␣!A0␣!PROCESSOR␣!FALSE",2)
(1,"?BUS_TRANS␣!M1␣!RWITM␣!A1␣!PROCESSOR␣!FALSE",2)
(1,"LMD_PUT␣!M0␣!A0␣!RCC_SH␣!FLAG␣(FALSE,␣TRUE)",3)
(1,"*",1)
ACCEPT␣3␣REFUSE␣2
```

The first line is the automaton descriptor. It indicates that the first state is 0, there are 11 transitions and 4 states. The first transition indicates that the processor P0 of M1 requests for a READ transaction on the block address A0. The statement REFUSE 2 on the last line indicates to TGV that the state 2 is a refusal state of the test purpose. The labels of transitions which lead to a refusal state are not considered by TGV while generating the test case. After the READ transaction requested by M1 on A0, we don't want to consider other READ transactions.

The statement ACCEPT 3 indicates to TGV that the state 3 is the acceptance state of the test purpose. When the Presence Cache status of Module M0 changes from Invalid to Shared, TGV should consider that the test purpose is reached. This is mentioned in the test purpose with the transition (1,"LMD_PUT !M0 !A0 !RCC_SH !FLAG (FALSE, TRUE)",3).

The label "*" stands for *otherwise.* With the transition (1,"*",1), TGV takes other intermediate observations into account until it observes the specified observations (from state 1).

4.3 Generated abstract test cases

We give here the test case generated by TGV starting from the test purpose of Table 1 and formally described above.

```
des (0, 28, 26)
(0,"!BUS_TRANS !M1 !READ !A0 !PROCESSOR !FALSE",1)
(1,"LOC_RESP ?M1 !ARESP_MODIF",2)
(2,"PACKET_TRANSFER ?M1 !M0 !READ !A0 !REQ_PACKET_TYPE !NIL_DATA
                          !NETRESP_NIL !OUTQIO !M1 !OUTQIO",3)
(3,"LMD_PUT ?M0 !A0 !RCC_SH !FLAG (FALSE, TRUE), (PASS)",4)
(3,"BUS_TRANS ?M0 !READ !A0 !RCC_INQ !FALSE",5)
(5,"LOC_RESP ?M0 !ARESP_RETRY",6)
(6,"LMD_PUT ?M0 !A0 !RCC_SH !FLAG (FALSE, TRUE), (PASS)",7)
(6,"BUS_TRANS ?M0 !READ !A0 !RCC_INQ !FALSE",8)
(8,"LOC_RESP ?M0 !ARESP_RETRY, INCONCLUSIVE",9)
(8,"LOC_RESP ?M0 !ARESP_NULL",10)
(10,"LMD_PUT ?M0 !A0 !RCC_SH !FLAG (FALSE, TRUE), (PASS)",11)
(10,"LOC_DATA_BUS_TRANS ?M0 !READ !A0 !D0 !SMC !RCC_INQ",12)
(12,"LMD_PUT ?M0 !A0 !RCC_SH !FLAG (FALSE, TRUE), (PASS)",13)
(12,"PACKET_TRANSFER ?M0 !M1 !RESP_DATA_PACKET_TYPE !D0 !NETRESP_DONE
                                                  !OUTQIO" 14)
(14,"LMD_PUT ?M0 !A0 !RCC_SH !FLAG (FALSE, TRUE), (PASS)",15)
(14,"RCT_PUT ?M1 !A0 !RCC_SH",16)
(16,"LMD_PUT ?M0 !A0 !RCC_SH !FLAG (FALSE, TRUE), (PASS)",17)
(16,"LOC_DATA_BUS_TRANS ?M1 !READ !A0 !D0 !RCC_OUTQ !PROCESSOR",18)
(18,"LMD_PUT ?M0 !A0 !RCC_SH !FLAG (FALSE, TRUE), (PASS)",19)
(18,"FREE_OUTQ ?M1",20)
(20,"LMD_PUT ?M0 !A0 !RCC_SH !FLAG (FALSE, TRUE), (PASS)",21)
(14,"LOC_DATA_BUS_TRANS ?M1 !READ !A0 !D0 !RCC_OUTQ !PROCESSOR",22)
(22,"LMD_PUT ?M0 !A0 !RCC_SH !FLAG (FALSE, TRUE), (PASS)",23)
(22,"RCT_PUT ?M1 !A0 !RCC_SH",24)
(24,"LMD_PUT ?M0 !A0 !RCC_SH !FLAG (FALSE, TRUE), (PASS)",25)
(24,"FREE_OUTQ ?M1",20)
(5,"LOC_RESP ?M0 !ARESP_NULL",10)
(1,"LOC_RESP ?M1 !ARESP_RETRY, INCONCLUSIVE",9)
```

We can recognize the reverse form ("!" rather than "?") of the first transition of the test purpose described above (because it is the tester's point of view): (0,"!BUS_TRANS !M1 !READ !A0 !PROCESSOR !FALSE",1). This is a stimuli of the tester. It consists of a READ transaction on the local bus of module M1. The target of this transaction is the address location A0 (local to M0). So, this is expected to be a remote operation. Let us now describe some important transitions of the test case:

- The transition (1,"LOC_RESP ?M1 !ARESP_MODIF",2) indicates that the remote cache controller recognizes the address as a non-local address and forces the arbiter of the node to give a modify response (ARESP_MODIF).
- The remote cache controller routes the request to the remote link, the request is directed to the home module M0: (2,"PACKET_TRANSFER ?M1 !M0 !READ !A0 !REQ_PACKET_TYPE !NIL_DATA !NETRESP_NIL !OUTQI0 !M1 !OUTQI0",3).
- (3,"LMD_PUT ?M0 !A0 !RCC_SH !FLAG (FALSE,TRUE),(PASS)",4):

Module M0 reads his Presence Cache to see the status of the cache line. After that, the entry is updated (the line is in a Shared (RCC_SH) status), which means that the line is present also in module M1; an array of booleans is used to represent the presence bits (FLAG (FALSE, TRUE)). The verdict (PASS) is then emited indicating that the test purpose is reached.

- The implementation is allowed to choose the order in which the operations are done. Thus, the three following transitions constitute an other way to reach the test purpose: (3,"BUS_TRANS ?M0 !READ !A0 !RCC_INQ !FALSE",5), the remote cache controller of module M0 requests localy the data. An agent on the bus can always decide to retry the transaction for any sort of reason (5,"LOC_RESP ?M0 !ARESP_RETRY",6). This means that the remote cache controller has to execute again the transaction. Then the Presence Cache changes to Shared: (6,"LMD_PUT ?M0 !A0 !RCC_SH !FLAG (FALSE, TRUE), (PASS)",7).
- (6,"BUS_TRANS ?M0 !READ !A0 !RCC_INQ !FALSE",8) and (8,"LOC_RESP ?M0 !ARESP_RETRY, INCONCLUSIVE",9) indicate that the remote cache controller of module M0 executes again the local operation. A second retry response should lead to an inconclusive verdict because in TGV we have chosen to cut cycles in order to generate finite test cases.
- All the other transitions of this test case can be easily interpreted as they correspond to different other orders of execution of the operations described in previous items.

4.4 Results on the abstract test cases generation

At this stage of the experimentation, we have formally specified all the test purposes described in the Test Groups 3 and 4 (see section 4.2) except those requiring an interactive behaviour of the system. For each test purpose, we have generated the corresponding abstract test case using TGV. The main problem here concerns with the time cost of the test generation with TGV. This is due to the complexity of the BULL's CC_NUMA architecture specification which required us sometimes to refine the test purposes in order to speed up the test generation with TGV.

5 IMPLEMENTATION OF THE GENERATED TEST CASES

The purpose of this section is to describe the techniques and tools we have developed in order to make the abstract test cases generated by TGV executable in the testing environment of BULL'S CC_NUMA architecture (called SIM1 environment). To do so, we start by describing the SIM1 environment and principally the testing methodology currently used in SIM1 environment. Then, we present the new testing architecture. Finally, we describe an example on using this architecture.

5.1 The current testing architecture

The SIM1 BULL'S CC_NUMA testing environment structure consists of three modules, connected on a Remote Interconnection Network. Each module is composed by Processor Behavioral Models (MPB Bus Model), Memory Array and Memory Control, Arbiter and I/O Block, Coherency Controller, Remote Cache Tag that contains the Tag of Remote Cache and the Presence Cache. Figure 3 shows the current testing environment. The simulation environment is composed of kernel event simulator (VSS kernel: VHDL System Simulator) and a front end human interface (VHDL Debugger). The VSS is in charge of down-loading the outputs which are issued by the PROBE lines into a file (PROBE.OUT file in Figure 3). A CPU#i (MPBi in Figure 3) expects an *input table* which contains the commands to the MPB model in an intermediate format. The MPBgen application is in charge of converting the MPB input commands format (*input files*) into the intermediate format (*input tables*). The first step in the testing methodology is to write the input files.

(a) Input files

An input file describes a sequence of transactions to be executed by one CPU. The input files are written according to the informal test purposes specified in the test plan document. This is currently done by hand. There is one input file per MPB, and the main difficulty in describing these files is the synchronization of the CPUs w.r.t the test purpose. In fact, there are two cases of synchronization:

Case 1 (Intra-CPU Synchronization): In the case where all the transactions have to be executed by one CPU, that is materialized by only one input file, the synchronization is achieved by using the SYNC_CYC trans-

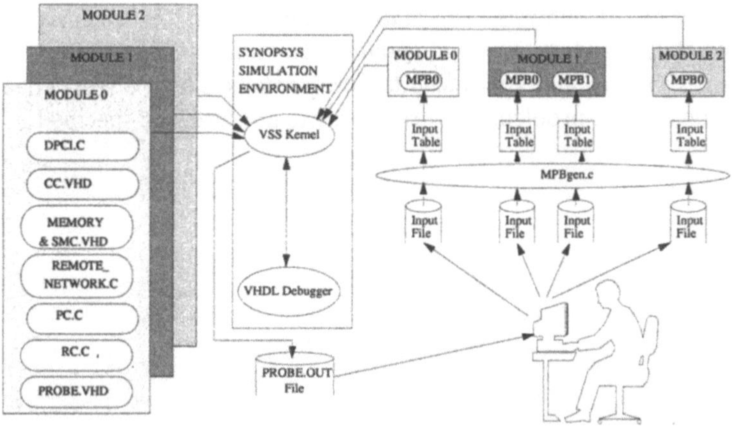

Figure 3 Current testing environment of the BULL'S CC_NUMA architecture

action. This transaction is a "barrier" for any subsequent operation issued by the same processor.

Case 2 (Inter-CPUs Synchronization): In the case where the transactions of the test purpose have to be executed by several CPUs, the problem is to achieve an inter-CPUs synchronization. The previously described SYNC_CYC transaction can also be used in this case. Another way to achieve this synchronization consists in submitting one input file to its corresponding CPU. Then after an estimated delay δ of the execution, the next input file is submitted to an other CPU. The difficulty of this synchronization mechanism lies in the estimation of δ.

(b) Output Analysis

Once the execution of the different input files has been completed, a PROBE output file is generated (see Figure 3). This file contains for each module the sequence of actions which has been effectively executed in the system together with the Local Memory Directory and Remote Cache status. Each action is associated with a stamp, that is the starting time of its execution. One line of this file has the following form:

```
PROBE #0 ---> L_Bus 620 burst rwitm A0 Tag 00 addr=014000AA00
                         Pos_Ack Resp_Rerun at time 660 NS
```

It means that the PROBE of Module#0 observes at time `660 NS` a `RWITM` transaction on the local bus 620. Currently, the analysis of the output file is done by hand using some empirical rules. It consists in comparing each line of the PROBE file with what was specified in the test purpose and what is informally described in the test plan document. Finally a verdict is emitted at the end of the analysis. *The main problem here is the analysis task which is completely based on informal specifications and informal notion of conformance which may sometimes lead to false verdicts.* The TGV approach brings a solution to that problem since the verdicts are formally specified in the test cases.

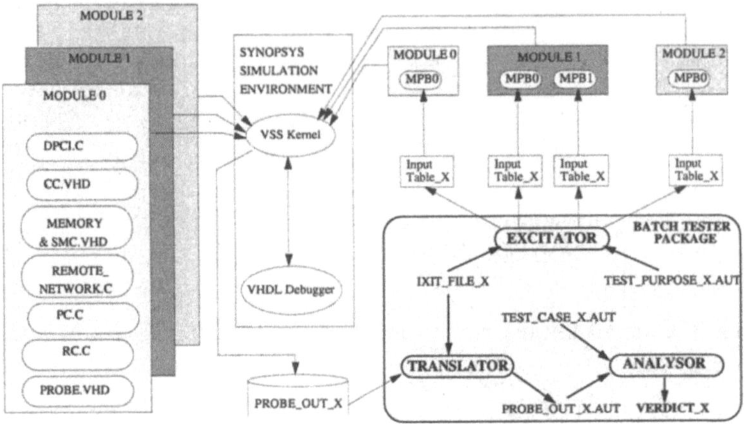

Figure 4 The BULL'S CC_NUMA architecture SIM1 batch testing environment proposal

5.2 The new testing architecture

The test cases generated by TGV are abstract in the sense that they are specified independently of the testing environment. There is one test case per test purpose (corresponding to one of the tables described in the test plan document). In this section, we present the tools we have developped to make this abstract test cases executable in SIM1 environment.

An abstract test case generated by TGV is a direct acyclic graph in which each branch describes a sequence of interactions between the tester and the system under test. This way of generating test cases is suitable to network protocols conformance testing where the testing activity is "interactive". We have seen in section 5.1 that the testing activity currently used in SIM1 environment is rather "batch". Indeed, it consists in three independent steps: (a) stimulating the system, (b) collecting all what has been observed (including the stimulus), and (c) analyzing and concluding with a verdict. So, the problem we have to tackle here is to implement interactive abstract test cases on top of batch testing environment. Basically, the solution we propose consists first in translating what has been observed during step (b) from the system into a trace in the specification. Then, this trace is analyzed according to what has been foreseen in the abstract test case.

Figure 4 shows the new testing environment and principally the overall structure of the batch tester package we have developed. The tester package consists of three applications.

The **EXCITATOR** application deals with the conversion of a test purpose (called TEST_PURPOSE_X.AUT in Figure 4) described in the format of TGV into a format readable by the MPBs. Once the conversion is done, the EXCITATOR proceeds to the stimulation of the MPBs. Then, the VSS kernel generates the probe output file (called PROBE_OUT_X). This file describes what has been effectively observed from the system under test. The **TRANSLATOR** application is in charge of translating the probe output file into a trace in the specification model. This translation is necessary to make possible the analysis of the observation according to what has been forseen in the specification. Both EXCITATOR and TRANSLATOR take into account some Implementation eXtra Information for Testing (called IXIT_FILE_X in Figure 4). These information describe the mapping between the abstract data values of the formal specification and the real data values of the system under test.

Finally, the **ANALYSOR** application proceeds to the analysis of the trace generated by the TRANSLATOR according to the given test case (called TEST_CASE_X.AUT) and delivers a verdict together with some diagnostic information. A correct trace must be a branch of the test case which leads to a PASS verdict. Since the TRANSLATOR and the EXCITATOR are automatically produced using compiler generators, the tester package can be reused to test other batch systems without major effort. In its current version, the tester package doesn't include the EXCITATOR application. Indeed, this application is quite easy to do by hand in the case of batch testing.

5.3 Example on using the tester package

We present in this section the results obtained using the tester package for the example of section 4.3.

Probe output file generation When the system under test is stimulated by the READ transaction, the VSS kernel produces the following trace. This trace describes all the operations performed by the system.

```
PROBE # 1 ---> L_Bus 620  burst  read_tt   A0 Tag 00 addr=0140042000 Pos_Ack
          Resp_Mod at time 660 NS
PROBE # 1 ---> BLINK SID=2 fm 0010 to home 0000 R_tag=07 (rq=0010 07)
          burst read_tt WIM=101 addr=000140042000 retry=00 at time 880 NS
PROBE # 0 ---> BLINK SID=0 fm 0010 to home 0000 R_tag=07  (rq=0010 07)
          burst read_tt WIM=101 addr=000140042000 retry=00 at time 1200 NS
PROBE # 0 ---> L_Bus LMD update : address = 0140042000 status =  SHD  1,
```

```
          at time 1560 NS
PROBE # 0 ---> L_Bus LMD update : address = 0000042000 status =  INV
          at time 1560 NS
PROBE # 0 ---> L_Bus LMD update : address = 0000042000 status =  INV
          at time 1560 NS
PROBE # 0 ---> L_Bus LMD update : address = 0000042000 status =  INV
          at time 1560 NS
PROBE # 0 ---> L_Bus RCC  burst  read_tt   A0 Tag 7C addr=0140042000 Pos_Ack
          Resp_Null at time 1800 NS
PROBE # 0 ---> L_Bus     Data Transaction for Tag 'L111'C data=DEADBEEFFEDCBA98
          at time 2260 NS
PROBE # 0 ---> BLINK SID=2 fm 0001 to part 0011 R_tag=07 R_Done
          data=DEADBEEFFEDCBA98 at time 3080 NS
PROBE # 1 ---> BLINK SID=1 fm 0001 to part 0011 R_tag=07 R_Done
          data=DEADBEEFFEDCBA98 at time 3640 NS
PROBE # 1 ---> L_Bus RCT update : address = 0140042000 status =  SHD
          at time 3700 NS
PROBE # 1 ---> L_Bus Intv. Data Xact. for Tag 00 data=DEADBEEFFEDCBA98
          at time 4020 NS
```

IXIT information The trace of the system is therefore submitted to the TRANSLATOR application with the following IXIT information. These information give the correspondence between abstract values and real values.

```
M0 = MODULE 0
M1 = MODULE 1
A0 = ADDRESS 0140042000
D0 = DATA DEADBEEFFEDCBA98
```

Trace of the specification The TRANSLATOR translates the trace of the system into a trace of the specification, the result is given as follow.

```
des(0,10,11)
(0,"BUS_TRANS !M1 !READ !A0 !PROCESSOR !FALSE",1)
(1,"LOC_RESP !M1 !ARESP_MODIF",2)
(2,"PACKET_TRANSFER !M1 !M0 !READ !A0 !REQ_PACKET_TYPE !NIL_DATA !NETRESP_NIL
                                                   !OUTQIO !M1 !OUTQIO",3)
(3,"LMD_PUT !M0 !A0 !RCC_SH !FLAG( FALSE, TRUE)",4)
(4,"BUS_TRANS !M0 !READ !A0 !RCC !FALSE",5)
(5,"LOC_RESP !M0 !ARESP_NULL",6)
(6,"LOC_DATA_BUS_TRANS !M0 !D0 !RCC",7)
(7,"PACKET_TRANSFER !M0 !M1 !RESP_DATA_PACKET_TYPE !D0 !NETRESP_DONE !OUTQIO",8)
(8,"RCT_PUT !M1 !A0 !RCC_SH",9)
(9,"LOC_DATA_BUS_TRANS !M1 !D0 !PROCESSOR",10)
```

Trace analysis Finally, the obtained trace is analysed according to the test case generated by TGV (see Section 4.3). This is done by the ANALYSOR application. The output of the ANALYSOR given below describes the traversed part of the test case during the analysis and the verdict which has been found. The pass verdict means that the system under test is conform to the specification w.r.t the given test purpose.

```
TC traversed part...
(0,"BUS_TRANS !M1 !READ !A0 !PROCESSOR !FALSE",1)
(1,"LOC_RESP !M1 !ARESP_MODIF",2)
(2,"PACKET_TRANSFER !M1 !M0 !READ !A0 !REQ_PACKET_TYPE !NIL_DATA !NETRESP_NIL
                                                        !OUTQIO !M1 !OUTQIO",3)
(3,"LMD_PUT !M0 !A0 !RCC_SH !FLAG (FALSE, TRUE), (PASS)",4)
IUT(3),TC(3) : PASS
```

5.4 Results on using the tester package

The main difficulty in executing the test cases was in the fact that the format of the test cases is different from the probe output format. The tester package brings solution to this problem. All the test cases generated by TGV have been executed in the testing SIM1 environment using the tester package. For each test case and the corresponding probe output file, the testing activity is almost instantaneous.

6 CONCLUSION

In this paper, we have presented an end-to-end industrial case-study concerning the automatic generation of executable tests suites for the Cache Coherency Protocol of the BULL'S CC_NUMA Architecture. Starting from the formal specification in Lotos language of this architecture and formalized test purposes, we have generated abstract test suites using the prototype TGV. The generated test cases have been experimented in the real testing environment of BULL'S CC_NUMA architecture using the tester package we have developped. At this stage of the experiment, we have covered all the test purposes described in the test plan, except those requiring an interactive approach. In order to cover all the test plan, some improvements are needed for both TGV and the tester package, such as:

- the introduction of cycles in the test cases in order to reduce the inconclusive cases; this should improve the quality of the generated test cases,
- some tests need to be executed in an interactive way; this requires to extend both the tester package and the testing environment.

The main benefit in using the TGV approach is that we only have to formally specify the system to test and the test purposes, then all the testing activity would be completely automated. The time spent in specifying the BULL'S CC_NUMA architecture, formalizing test purposes and generating the test cases with TGV is completely paid by the better correctness and the confidence to put in the implementation.

This industrial experiment also demonstrates that the prototype TGV which was developped for conformance testing of communication protocols can also be efficiently used to generate tests for hardware architectures.

Acknowledgements : This work has been done in the framework of DYADE, the Bull-Inria Research Joint Venture. It has been supported by the Bull R&D PowerPC(TM) technology Platforms Division, headed by Angelo Ramolini. Special thanks to Pierpaolo Maggi and Paolo Coerezza and all the methodology office in Pregnana for their help in providing the inputs for our case study. We also want to thank all the member of the TGV-TEAM of Irisa-Rennes/Pampa & Verimag-Grenoble/Spectre, and particularly Pierre Morel, Thierry Jéron & Claude Jard, for the time they spent in improving "on-the-fly" TGV during this experiment.

REFERENCES

[1] J.-C. Fernandez, C. Jard, T. Jéron, and C. Viho. Using on-the-fly verification techniques for the generation of test suites. In A. Alur and T. Henzinger, editors, *Conference on Computer-Aided Verification (CAV '96), New Brunswick, New Jersey, USA*, LNCS 1102. Springer, July 1996.

[2] J.-C. Fernandez, C. Jard, T. Jéron, and C. Viho. An experiment in automatic generation of test suites for protocols with verification technology. *Science of Computer Programming - Special Issue on Industrial Relevant Applications of Formal Analysis Techniques*, 29, p. 123-146, 1997.

[3] L. Doldi, V. Encontre, J.-C. Fernandez, T. Jéron, S. Le Bricquir, N. Texier, and M. Phalippou. Assessment of automatic generation methods of conformance test suites in an industrial context. In B. Baumgarten, H.-J. Burkhardt, and A. Giessler, editors, *IFIP TC6 9th International Workshop on Testing of Communicating Systems.* Chapman & Hall, September 1996.

[4] M. Faci and L. Logrippo. Specifying Hardware in LOTOS. In D. Agnew, L. Claesen, and R. Camposano, editors, *Proceedings of the 11th International Conference on Computer Hardware Description Languages and their Applications*, pages 305–312, Ottawa, Ontario, Canada, April 1993.

[5] G. Chehaibar, H. Garavel, L. Mounier, N. Tawbi, and F. Zulian. Specification and Verification of the PowerScale^{TM} Bus Arbitration Protocol : An Industrial Experiment with LOTOS. In R. Gotzhein and J. Bredereke, editors, *Proceedings of the Joint International Conference on Formal Description Techniques for Distributed Systems and Protocols, and Protocol Specification, Testing, and Verification FORTE/PSTV'96*, Kaiserslautern, Germany, October 1996.

[6] J.-C. Fernandez, H. Garavel, L. Mounier, A. Rasse, C. Rodriguez, and J. Sifakis. A Tool Box for the Verification of Lotos Programs. In *14th International Conference on Software Engineering*, Melbourne, Australia, May 1992.

[7] H. Garavel and J. Sifakis. Compilation and Verification of LOTOS Specification. In L. Logrippo, R. Probert, and H. Ural, editors, *Proceedings of the 10th International Symposium on Protocol Specification, Testing and Verification*, pages 379–394, Amsterdam, North-Holland, June 1990.

[8] T. Jéron and P. Morel. Abstraction et déterminisation à la volée : application à la génération de test. In G. Leduc, editor, *CFIP'97 : Colloque Francophone sur l'Ingénierie des Protocoles*, pages 255–270. Hermes, September 1997.

[9] D. Lenoski, J. Laudon, K. Gharachorloo, and J. Hennessy. The Directory-Based Cache Coherency Protocol for the DASH Multiprocessor. Technical Report CSL-TR-89-403, Stanford University, CA 94305+, 1989.

Autolink - Putting SDL-based test generation into practice

M. Schmitt[†], *A. Ek*[*], *J. Grabowski*[†], *D. Hogrefe*[†], *B. Koch*[†]
[†] *Medical University of Lübeck, Institute for Telematics*
Ratzeburger Allee 160, 23538 Lübeck, Germany
email: {schmitt, koch, grabowsk, hogrefe}@itm.mu-luebeck.de
[*] *Telelogic AB, P.O. Box 4128, S-20312 Malmö, Sweden*
email: anders.ek@telelogic.com

Abstract

AUTOLINK is a tool for automatic test generation. It allows to generate TTCN test suites based on a given SDL specification and MSC requirements. The first big challenge for AUTOLINK was the creation of a test suite for the *Intelligent Network Application Protocol* at ETSI. In this paper we discuss our experience in applying AUTOLINK to a real-life protocol and the improvements of AUTOLINK which were developed during this project. We also present future enhancements which will further ease the work of test suite developers.

Keywords

Autolink, SDL, MSC, TTCN, SDT, ITEX, Test generation

1 INTRODUCTION

In recent years, several formal methods have been developed for automatic test generation. However, when putting these methods into practice, many test generation tools fail due to implementation-specific restrictions. Only a few promising reports are presented in literature for real-life protocols, see e.g. [5, 6, 13, 17].

AUTOLINK is a research and development project which aims at tackling this problem. It has been started in 1996 by the Institute for Telematics in Lübeck (Germany) and Telelogic AB in Malmö (Sweden). AUTOLINK is part of Telelogic's TAU development environment. TAU provides tools for the design, analysis and compilation of systems and protocols specified in *SDL* (*Specification and Description Language*) [2, 8], *MSC* (*Message Sequence Chart*) [3] and *TTCN* (*Tree and Tabular Combined Notation*) [16]. It supports the object-oriented features of SDL'96 and also allows the combined use of SDL with *ASN.1* (*Abstract Syntax Notation One*) as defined in ITU-T Recommendation Z.105 [1].

An SDL specification which makes use of both the object oriented features and ASN.1 data descriptions was developed for the *Intelligent Network Application Protocol* (*INAP*). Based on the SDL specification and by using AUTOLINK, a TTCN test suite was created by a project team at the *European Telecommunications Standards Institute* (*ETSI*). Due to its complexity, INAP was a good example to demonstrate and verify the applicability of AUTOLINK to real-life systems. Feedback from the project has also directly influenced the development of AUTOLINK.

The rest of this paper is structured as follows: In Section 2, a short introduction is given to the general concepts of AUTOLINK and its embedding in the TAU tool environment. Sections 3 to 5 describe some aspects of AUTOLINK that have been of particular relevance for the construction of the INAP test suite. Section 3 discusses the influence of state space exploration heuristics on test case generation. A direct translation from MSC to TTCN which does not perform a state space search is motivated in Section 4. Section 5 introduces a language which allows to describe constraint naming conventions and parameterization. INAP and some test generation results are presented in Section 6. Finally, a summary and outlook is given in Section 7.

2 THE AUTOLINK TOOL

AUTOLINK is a tool which supports the automatic generation of TTCN test suites based on SDL specifications. Its basic concepts have already been documented in [7] and [18].

AUTOLINK has been influenced by the SAMSTAG method and tool [12, 19]. SAMSTAG was an experimental system, developed at the University of Berne together with Swisscom. It was applied successfully to large scale protocols, e. g. [13].

2.1 Integration into the Tau tool set

AUTOLINK is tightly integrated within the TAU tool family which comprises the well-known SDT and ITEX tools. AUTOLINK is a component of the SDT Validator. The Validator is based on state space exploration techniques and can be used to find dynamic errors and inconsistencies in SDL specifications. Additionally, it allows to verify an SDL system against requirements described by MSCs. AUTOLINK makes use of the core functionalities provided by the Validator and extends it with respect to test generation facilities. Besides AUTOLINK, some other TAU tools are involved in the generation of a complete TTCN test suite. Figure 1 shows the relations between all tools and information involved in the test generation process.

Based on an SDL system which is specified by the user (Task 1 in Figure 1)

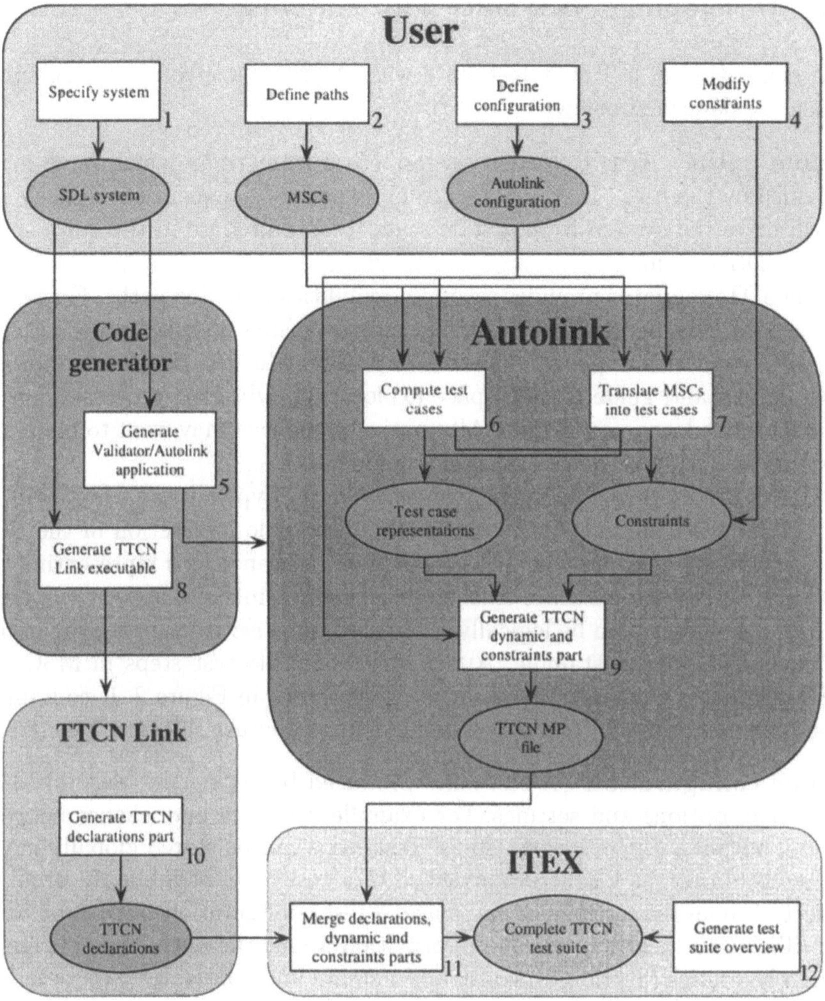

Figure 1 AUTOLINK and its integration into the TAU tool family

the code generator produces both an AUTOLINK/Validator and a TTCN Link application (Task 5 and Task 8).

Using AUTOLINK, a TTCN test suite can be generated which contains constraint and dynamic behavior tables (Task 9). This test suite can be completed and refined in ITEX, a development environment for test suites specified in TTCN. The TTCN Link application derived from the SDL specification is able to generate all static TTCN declarations (Task 10). These declarations can be merged with the AUTOLINK test suite (Task 11). Finally, the test suite overview can be generated automatically by ITEX (Task 12).

2.2 Developing a test suite with Autolink

The generation of a TTCN test suite with AUTOLINK involves several steps which are described below:

Define paths AUTOLINK derives test cases from *paths* which have to be provided by the user (Task 4 in Figure 1). A path is a sequence of SDL events which drive the system from a start state to an end state in the state space of the SDL system.

The SDT Validator provides several possibilities to define paths. For example, a path may be generated automatically by using observer processes. An observer process is a special kind of SDL process which is able to monitor the SDL system and guide a state space exploration. Observer processes can be used to define large sets of tests. Alternatively, the user may want to manually navigate in the state space and select single paths.

A path is stored as a Message Sequence Chart. Typically, an MSC used by AUTOLINK may only show the externally observable interaction of the SDL system with its environment. It consists of one instance axis representing the SDL system and one instance axis for each channel linked to the environment.

TTCN test cases can be logically structured into *test steps*, e. g. a preamble, a test body and a postamble. AUTOLINK represents test steps in MSCs by MSC references. A typical MSC for INAP is shown in Figure 2. It contains a preamble named *O_OS* and a postamble named *ReleaseCallAB_Cause_00.*

Define configuration A test suite produced by AUTOLINK depends on a number of options and settings. For example, the user may choose between several output formats for test steps. Test steps can be stored globally in the test step library, as local trees attached to a test case, or inline. In order to collect all relevant settings in one place, a test configuration file can be written (Task 3) which may include information about the output of test steps, search heuristics (Section 3), constraint naming and constraint parameterization (Section 5).

Process test cases Based on the MSCs and the configuration file provided by the user, AUTOLINK computes an internal representation for each single test case. The representations contain all sequences of send and receive events that lead to a *pass* or an *inconclusive* verdict. Additionally, it keeps track of the test case structure, i. e. the embedding of test steps.

There are two different approaches to generate test cases from MSCs. Normally, a state space exploration is started which simulates both a given MSC and the SDL system (Task 6). In this case, alternative receive events which violate the MSC but are valid according to the SDL specification are added to the test case representation with a TTCN *inconclusive* verdict. If a state space exploration is not applicable, MSCs have to be translated directly into test cases (Task 7; Section 4).

Send and receive events in a test case are associated with constraints codifying the signal parameters. Since constraints can be shared among several events in different test cases, they are stored separately from the test case representations. AUTOLINK merges identical constraints automatically and resolves naming conflicts. In addition, the user is allowed to define new constraints and to rename, merge and remove existing constraints (Task 4).

Generate TTCN test suite Based on the internal test case representations and the list of constraints, a TTCN test suite in MP format can be generated (Task 9). The appearance of the dynamic behavior and constraint tables can be controlled by various options defined in the configuration file. For example, constraints can either be stored as ASN.1 PDU or as ASN.1 ASP constraints. Before writing a test suite in a file, AUTOLINK checks the consistency of the test cases. For example, a postamble which is represented by an MSC and which is used for more than one test case description does not necessarily result in identical test steps. In this case, AUTOLINK has to distinguish the test steps by renaming them.

With respect to the *Framework of Formal Methods in Conformance Testing* [4], AUTOLINK uses trace preorder for relating a TTCN test suite to an SDL specification. This means that in the best case, AUTOLINK produces all possible traces from a specification and transforms them into test cases in TTCN. However, this is only an ideal scenario. In practice, the number of possible traces is much too large to be seriously considered. The MSCs are used to constrain the test cases to those which are considered relevant for testing the most important functions and to those which exhibit most likely an error.

3 THE STATE SPACE EXPLORATION

For test case generation, AUTOLINK performs a state space exploration based on the well-known bit-state algorithm [15]. Therefore it also has to cope with the state space explosion problem. In [14], Grabowski et al. list several *heuristics* to deal with the complexity of state space exploration algorithms. Heuristics make assumptions about the system. They avoid the analysis of system traces which do not comply to these assumptions.

The SDT Validator and hence also AUTOLINK allow to set several state space exploration options which can be related to heuristics. However, there are some options which definitely prevent AUTOLINK from generating all sequences of test events which lead to a *pass* verdict. Therefore some options are fixed, while others may be changed. Some relevant options are listed below:

- **Channel queues**
  Channel queues drastically contribute to the state space explosion. Therefore, deviating from the SDL standard, the SDT Validator and AUTOLINK allow to disable channel queues. However, queues should always be acti-

vated for all channels linked with the environment of the SDL system. Otherwise, most likely possible sequences of test events get lost.

- **Priorities of classes of SDL events**
  AUTOLINK allows to define priorities for five classes of SDL events: Internal events, input from the system environment, timeouts, channel outputs and spontaneous transitions. In the context of conformance testing, we assume that the tester is faster than the System Under Test (SUT). Therefore, when simulating the SDL system, input from the environment has highest priority. Since the number of inputs from the environment is limited by the MSC, this indeed reduces the complexity. Usually, SDL timers are used for exception handling. Therefore timeouts may be assigned a low priority at the risk of not finding all test events leading to an *inconclusive* verdict.
- **Process scheduling**
  In each system state, either all process instances in the ready queue are allowed to execute or only the first process instance. By using the second alternative, the state space is strongly reduced at the cost of not detecting any signal races.

Several other parameters can be adjusted, e. g. the maximum length of channel queues, the maximum search depth or the (in-)divisibility of SDL state transitions. Our experience with INAP has shown that it is essential to use the restricted process scheduling for complex SDL specifications. Additionally, internal channel queues have to be disabled.

4 TRANSLATING MSCS INTO TTCN TEST CASES

If a test purpose covers certain aspects of a protocol specification which are not represented in the corresponding SDL model, it is obviously not possible to generate a test case by starting a state space exploration. However, for a uniform test suite development process, it is desirable to formalize *all* test purposes as MSCs. Those MSCs which cannot be handled by a state space exploration should be converted directly into TTCN test cases.

AUTOLINK provides a function which performs a direct translation from MSC into TTCN. Figure 2 shows an MSC which has been constructed for INAP. The resulting TTCN test case generated by AUTOLINK is presented in Figure 3.

Although AUTOLINK does not need to perform a state space exploration, it requires some information about the interface of the specification. Therefore an SDL system has to be provided which at least defines the channels to the system environment and the signals sent via these channels. Using this SDL system, AUTOLINK can find out which MSC instances represent PCOs. Additionally, it can check whether the MSC is syntactically correct with regard to signals and signal parameters.

Direct translation of MSCs into TTCN test cases has to be applied with

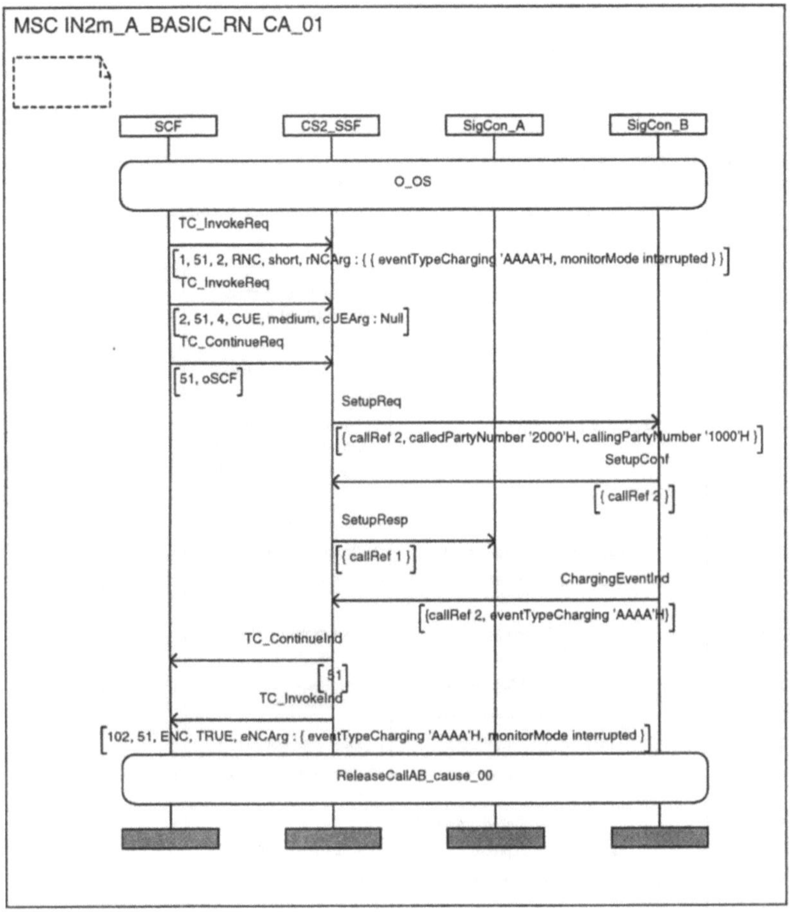

Figure 2 MSC *IN2m_A_BASIC_RN_CA_01* is translated ...

caution. There is no guarantee that the MSCs and hence the test cases describe valid traces of the specification or the implementation, respectively. Instead, AUTOLINK relies on the developer that the test cases are valid. Furthermore, is is not possible to compute test events which lead to an *inconclusive* verdict, meaning any deviation from the behavior described in the MSC is considered false.

On the other hand, there are good reasons to use MSCs instead of directly writing TTCN test cases. First, test cases typically span trees with several tree leaves because of the partial order of test events. For example, the test case in Figure 3 contains three valid sequences of test events. In MSCs the partial order is expressed inherently due to the semantics of MSC. While it is arduous for a test suite developer to write down a complete TTCN test case, AUTOLINK automatically computes all valid permutations of test events for a given MSC. Second, since AUTOLINK always translates MSCs into an intermediate internal

Test Case Dynamic Behaviour					
Test Case Name : IN2m_A_BASIC_RN_CA_01 Group : Purpose : Configuration : Default : OtherwiseFail Comments :					
Nr	Label	Behaviour Description	Constraints Ref	Verdict	Comments
1		+O_OS			
2		SCF ! TC_InvokeReq	CIR_RequestNotificationCharging_002(1 , 51)		
3		SCF ! TC_InvokeReq	CIR_Continue_004(2 , 51)		
4		SCF ! TC_ContinueReq	C_TC_ContinueReq_001(51)		
5		SigCon_B ? SetupReq	C_SetupReq({ callRef 2, calledPartyNumber '2000'H, callingPartyNumber '1000'H })		
6		SigCon_B ! SetupConf	C_SetupConf({ callRef 2 })		
7		SigCon_B ! ChargingEventInd	C_ChargingEventInd_002		
8		SCF ? TC_ContinueInd	C_TC_ContinueInd_003(51)		
9		SCF ? TC_InvokeInd	CII_EventNotificationCharging_001(102 , 51)		
10		SigCon_A ? SetupResp	C_SetupResp({ callRef 1 })	(PASS)	
11		+ReleaseCallAB_cause_00			
12		SigCon_A ? SetupResp	C_SetupResp({ callRef 1 })		
13		SCF ? TC_InvokeInd	CII_EventNotificationCharging_001(102 , 51)	(PASS)	
14		+ReleaseCallAB_cause_00			
15		SigCon_A ? SetupResp	C_SetupResp({ callRef 1 })		
16		SCF ? TC_ContinueInd	C_TC_ContinueInd_003(51)		
17		SCF ? TC_InvokeInd	CII_EventNotificationCharging_001(102 , 51)	(PASS)	
18		+ReleaseCallAB_cause_00			
Detailed Comments:					

Figure 3 ... into a TTCN test case.

test case representation, test cases generated by an MSC-TTCN translation can be merged with test cases generated by state space exploration. This leads to uniform and compact test suites with a reduced number of constraints.

5 CONSTRAINT RULES

Early tests with AUTOLINK have shown that the readability of automatically generated TTCN test suites is not very good. One particular problem is the naming of constraints. Due to the lack of information about the meaning of constraints, their names have to be created generically. For instance, a constraint may be named after its signal or the test case in which it is used. If there are different constraints with the same name, they might be distinguished by appending a sequence number.

In practice this naming scheme is not acceptable, even though AUTOLINK provides functions for the subsequent manipulation of constraints. Especially, if a test suite has to be regenerated due to a modification of the underlying SDL specification, a lot of manual work has to be repeated in order to assign meaningful constraint names.

Another important aspect is the parameterization of constraints. Without parameterization, a vast number of similar constraints is generated. This also makes the naming problem worse, since all these constraints have to get unique names.

For these reasons, AUTOLINK allows the user to specify rules which tell the tool how to map SDL signals onto TTCN constraints during the test generation process. Both the names of constraints and their parameterization can be controlled by these rules. The rules have to be provided in advance, i. e. before the test generation starts, as part of the configuration file.

A typical constraint rule looks like this:

Example 1

```
TRANSLATE
  FROM       TC_ContinueReq
  TO         "C_TC_ContinueReq"
  PARAMETERS $1="Dialog_ID"
END
```

Constraint rules can be considered as mapping rules: AUTOLINK translates *from* signals *to* constraints. Example 1 instructs AUTOLINK to map *TC_ContinueReq* signals onto constraints whose name is *C_TC_ContinueReq*. If more than one constraint is built, all constraints are distinguished by an additional sequence number. Moreover, the first parameter of each concrete signal (referred to by *$1*) becomes a parameter of the resulting constraint. The name of the formal parameter used in the constraint declaration table is *Dialog_ID*.

Constraint names may not only be composed of plain texts, they can also depend on signal parameters. However, in some cases it is not desirable to take the textual representation of a parameter value directly as part of a constraint name. E. g., a protocol engineer might use abbreviations as signal parameter values. But for the TTCN test suite, these abbreviations are intended to be mapped onto extended names.

In Example 2, the fourth parameter of signal *TC_InvokeReq* is taken as input for function *OpName*. Depending on its input value, the function returns a text which forms the second half of the constraint name. As a consequence, *TC_InvokeReq* signals with different fourth parameter are automat-

Example 2

```
TRANSLATE
  FROM        TC_InvokeReq
  TO          "CIR_" + OpName($4)
  PARAMETERS $1="Invoke_ID", $2="Dialog_ID"
END

FUNCTION OpName
     $1 == "ASF"    : "ActivateServiceFiltering"
  ...
   | $1 == "RC"     : "ReleaseCall"
  ...
   | $1 == "SL_R"   : "SplitLegResult"
   | TRUE           : "OperatorTypeNameUndefined"
END
```

ASN.1 ASP Constraint Declaration
Constraint Name: CIR_ReleaseCall(Invoke_ID : InvokeIDtype; Dialog_ID : DialogIDtype) ASP Type : TC_InvokeReq Derivation Path: Comments :
Constraint Value
{ invokeIDtype1 Invoke_ID, dialogIDtype2 Dialog_ID, opClassType3 4, opCodeType4 RC, timeoutValType5 short, argType6 rCArg : initialCallSegment : '00'H }
Detailed Comments:

Figure 4 A parameterized constraint

ically mapped onto constraints with different names. A possible constraint declaration table for a *TC_InvokeReq* signal is shown in Figure 4.

Besides the constructs outlined above, AUTOLINK's constraint description language allows to define conditional rules. By using conditions in a TRANSLATE statement, constraints can be customized to specific requirements. For example, constraint parameterization can be guided by signal parameters. Moreover, it is possible to combine constraint rules for several signals by the use of regular expressions.

One goal for the design of the constraint description language was simplicity. Even an unexperienced user should be able to understand and define constraint rules. The syntax is not very strict in the sense that, for example, function parameters do not have to be declared. Instead, potential inconsistencies are checked and resolved at run-time.

The only data type used in the constraint description language is *text*. Any reference to a signal parameter returns a text. The same is true for function calls. Conditions are also evaluated on a textual basis.

Despite its simplicity, the language has proven to be sufficiently powerful. Nevertheless, it can be easily extended by additional built-in functions. One restriction of the current implementation is that AUTOLINK can only refer to top-level signal parameters, i. e. it is not possible to address nested parameters. We plan to remove this limitation in a future release.

6 THE INAP EXAMPLE

The *Intelligent Network Application Protocol* [9] is the first protocol specified by ETSI for which a machine-processable SDL model is available. The SDL model was developed by ETSI Sub-Technical Committee SPS3 with support of the Protocol Expert Group and the Technical Committee 'Methods for Testing and Specification'.

The specification of INAP Capability Set 2 (CS–2) makes use of the object-oriented features of SDL'96 by inheriting CS–1. Data types are defined in ASN.1.

The SDL specification of ETSI's INAP CS–2 is voluminous. It comprises more than 450 pages in printed form. The phrase representation is about 1.6 MByte large (approximately 570 KByte without comments). When translating the specification into C with SDT's code generator, about 350 000 lines or 13.6 MByte of source code are generated.

6.1 Test suite generation

TTCN test suites for INAP CS–2 are developed by ETSI Specialists Task Force STF 100. A first test suite which covers the basic capability set, i. e. the CS–1 operations with CS–2 additions, has already been published in [11]. Another test suite covering the CS–2 operations is currently under development.

With respect to the CS–1 operations, test purposes were defined with textual descriptions and rough MSCs, first. Next, these test purposes were formalized as detailed MSCs using the SDT Simulator. In total, 126 test purposes were specified [10]. For 67 test purposes the MSCs could be simulated in order to produce the corresponding test cases. The remaining 59 test purposes had to be translated directly into TTCN due to unspecified parts in the SDL model.

The test suite resulted from a repetitive process of SDL/MSC refinements and modifications, MSC verifications and test generation runs. Whenever a modification of the SDL model was made, all MSCs were verified with the SDT Validator. If errors emerged, the SDL model or the MSCs were modified again until all MSCs passed the verification. Thereafter the test case generation using AUTOLINK was started.

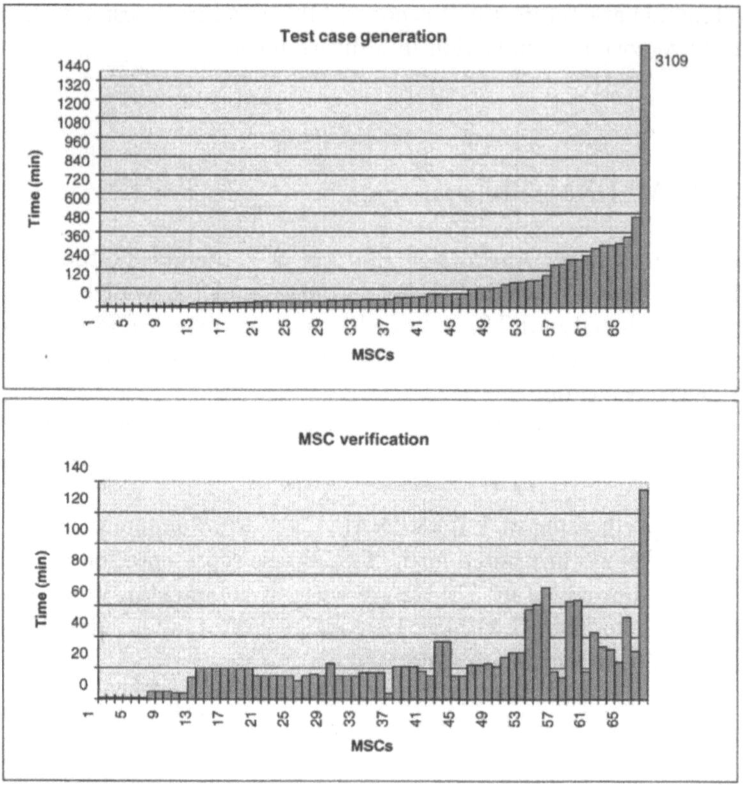

Figure 5 Computation time for MSC verifications and test generations

6.2 Statistics

Both the MSC verification and the test generation runs were executed at the Institute for Telematics in Lübeck. The test results discussed below were obtained on SUN ULTRA 2 workstations with 300 MHz processors.

Figure 5 shows the computation time for both the MSC verification and the test generation with AUTOLINK. The time needed for the verification of an MSC ranged from 1 min 24 sec to 2 h 15 min. It took between 6 min 44 sec and 51 h 49 min (= 3109 min) to generate a test case.

The larger amount of time needed for test generation is not surprising: During MSC verification, a path in the state space graph is truncated as soon as an event in an SDL transition conflicts with the MSC. On the other hand during test generation, the path needs to be extended until an observable event occurs.

Interestingly, there is no general correlation between the computation time for MSC verification and test generation. For example, MSC no. 57 in Figure 5 can be verified comparably fast, whereas its test case generation takes about 5 hours.

Due to the large number of data used in the SDL system, on average only 22 states per minute could be explored by AUTOLINK. Much time was spent for the computation of the hash values of each state needed for the bit-state algorithm.

6.3 Distributed test case processing

Verification of all MSCs on a single machine would have taken about a day; generation of all test cases would have taken about a week. Therefore, the processing of test purposes was distributed among up to fifteen workstations.

As described in Section 2, AUTOLINK does not directly write a generated test case into a TTCN MP file. Instead, it stores each test case in an internal representation in memory. This representation and the corresponding constraints can be saved on disk and reloaded later. This feature was used to compute each test case separately. After the computation finished, all test cases were reloaded and combined into a single test suite. Identical constraints were merged automatically during this process.

With the help of shell scripts, test generation runs were executed in batch mode, so no manual intervention was needed to start the generation of each single test case. This way, test cases could be generated overnight. In addition, information about previous test generation runs could be used in order to minimize computation time by placing time-intensive test cases on fast machines first.

6.4 Test suite post-processing

Even though AUTOLINK (in combination with TTCN Link) produced a complete, readable test suite, some manual steps were still needed to enhance the result:

1. Preambles which were generated during direct translation from MSC into TTCN were replaced by the ones generated with state space exploration.
2. Using parameterization on the level of test steps, the number of postambles was reduced significantly.
3. PIXIT information was added.
4. Test group information was added.
5. The test suite was converted to concurrent TTCN.

7 SUMMARY AND OUTLOOK

AUTOLINK makes it much easier to generate TTCN test suites based on an SDL specification. In particular, constraint rules save a lot of time. Additionally, the integration of MSCs for which no state space search can be performed allows a uniform development process. AUTOLINK is used by project STF 100 at ETSI whose goal is the development of test suites for INAP CS–2.

Due to the complexity of modern protocol specifications, it seems to be almost impossible to create correct test suites by hand. AUTOLINK allows to check several properties of the test suite which would otherwise have been overlooked. For instance, AUTOLINK checks whether a test step can be shared among several test cases.

The time needed for test generation depends both on the complexity of the SDL model, the size of the MSCs describing the test purposes and the heuristics for the state space search. While the first two factors cannot be altered, the state space options have to be chosen carefully in order to find a good compromise between computation time and the chance to find all events with *inconclusive* verdict. More tool assistance is needed by the user to choose appropriate options.

With regard to the whole development process, the time effort for the actual test generation is not relevant (if some restrictions of the state space are accepted). Most time is spent for refinements of the SDL specification and the test purposes. ETSI estimates that about 20% of the expenses for the development of the first INAP test suite could be saved by tool support in comparison with a manual test suite development.

However, experience has shown that the amount of time spent for manual post-processing of a generated test suite can be further decreased. Therefore, improvements of AUTOLINK will focus on the readability of the generated TTCN code. In particular, we plan to implement the following extensions:

- **Support of concurrent TTCN**
  In order to generate concurrent TTCN, AUTOLINK needs further information that cannot be automatically retrieved from the SDL specification. E. g. information is needed about the assignment of PCOs to channels and the relation between PTCs and PCOs. Therefore the user will have to provide this information as part of the configuration file. Several strategies for the coordination of PTCs will be implemented, e. g. a strict synchronization or synchronization only when indicated in the MSC.
- **Parameterization of test steps**
  MSC references can be used in test purpose descriptions to refer to test steps. If an MSC test step is used in several test cases, it may lead to several test steps which only differ in a few signal parameters. Parameterization of test steps can be handled similarly to constraint parameterization and should include parameterization of signals, PCOs and signal parameters.

- **Support of timer events**
  TTCN timers could be created automatically either once for a complete test case or for each event separately. There should also be a possibility to explicitly specify timers in the MSCs which are transformed to TTCN timer events.
- **Test suite structure**
  A test suite is typically structured into test groups, i. e. sets of test cases which test a specific aspect of the specification. Since the test suite structure is often reflected in the names of the test purposes, a mechanism will be implemented that groups test cases based on their names.
- **PICS/PIXIT parameterization**
  Since SDL does not allow the use of symbolic values, PICS/PIXIT parameters have to be encoded as concrete values for test generation and replaced by symbolic values in a post-processing step. This time-consuming task can be automatized.
- **Automatic constraint parameterization**
  Automatic constraint parameterization is a way to minimize the number of constraints. However, it is not yet clear whether it will also enhance the readability of a test suite.

REFERENCES

[1] ITU Telecommunication Standards Sector SG 10. ITU-T Recommendation Z.105: Specification and Description Language (SDL) combined with Abstract Syntax Notation One (ASN.1). ITU, Geneva, 1995.

[2] ITU Telecommunication Standards Sector SG 10. ITU-T Recommendation Z.100: Specification and Description Language (SDL). ITU, Geneva, 1996.

[3] ITU Telecommunication Standards Sector SG 10. ITU-T Recommendation Z.120: Message Sequence Chart (MSC). ITU, Geneva, 1996.

[4] ITU Telecommunication Standards Sector SG 10. ITU-T Recommendation Z.500: Framework of Formal Methods in Conformance Testing. ITU, Geneva, 1998.

[5] R. Anido, A. Cavalli, T. Macavei, L. P. Lima, M. Clatin, and M. Phalippou. Testing a real protocol with the aid of verification techniques. In *XXIII Seminario Integrado de Software e Hardware (SEMISH'96)*, pages 237–248, Brazil, August 1996.

[6] A. Cavalli, B.-H. Lee, and T. Macavei. Test generation for the SSCOP-ATM networks protocol. In *SDL '97 Time for Testing – Proceedings of the Eighth SDL Forum*, Evry, France, September 1997.

[7] A. Ek, J. Grabowski, D. Hogrefe, R. Jerome, B. Koch, and M. Schmitt. Towards the Industrial Use of Validation Techniques and Automatic Test Generation Methods for SDL Specifications. In *SDL '97 Time for Testing – Proceedings of the Eighth SDL Forum*, Evry, France, Septem-

ber 1997.

[8] J. Ellsberger, D. Hogrefe, and A. Sarma. *SDL – Formal Object-oriented Language for Communicating Systems.* Prentice Hall, 1997.

[9] European Telecommunications Standards Institute. *DEN 03038-1. ETSI Core INAP CS-2; Part 1: Protocol Specification*, 1997.

[10] European Telecommunications Standards Institute. *DEN 03038-3. ETSI Core INAP CS-2; Part 3: Test Suite Structure and Test Purposes specification for Service Switching Function (SSF), Specialized Resource Function (SRF) and Service Control Function (SCF)*, 1998.

[11] European Telecommunications Standards Institute. *DEN 03038-4. ETSI Core INAP CS-2; Part 4: Abstract Test Suite (ATS) for Service Switching Function (SSF), Specialized Resource Function (SFR) and Service Control Function (SCF)*, 1998.

[12] J. Grabowski, D. Hogrefe, and Nahm. R. Test Case Generation with Test Purpose Specification by MSCs. In *SDL '93 Using Objects – Proceedings of the Sixth SDL Forum*, Darmstadt, Germany, October 1993. North-Holland.

[13] J. Grabowski, D. Hogrefe, R. Scheurer, and Z. Dai. Applying SAMSTAG to the B-ISDN Protocol SSCOP - Technical Description and TTCN Testsuite. In *Testing of Communicating Systems*, volume 10, Cheju Islands, Korea, September 1997.

[14] J. Grabowski, R. Scheurer, D. Toggweiler, and D. Hogrefe. Dealing with the Complexity of State Space Exploration Algorithms. In *Proceedings of the Sixth GI/ITG technical meeting on 'Formal Description Techniques for Distributed Systems'*, University of Erlangen, June 1996.

[15] G. J. Holzmann. *Design and Validation of Computer Protocols.* Prentice-Hall International, Inc., 1991.

[16] ISO/IEC JTC 1/SC21. Information Technology - Open Systems Interconnection - Conformance Testing Methodology and Framework - Part 3: The Tree and Tabular Combined Notation. International Standard 9646-3, ISO/IEC, 1992.

[17] E. Perez, E. Algaba, and M. Monedero. A pragmatic approach to test generation. In *Testing of Communicating Systems*, volume 10, Cheju Islands, Korea, September 1997.

[18] M. Schmitt, B. Koch, J. Grabowski, and D. Hogrefe. Autolink - A Tool for the Automatic and Semi-Automatic Test Generation. In *Proceedings of the Seventh GI/ITG Technical Meeting on Formal Description Techniques for Distributed Systems*, Berlin, June 1997.

[19] D. Toggweiler, J. Grabowski, and D. Hogrefe. Partial order simulation of SDL specifications. In *SDL '95 with MSC in CASE – Proceedings of the Seventh SDL Forum*, Oslo, Norway, September 1995. Elsevier.

Publications concerning AUTOLINK can be downloaded from
http://www.itm.mu-luebeck.de/research/autolink/

8 BIOGRAPHY

Michael Schmitt studied computer science with focus on computational linguistics at the University of Koblenz, Germany, where he graduated with a diploma degree in September 1996. Since October 1996 Michael Schmitt is a research assistant at the Institute for Telematics at the Medical University of Lübeck, Germany.

Anders Ek studied computer science and engineering at Lund Institute of Technology in Sweden. After his graduation 1986 he started working at Telia, the national Swedish telecom operator. At Telia he did research on formal methods and associated development methodology. Since 1992 he has been working with development of tools and methods at Telelogic, a commercial CASE tool vendor providing tools based on formal description techniques.

Jens Grabowski studied computer science and chemistry at the University of Hamburg, Germany, where he graduated with a diploma degree. From 1990 to October 1995 he was a research scientist at the University of Berne, Switzerland, where he received his Ph.D. degree in 1994. Since October 1995 Jens Grabowski is a researcher and lecturer at the Institute for Telematics.

Dieter Hogrefe studied computer science and mathematics at the University of Hannover, Germany, where he graduated with a diploma degree and later received his PhD. From 1983 to 1995 he worked at various positions in the industry and at universities. Since 1996 he is director of the Institute for Telematics and full professor at the Medical University of Lübeck.

Beat Koch studied computer science at the University of Bern, Switzerland. After his graduation in 1994, he worked in the industry for two years. Since 1996, he is a research assistant at the Institute for Telematics.

PART SEVEN

Testing of Distributed Systems

Testing Temporal Logic Properties in Distributed Systems

F. Dietrich, X. Logean, S. Koppenhoefer, J.-P. Hubaux
Institute for computer Communications and Applications (ICA)
Swiss Federal Institute of Technology (EPFL)
Lausanne, Switzerland
telephone: +41 21 693 52 57, fax: +41 21 693 6610
e-mail: {falk.dietrich, xavier.logean, shawn.koppenhoefer, jean-pierre.hubaux}@epfl.ch

Abstract

Based on the notion of event-based behavioral abstraction (EBBA) we specify properties of object-oriented distributed systems in linear time temporal logic. These properties are then observed at system run-time and it is checked whether or not the system violates the specified behavioral constraints. In our approach, several steps in the testing process can be automized: instrumenting the source code, constructing test-oracles and generating an observer. Taking an industrial example as basis, we discuss how our proposal can be integrated into the software design- and testing process.

Keywords

Event-based behavioral abstraction, Linear-time Temporal Logic, testing

1 INTRODUCTION

We describe a way to ***automatically*** generate an implementation that observes the dynamic behavior of an object-oriented distributed system, maintaining a notion of whether or not that behavior violates some predefined properties. Therefore, we are concentrating on the twofold problem of specification and testing of object-oriented distributed services; what behavior needs to be observed at runtime, how is that behavior specified and how is the automatization based on that specification to be achieved?

Event-based behavioral abstraction is being frequently used during testing (Dillon & Yu 1994) and debugging (Bates 1995). However, when the events generated by a system are to be observed and analyzed, most proposals rely on ***manual*** source code annotations for event generation. Some proposals, e.g. (Bates 1995), allow for the definition of arbitrary events using an event description language.

In this paper, we follow another avenue. We provide a set of *predefined* events that is appropriate for expressing properties of object-oriented distributed systems. This set has been determined by collaborating with several industrial players and by taking into account the tradeoffs between flexibility and complexity of the property language. The set of events is chosen very carefully, often making it possible to perform source code annotation for event generation in an *automated* manner.

In our framework, behavioral constraints (also called properties) are to be expressed using these predefined events and Linear-time Temporal Logic (LTL) (Manna & Pnueli 1991*a*). By using LTL as property language we can benefit from the well-known solutions for constructing test oracles.

Testing is still a human intensive activity, thus error-prone. This is to a certain extent due to a lack of formality which would otherwise allow testers to do their job efficiently and to take advantage of powerful tools. To make the use of formal approaches more appealing to software designers and testers, it is beneficial to encapsulate formal methods concepts and/or algorithms: users do not have to know how it works, or even that it is there (Goguen & Luqi 1995). By using the approach advocated in this paper several important steps in the testing process, namely the source code annotation, the generation of test oracles, the test trace collection and test trace analysis can be automated and hidden from the service designer and tester.

We informally describe the set of events and, taking an industrial service as an example, demonstrate how properties can be expressed with these events and LTL. We show how the automatization of parts of the testing process can be achieved and briefly describe MOTEL, a MOnitoring and TEsting tooL, being developed by our institute.

The remainder of this paper is structured as follows: In Section 2 we describe the set of events that is appropriate for expressing behavioral properties of object-oriented distributed systems. In Section 3 we show, on a concrete example, how properties can be formally specified with our approach. In Section 4 we analyze how the automatization of several important steps in the testing process can be achieved and briefly describe MOTEL. Finally, our conclusions are presented.

2 THE SET OF EVENTS

The question we answer in this section is the following: How can we, using event-based behavioral abstraction, faithfully represent the behavior of object-oriented distributed systems such that (i) the chosen events reflect to a large extent the abstraction level found in today's industrial implementations of distributed systems and (ii) we can perform source code annotation in an automatic manner.

To respond to this question we derive a set of twenty events that is appropriate for modelling object-oriented distributed systems. These twenty events satisfy the requirement of being easily observable. We consider observable events at four different levels: the object-, thread-, process- and system level. The classification of events into these four groups is mainly intended to facilitate the presentation. The events are summarized in Table 1. For a more detailed (and formal) description of these events we refer the interested reader to (Dietrich, Logean, Koppenhöfer & Hubaux 1998).

The notion of observable event can be seen as a filter that screens out all events that are irrelevant at the given level of abstraction. We consider observable events to occur instantaneously and to be atomic.

In this paper, we will concentrate on the four observable events at the object level. For that purpose we will briefly describe the formal notation we use for observable events at this level.

Let *OID* be the set of object identifiers for all objects in the system. Each observable event at the object level is represented as a pair (*o_type*, *op_req*) where *o_type* is an element of the set of object events types *o_type* $\in$ {*o_outReq*, *o_inReq*, *o_outRep*, *o_inRep*} and *op_req* is an operation request. An event of type *o_outReq* occurs when an object is sending a request to execute an operation on another object. An event of type *o_inReq* occurs when an object starts executing an operation as requested by another object. An event of type *o_outRep* occurs when an object is sending the result of an operation back to the object that requested the execution of the operation. An event of type *o_inRep* occurs when an object receives the reply for the execution of an operation from the called object. An operation request is a quadruple (*src*, *tgt*, *oper*, *param_list*) where *src* $\in$ *OID* is the object identifier for the source object, i.e. the object that requests the execution of an operation on another object, *tgt* $\in$ *OID* is the object identifier for the target object, i.e. the object that executes the operation, *oper* is the name of the called operation and *param_list* is a list of parameter values of the operation where each item has to be in the domain of its corresponding parameter type.

For illustration consider Figure 1. The execution of an operation (as seen at the object level) involves four observable events, each of these four events describing a different stage during the execution. The numbers in Figure 1 indicate the order in which these events occur during the execution of the operation offered by object o_2 and invoked by object o_1.

For example, an event described as $(o\_inReq, (o_1, o_2, oper, *))$ occurs when object o_2 starts executing the operation *oper* that has been called by object o_1. Throughout this paper we use "*" to denote that a value is unrestricted

Name	Description
o_outReq	An event of type *o_outReq* occurs when an object is sending a request to execute an operation on another object.
o_inReq	An event of type *o_inReq* occurs when an object starts executing an operation as requested by another object.
o_outRep	An event of type *o_outRep* occurs when an object is sending the result of an operation back to the object that requested the execution of the operation.
o_inRep	An event of type *o_inRep* occurs when an object receives the reply for the execution of an operation from the called object.
t_assThr	An event of type *t_assThr* occurs when an operation request is assigned to a thread.
t_relThr	An event of type *t_relThr* occurs when a thread becomes idle after processing an operation request to completion.
t_outReq	An event of type *t_outReq* occurs when, during the execution of an operation request, a request to invoke another operation on another object is being sent.
t_outRep	An event of type *t_outRep* occurs when a thread completes the execution of an operation, i.e. when the result of the operation is being sent back to the calling object.
t_inRep	An event of type *t_inRep* occurs when the response for a previous *t_outReq* arrives and the thread continues to execute the original operation.
p_inReq	An event of type *p_inReq* indicates the arrival of an operation request at a process.
p_oReg	Occurs when an object is registered in the system thereby making it possible for other objects to invoke operations on it.
p_oDereg	Occurs when an object is de-registered.
p_newO	Occurs when the creation of an object takes place.
p_delO	Occurs when an object is deleted.
p_newT	Occurs when the creation of a thread takes place.
p_delT	Occurs when a thread is deleted.
p_reqRef	Occurs when an object reference is requested.
p_recRef	Occurs when an object reference is received.
s_newP	Occurs when the creation of a process takes place.
s_delP	Occurs when the deletion of a process takes place.

Table 1 Observable events: Summary

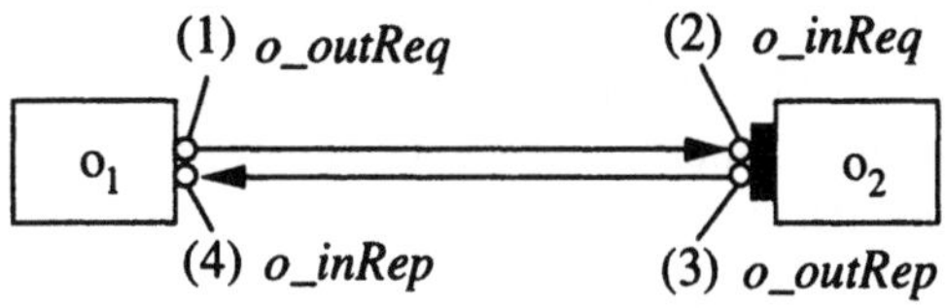

Figure 1 Observable object events

or irrelevant. In the above example, the parameters of the operation are of no interest.

An observable event *occurrence* is an instance of an observable event. We assume that each event occurrence can be distinguished from other event occurrences of the same event. This can be done by using a unique event occurrence identifier. However, an event occurrence identifier is not part of the event's tuple notation. Distinct event occurrences can obviously have the same event tuple.

3 EXPRESSING PROPERTIES

In this section, using an industrial example, we will demonstrate how behavioral properties can be expressed using the events described in the previous section and LTL.

LTL formulae are interpreted over an infinite sequence of states $\sigma = s_0, s_1, \ldots$. Given a state sequence σ and a temporal formula p, $(\sigma, j) \models p$ denotes that p holds at position $j \geq 0$ in σ. In this paper we will use the notation $\odot e$ to denote that an event e just happened, i.e. $(\sigma, j) \models \odot e$ iff event e just happened. We restrict ourselves to the use of the following future temporal operators: $\Box$ (always), $\Diamond$ (eventually) and $\mathcal{U}$ (Until) which are defined as follows: $(\sigma, j) \models \Box p \Longleftrightarrow \forall k \geq j, (\sigma, k) \models p$; $(\sigma, j) \models \Diamond p \Longleftrightarrow \exists k \geq j, (\sigma, k) \models p$ and finally $(\sigma, j) \models p \,\mathcal{U}\, q \Longleftrightarrow \exists k \geq j, (\sigma, k) \models q$ *and* $\forall i, j \leq i < k, (\sigma, i) \models p$.

In the following we will show on an example how formal properties can be derived from informal service specifications. The application chosen to validate our approach was selected independently of the approach. The target application, a Desktop Video Conference (DVC) System built according to the TINA architecture (Chapman & Montesi 1995) on top of CORBA, was provided by Swisscom.

CORBA is a standardized architecture for object-oriented distributed systems with transparent distribution and easy access to components. CORBA requires that every object's interface be expressed in the Interface Definition Language (IDL). Clients only see the object's interface but never any of the implementation details. Every invocation of a CORBA object is passed to the Object Request Broker (ORB); even when the object is local. All distribution issues like parameter transfer to the remote object, are handled by the ORB.

IDL provides an implementation language independent representation of the system, more specifically, of the interface templates that the objects in the distributed system support. There exist several well-defined and standardized mappings from IDL to implementation languages like C++ and JAVA.

We were given the informal service specification documents and the implementation code once the service had been developed by Swisscom. In contrast to many other formal methods projects from the literature, we had therefore to cope with two major handicaps: (i) The persons formally specifying the properties on the service had not been involved in designing and implementing the service. The properties had to be expressed purely on the information given in the informal service specifications. (ii) The service had been designed and implemented without paying any attention to formality.

Linear-time temporal logic has already been used in several industrial projects to express properties that the software under construction should satisfy (Holzmann 1994) (Jagadeesan, Puchol & Olnhausen 1995). However, there is only limited information in the literature about the complexity of the properties as they arise from industrial software development. In most papers, the complexity of the properties expressed in real systems remains unclear.

In (Manna & Pnueli 1991*b*), Manna and Pnueli give three classes of properties that are believed to cover the majority of properties one would ever wish to verify: invariance ($\Box p$), response ($\Box(p \rightarrow \Diamond q)$) and precedence ($\Box(p \rightarrow q \, \mathcal{U} \, r)$).

Holzmann (Holzmann 1994) followed the argumentation of Manna and Pnueli and considers only the three above-mentioned classes. In a similar project (Jagadeesan et al. 1995), only safety properties (invariance properties) were considered.

In our work it turned out that safety and precedence properties cover a multitude of properties as they are stated upon industrial systems. However, the complexity of the system we had to deal with was such that some properties we needed to express, could not fall into any of the three property classes.

Let us look at an extract from the informal DVC specification (Figure 2) to illustrate the expression of properties and the problems encountered during the property specification process.

The four informal properties in Figure 2 can be expressed as simple safety properties.

We pick the second property as example. To formally specify this property we first need to identify the event that denotes the addition of a party to a session. The informal specification (and the IDL specification of the service components) reveals that objects of class DVC_UAPSessionReq offer an operation add_dvc_parties which takes the user id of the user to add to the session

It is the responsibility of the DVC_GUI to check the consistency of a number of end-user requirements, such as: 1. don't use invalid userIds (userIds can only be provided by selecting them from a list of valid userIds) 2. don't add the same party twice to the same session 3. don't add more users than the predefined maximum 4. don't select a video QoS which exceeds the maximum session QoS

Figure 2 DVC Specification

as parameter. Based on the syntax described earlier this event can therefore be described as:

$$(o\_inReq, (*, oid, add\_dvc\_parties, (uid)))$$

We refer to the number of event occurrences of type ϕ by writing $\#[\phi]$ which is defined as follows:

$$\#[\phi]_{(\sigma,n)} \stackrel{def}{=} \begin{cases} 0 & if\ n=0 \\ \#[\phi]_{(\sigma,n-1)} & if\ n>0 \wedge (\sigma,n) \not\models \odot\phi \\ \#[\phi]_{(\sigma,n-1)}+1 & if\ n>0 \wedge (\sigma,n) \models \odot\phi \end{cases}$$

A first representation of the property looks as follows:

$$\forall o \in DVC\_UAPSessionReq\ .$$
$$\Box(\#[(o\_inReq, (*, o, add\_dvc\_parties, (uid)))] < 2))$$

However, even though this property seems to give a formal representation of the informal property at the first glance, deeper investigation reveals that it is not the property we intended to specify.

If a user joins the session, leaves it and joins it again, the number of join-operations is equal to two and the property is violated. However, the formal property exactly expresses what the informal property states which means that the informal property is not free of ambiguity. To rectify the property we have to change it to:

$$\forall o \in DVC\_UAPSessionReq\ .$$
$$\Box(\#[(o\_inReq, (*, o, add\_dvc\_parties, (uid)))] -$$
$$\#[(o\_inReq, (*, o, remove\_dvc\_parties, (uid)))] < 2))$$

This relatively simple example shows already that it is not always easy to identify the ambiguities of the informal specifications when deriving formal properties from it.

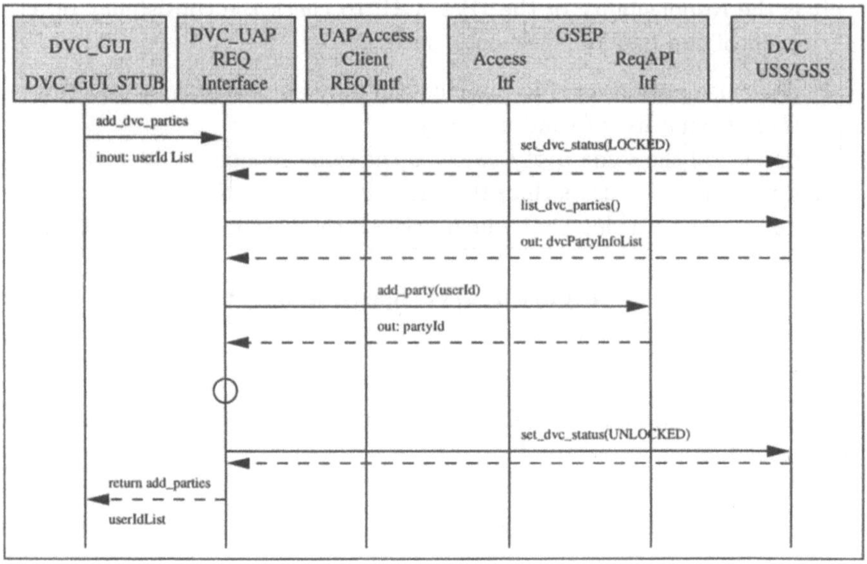

Figure 3 Add Parties Scenario

Scenarios are frequently used in informal specifications to illustrate certain behavior aspects. Behavioral constraints as they can be derived from scenarios, can frequently be expressed by using precedence properties. Consider Figure 3 for a scenario for adding parties to a video conference session. It is relatively straightforward to derive LTL properties from such scenarios. The entire scenario can be expressed using LTL. Let us consider one part of this scenario which requires that when DVC parties are added, the DVC status has to be set to LOCKED before any other action can be taken. A first property that one might express is that we always have to set the DVC status to LOCKED before we can call the list_dvc_parties-operation: Each time we call add_dvc_parties, we will not call list_dvc_parties unless we have set the DVC status to LOCKED before.

$$
\begin{array}{l}
\forall o \in \mathit{DVC\_UAP\_REQ}\ . \\
\Box(\odot(\mathit{o\_inReq}, (*, o, \mathit{add\_dvc\_parties}, *)) \rightarrow \\
\;\neg \odot (\mathit{o\_inReq}, (o, *, \mathit{list\_dvc\_parties}, *))\ \mathcal{U} \\
\quad \odot(\mathit{o\_inReq}, (o, *, \mathit{set\_dvc\_status}, (\mathit{LOCKED}))))
\end{array}
$$

Let us finally consider a more complicated property. It states that a chairperson (the owner) of a DVC session is not allowed to exit the session (by calling the exit_dvc_session operation) unless he/she has transfered the session ownership to another person. At first glance this seems to be straightforward. However, finding a correct formal representation turns out to be quite difficult. A person is automatically chairman of a session if he has requested the session creation by calling the request_dvc_service operation. We now identify the events that we need for expressing the property.

$$e_1 = (o\_inReq, (*, oid, request\_dvc\_service, (*, uid_1, *, *, *)))$$

The first event denotes the invocation of the request_dvc_service-operation. We skip the details of the operation and only note that it takes five parameters, only the second parameter is of interest for the specification of the property. This parameter specifies the user id for the user requesting the service.

$$e_2 = (o\_outRep, (*, oid, request\_dvc\_service, (*, *, *, *, i\_req)))$$

The second event denotes the termination of the *request_dvc_service*-operation. This operation returns an interface reference (object reference) as out parameter. The object reference returned by this operation identifies the interface that the user can use to add parties to the requested session, to transfer the ownership of a session etc.

$$e_3 = (o\_inReq, (*, i\_req, exit\_dvc\_session, *))$$

The third event describes the invocation of the *exit_dvc_session* operation, i.e. the operation that the chairman is not allowed to call.

$$e_4 = (o\_inReq, (*, i\_req, transfer\_dvc\_ownership, (uid_2))) \wedge uid_2 \neq uid_1$$
$$e_5 = (o\_inReq, (*, i\_req, transfer\_dvc\_ownership, (uid_2))) \wedge uid_2 = uid_1$$

The fourth event describes the transfer of the session ownership from one user to another user while the fifth event describes the case where the ownership is not changed (it is transfered from a user to that user).

Having described these five events we are now ready to give the formal representation of our property:

$$\Box((\odot e_1 \rightarrow \Diamond \odot e_2) \rightarrow ((\neg \odot e_3 \ \mathcal{U} \ \odot e_4) \wedge (\Box(\odot e_5 \rightarrow \neg \odot e_3 \ \mathcal{U} \ \odot e_4))))$$

In contrast to DisCo (Järvinen, Kruki-Suonio, Sakkinen & Systä 1990) and TLA (Lamport 1995), we only use a limited set of predefined events to specify behavior, no internal states or internal transitions are used to express behavior. We agree with Lamport (Lamport 1988) that purely temporal specifications are often hard to understand. However, in our approach these difficulties are compensated by the possibility to automatize several steps in the testing process as we will show in the next section.

Furthermore, it turns out that many properties as they are derived from industrial specifications, can be classified and there is a set of property structures that occur frequently. Based on this observation it is possible to offer a graphical user interface to the property specifier where he only has to select a property class from a list and then fill out the missing artifacts.

4 AUTOMATIZATION

In this section we will demonstrate some benefits of having these formally expressed properties. We show how the automatization of several testing steps can be achieved thereby illustrating how the combination of EBBA and LTL can be used to facilitate the testing process.

Consider Figure 4 for an overview about the development process of dis-

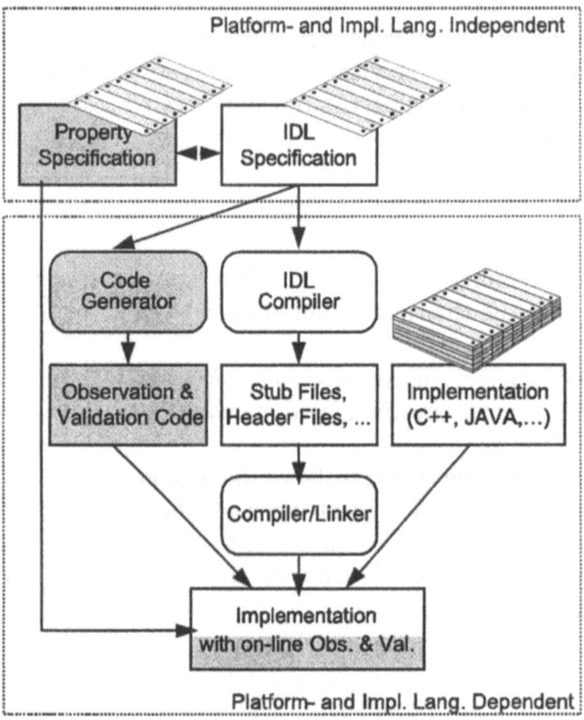

Figure 4 General Framework

tributed applications in the CORBA framework. The white boxes depict the normal development process of distributed applications; the gray boxes describe the extensions proposed in this paper. Rounded boxes denote tools.

The IDL specification of the interfaces is passed to an IDL compiler which generates stub code and header files which are then linked to the actual implementation code thereby shielding the developer of the distributed application from the difficult task of handling the distribution issues. Up to this point, the process is straightforward and mostly well-understood by today's software industry. But here is where we propose the innovative part developed in our work: In addition to passing the IDL specifications to the IDL compiler we also feed a code generator with the IDL specifications. This code generator tool generates some generic observation- and validation code which can then be linked to the actual implementation, thereby providing an on-line observer and -validator.

When running the distributed application we can pass our formally specified properties to the on-line validator which will then compare them to the observed behavior of the system and report all property violations.

When expressing properties it is not necessary, but certainly possible, to give a more detailed behavior specification. When expressing properties we can concentrate on a selected set of properties that we wish to be exhibited by the system.

The abstraction level that is provided by an IDL specification makes it also an excellent place for expressing properties that can later be tested at run-time.

The generic code generated by our code generation tool is comprised of two major parts: one part deals with the observation of the distributed system and the collection of traces, the other part is responsible for the analysis and interpretation of the traces.

The notion embodied in the observation part of our approach is not new: Many distributed platforms have been implemented so that run-time observation can be exploited. For example, CORBA compliant Orbix from IONA provides the filtering mechanism; the CHORUS Cool distributed platform offers a mechanism similar to Orbix filters, termed interceptor. We make the assumption that a run-time observation mechanism, is provided by the distributed platform. This assumption is not restrictive. In the case that such a mechanism is not offered by the distributed platform, a proxy object can be added for each object within the system playing the role of the observation filter.

Using the filter mechanism provided by IONA's Orbix CORBA platform (IONA Technologies PLC 1997) we can spy on the distributed system. Orbix offers two kinds of filters: process filters and object filters. Filters allow the execution of additional code for each filtered event. A process filter intercepts all incoming and outgoing operation requests for a given process. When objects inside a process invoke an operation on an object in the same process, then these invocations are also fully visible to the process filters. Object filters are executed before and after each operation invocation on an object. Orbix process filters also allow the possibility to piggy-back data to operation request as long as the receiving process removes the added data before passing it on to the object.

The way of an operation request from one object to another object is depicted

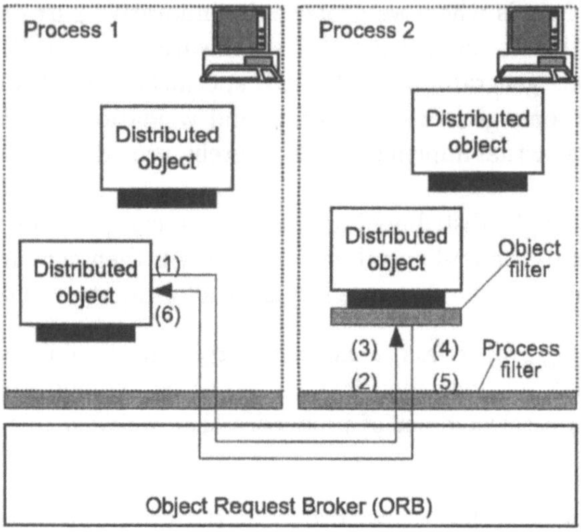

Figure 5 Orbix Filters

in Figure 5. The Orbix filters that are used on the way are numbered in the order they are executed. As shown in this Figure, we have six filters which map to our observable events as indicted in Table 2. It can be seen that our framework captures the abstraction level that is useful for filling the needs of todays industrial software development.

#	Filter level	Event type
1	process	*o_outReq*
2	process	N/A
3	object	*o_inReq*
4	object	N/A
5	process	*o_outRep*
6	process	*o_inRep*

Table 2 Mapping Orbix filters to observable events

The generation of test oracles from properties specified in LTL is also a well-understood problem and can be automized.

When running the system we need to collect the test traces and to reorder them at the observer side. As the observation mechanism can be dynamically activated and deactivated – filters can be dynamically attached to objects and detached – the impact of the validator on the system is marginal if no properties are to be tested. When feeding the observer with a property, the observation mechanism for the corresponding events would be activated and thereafter the validator would receive notifications about these two operations from the objects. As soon as a property gets violated, the on-line validator will report a property violation.

A screen dump of MOTEL, the MOnitoring and TEsting tool is given in Figure 6: LTL properties can be specified and activated (Window entitled "Properties") and relevant events can be observed (window entitled "MOTEL"). Test oracles for the properties are automatically generated (bottom window). The observed events are analyzed and property violations are reported to the user (window entitled "Property violation"). For a detailed description of MOTEL we refer the interested reader to (Logean 1998).

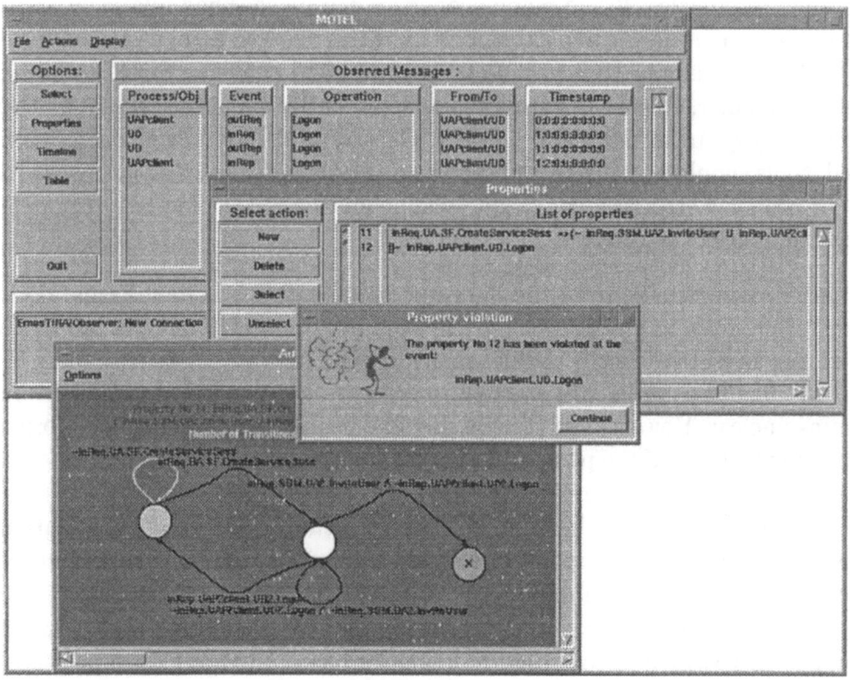

Figure 6 MOTEL screen dump

5 DISCUSSION

The set of observable events turned out to be largely sufficient for specifying the properties we derived from the informal documentations. Most properties could be expressed at the object level using the two event types ***o_inReq*** and ***o_outRep***.

The abstraction level we achieve through event–based behavioral abstraction with our events matches the abstraction level that the properties in the documentation are expressed at.

Since its inception we have identified several weaknesses of our property language. Firstly, the property language does not allow for expressing properties on complex data structures like lists and various records that are somewhere defined in the program and later used as parameters. Since most operations use these complex data structures, expressing properties on parameters is hardly possible with our property language which considers only simple data types like integers.

Secondly, it is not always easy to come up with a temporal logic formula for complex properties. While many properties can be specified relatively easily, there are some more complex properties which require a good deal of experience in developing LTL formulas. This problem could be fixed in the following manner. Deeper investigation would identify the property classes that frequently occur in industrial services. Once several classes have been identified, tools should be constructed that help the property specifier to choose the right property structure. He might even be unaware of the fact that temporal logic is behind the property he expresses.

Other problems arise from the informal documentation. The informal documentation gives in many cases only limited information about the properties that are useful to specify. While many properties can be derived from the SPOT documentation, the practical relevance of the specified properties remains unclear. However, we assume that the persons writing such an informal documentation and the persons designing and implementing the service could derive useful properties relatively easily.

Another problem results from the use of scenarios in the documentation. Since they are supposed to reflect a single system run, they do not, in general, give enough information about special cases that might be encountered. In such a case, a formal property which requires that something always has to happen as specified in the scenario might be based on wrong assumptions.

We are currently investigating several other issues: The observable events we were considering are primitive events (as opposed to aggregate events). The specification of aggregate events could be used to facilitate the specification of more complicated properties.

We are extending the observer tool that we have developed for CORBA-based applications. The basic observation mechanism has already been implemented (including dynamic activation/deactivation of event-generating code fragments, test oracle generation, time-stamping- and reordering mechanism etc.). To better address the problem of scalability we are also investigating *distributed* observers.

In order to tackle specific problems in distributed applications we are currently tailoring and extending our model for the two areas of fault tolerance and security. For example, additional events, e.g., for check-pointing and node crashes, will make our model applicable for the specification of a large number of properties related to fault tolerance.

6 CONCLUSIONS

There are many solid theoretical foundations related to testing and formal approaches but there seems to be a lack of assimilation of this work into the mainstream testing process. In particular, formality is difficult to justify in industrial projects. Hiding part of the formality and automizing parts of the testing process can break some barriers currently present.

In our approach, only the property specification must be derived manually. The observation- and validation code, the selection of filters to activate, the examination of the observation messages and the property checking are all derived automatically.

The landmark characteristics of our approach are the expression of properties in an implementation language independent manner and the verification those properties at system run-time without requiring any help from the programmer/tester to map the properties to the implementation level. The observation of the distributed system and the analysis of the test traces is also completely hidden from the service tester.

An IDL specification is written at a level of abstraction that makes to particularly suitable for providing a basis on which to express behavioral properties.

We have outlined how some properties can be specified in this framework. We have shown how the property relevant information can be collected in such systems.

7 ACKNOWLEDGEMENTS

This work is being partially supported by Swisscom. We would like to thank C. Delcourt and S. Grisouard at Alcatel Alsthom Research, Paris, and P.-A. Etique at Swisscom, Bern, for many interesting discussions. We thank H. Karamyan and F. Pont for their work on the CORBA observer implementation.

REFERENCES

Bates, P. (1995), 'Debugging heterogeneous distributed systems using event-based models of behavior', *ACM Transactions on Computer Systems* **13**(1), 1–31.

Chapman, M. & Montesi, S. (1995), *Overall Concepts and Principles of TINA, Version 1.0*, TINA-C.

Dietrich, F., Logean, X., Koppenhöfer, S. & Hubaux, J.-P. (1998), Modelling and testing object-oriented distributed systems with linear-time temporal logic, Technical Report SSC/1998/011, Swiss Federal Institute of Technology, Lausanne. Available at http://sscwww.epfl.ch.

Dillon, L. & Yu, Q. (1994), Oracles for checking temporal properties of concurrent systems, *in* 'Proceedings of the 2nd ACM SIGSOFT Symposium on Foundations of Software Engineering', Vol. 19, pp. 140–153.

Goguen, J. & Luqi (1995), Formal methods and social context in software development, *in* 'TAP-SOFT'95: 6th International Conference on Theory and Practice of Software Development', number 915 *in* 'Lecture Notes in Computer Science', Springer-Verlag, pp. 62–81.

Holzmann, G. (1994), The theory and practice of a formal method: NewCoRe, *in* 'Proceedings of the IFIP World Computer Congress', Vol. I, North-Holland Publ., Amsterdam, The Netherlands, Hamburg, Germany, pp. 35–44.

IONA Technologies PLC (1997), *Orbix 2: Programming guide, Version 2.2.*

Jagadeesan, L., Puchol, C. & Olnhausen, J. (1995), 'A formal approach to reactive systems software: A telecommunications application in ESTEREL', *Journal of Formal Methods in System Design* .

Järvinen, H.-M., Kruki-Suonio, R., Sakkinen, M. & Systä, K. (1990), Object-oriented specification of reactive systems, *in* 'Proceedings of the 12th International Conference on Software Engineering', IEEE Computer Society Press, pp. 63–71.

Lamport, L. (1988), A simple approach to specifying concurrent systems, Technical report, Digital Equipment Corporation, SRC.

Lamport, L. (1995), 'TLA in pictures', *IEEE Transactions on Software Engineering* pp. 768–775.

Logean, X. (1998), MOTEL – MOnitoring and TEsting tooL for distributed applications, Technical report, Swiss Federal Institute of Technology. Available from the authors.

Manna, Z. & Pnueli, A. (1991*a*), *The Temporal Logic of Reactive and Concurrent Systems*, Springer-Verlag.

Manna, Z. & Pnueli, A. (1991*b*), Tools and rules for the practicing verifier, Technical report, Stanford University.

Programming service tests with TTCN

Martin Anlauf
Bosch Telecom GmbH
Kleyerstraße 94, D-60326 Frankfurt am Main, Germany
martin.anlauf@fr.bosch.de
tel: +49 69 7505 6148   fax: +49 69 7505 6168

Abstract

This paper describes an experiment in which TTCN is used to program service test cases for a private switch. One goal of the experiment was to evaluate the suitability of TTCN for another context than conformance testing. The requirements of the experiment and the nature of the test cases under consideration are described. The importance of concurrency is emphasized. The methodology of development is outlined, including the derivation of test purposes from verbal descriptions. The test method is indicated. Concepts directing the design of the test suite are introduced. Experiences and conclusions are presented.

Keywords

Service Test, TTCN, ISDN

1 INTRODUCTION

The experiences described in the paper result from an experiment run at Bosch Telecom GmbH. The suitability of TTCN is evaluated to improve the testing phase in the development process for private communication switches. The functionality under focus is that of ISDN layer 3 protocol DSS1 (ETS 95).

The introduction describes requirements and goals of the experiment. Section 2 discusses the nature of the considered test cases. Section 3 presents the test method as well as concepts which have directed the design of the test suite. Section 4 describes experiences. Section 5 presents concluding results.

1.1 Current test practise and requirements

The peculiarities of the described TTCN application in contrast to pure conformance test shall be clarified first. To this purpose, the current test practise for private switches is outlined and emerging requirements are sketched.

When a new software release is integrated on the target hardware a regression test phase is entered. In this phase, a number of selected test cases is executed to make sure that the functionality of the previous versions is still valid. Only when the release has passed the regression test, more extensive tests are executed. If the regression test is not passed, it has to be repeated after repair of the release.

Regression test cases are carried out by hand once and recorded. Basically they can be executed automatically the next time. In an automatic execution, the test machine sends recorded stimuli to the switch and compares the received reactions to the recorded reactions. If theses reactions are identical then the verdict pass is given. Else the test case is regarded as failed.

A new software release may cause some changes in the behaviour of the switch, i.e. in reactions to given stimuli. Such changes range from additional information elements inside of protocol messages to the sending of different protocol messages. In such cases, the automatic execution of test cases may lead to the verdict fail, even if these changes are conformant to the standard. This means that the manual execution and the recording of a large number of test cases may have to be repeated, which costs time and effort.

1.2 The experiment

The goal in programming test cases is to save time and effort in the regression test phase. The test cases should be automatically executable but more robust with regard to those changes in behaviour which remain within the limits of the standard. If repeated manual execution of test cases can be avoided, the effort spent to develop a test suite should be regained.

The considered test cases are estimated to belong to service testing, in contrast to conformance testing. This is because not the protocol conformance of the switch is focused. The functioning and interaction of integral communication services are checked. However, in our approach of service testing the exchanged protocol messages are examined, and it is difficult to define a clear distinction to conformance testing.

TTCN is a standardized notation (ISO 97). It is designed to be used in conformance testing (ISO 94). The suitability of TTCN for other testing contexts remains to be evaluated. This paper describes an application in service testing.

The evaluation of this TTCN application was done in the RESTATE project. RESTATE was funded by the European Systems and Software Initiative (ESSI) as a Process Improvement Experiment (PIE), ESSI project no. 23978. RESTATE started in March 1997 for a duration of 18 months.

The project was accompanied by a measurement program. The Goal-Question-Metrics method (GQM) was applied to measure efforts and benefits of the TTCN employment. Since at the time of edition of the paper there were no representative results of the measurement program available, they will be presented on slides.

The paper does concentrate on design aspects. For a more complete description of the RESTATE project (Münzel 97).

2 THE TEST CASES TO CODE

The first subsection gives an example of a service test case. The second outlines some characteristics of the test cases. The following subsections discuss the task to do and sketch the test environment.

2.1 An example

The test cases of the regression test are currently executed manually by humans, and also the described actions and reactions belong to the level of perception of a human. The test cases are described in an informal and verbal way. To give a flavour of such a description, an example is presented.

Test Conditions:

Subscriber B has no knocking protection

Actions:

a) Subscriber A takes the handset off hook
Subscriber A dials all digits
b) Subscriber B takes the handset off hook
c) Subscriber A puts the handset on hook
d) Subscriber B puts the handset on hook

Reactions:

a) Terminal B is ringing
b) Subscribers A and B have connection
c) Subscriber B hears busy tone
Subscriber A has initial state
d) Subscriber B has initial state

There are also more complex test cases involving more subscribers, some subscribers handling several calls.

2.2 Characteristics

Some characteristics of the test cases under consideration are pointed out.

- **Concurrency** Several parties are acting concurrently.

- **Dependencies** There are dependencies between the actions of the different parties. Some dependencies are synchronized implicitly by the switch and do not have to be modelled explicitly. For example it is supposed that terminal B rings only after subscriber A has dialed the number of B. Other dependencies have to be synchronized explicitly. An example would be when in the presented test case a third subscriber C would also want to call B, but only after A and B have established a connection.
- **Configuration** A certain configuration of the switch may be a precondition to execute a test case. For example a subscriber is authorised to use a certain service or not. If several test cases are executed in a series automatically, then such configurations of the switch may have to be done automatically as well.
- **Internal values** For some test cases it is necessary to check internal parameter values of the switch. For example a test case could have to check the correct accounting of charges. Such checks of internal values should be automated as well.

It is to be noted that the last two points leave the realm of pure black box testing. Further the configuration of the switch and the check of internal values can be done via a proprierary protocol. The aspects concerning the configuration of the switch and the check of internal values are not covered in the paper.

2.3 The task

In the experiment there are considered 33 cases to test the basic functionality "basic call" of the switch. Additionally the functionality of supplementary service "Advice of charge" is checked by 17 test cases. These 50 test cases represent together only a small part of the regression test.

The task is to take the given verbal descriptions of the test cases and to code them in TTCN. There is work to be done in two respects.

- **Bridging the gap between the levels of abstraction** The verbal descriptions represent the human point of view. The message flow on the protocol interface has to be derived from the sequences of actions and reactions. For example the action *take handset off hoo*k may correspond to the sending of a *Setup* message and the receipt of a *Setup Acknowledge.* The flow of the messages represents the "protocol point of view".
- **Refinement of the descriptions** To obtain a complete TTCN test suite, there is more information to be added. Declarations and constraints are needed. The Messages to be sent or received have to be specified more precisely, up to the values of contained information elements.

2.4 The test environment

The test cases are executed on a A8619 from Alcatel/Sel. It is connected to the switch with 2 S_0 and 2 UP_0 ports. The 4 ports can be used either as 4 subscribers or as 3 subscribers and 1 system console. The system console is used to configure the switch and to check internal values.

The test machine is running the TTCN Professional software package from Expert Telecoms. The package includes an editor, a compiler, a test campaign manager, a simulator for the lower layers as well as a tool supporting the analysis of recorded test runs.

The development of the test suites is done with the Concerto TTCN editor from Sema Group, which is running on Sun workstations. The test environment is sketched in Figure 1.

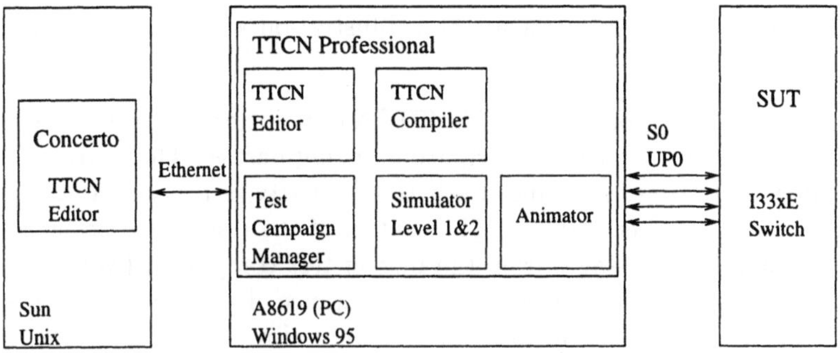

Figure 1 Test Environment

3 CONCEPTS

This section introduces concepts and design decisions. It presents the applied test method and sketches the test case configurations. A model of the user's behaviour is introduced which directs the structure of the test step library. Some of the conventions applied are mentioned and the use of message flow diagrams is illustrated.

3.1 The test method

A multi party test method is applied which is sketched in Figure 2. Among the four classical single party test methods, it has the closest resemblance to the remote test method (ISO 94).

There are several testing parties involved, all lower testers running on the same test machine. The exact number of testing parties depends on the test

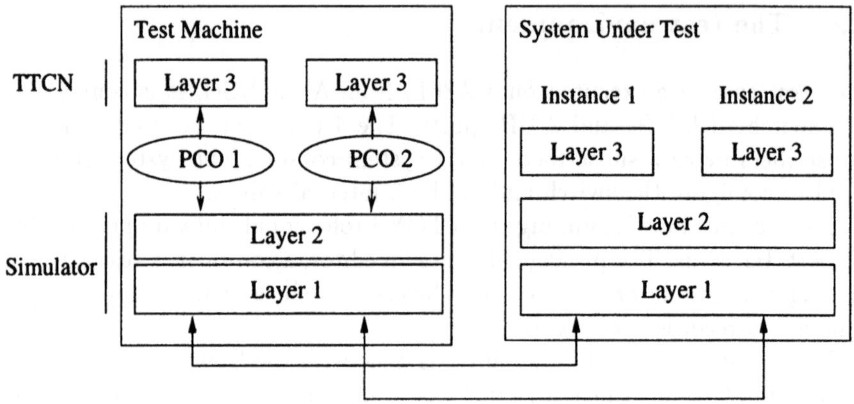

Figure 2 Multi party test method. Illustration with 2 testing parties.

case. There are no upper testers. The TTCN description of a test case specifies the layer 3 behaviour of all participating testing parties. The lower levels are simulated by the test machine. Each testing party is realized as a separate parallel test component (PTC). The control of the testing parties is realized with coordination messages from the master test component (MTC).

Each testing party has a corresponding instance of a protocol state machine on the switch. The test purposes do not only aim at the correct behaviour of such a single instance. The interaction between several instances is also important to assure the services of the switch.

3.2 The test case configurations

The configurations used by the test cases all have the same structure. They only differ in the number of participating PTCs. Figure 3 shows an example.

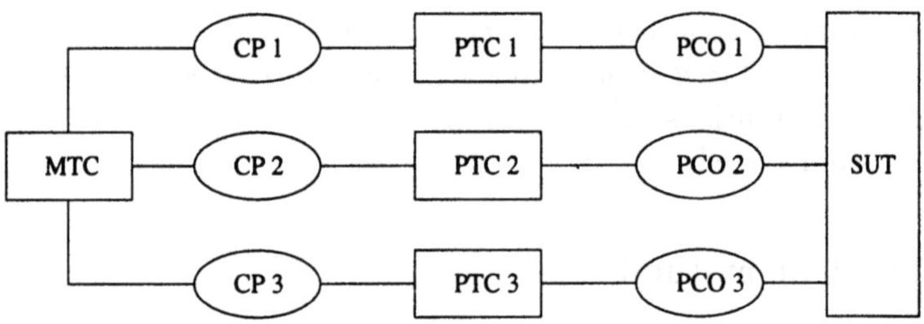

Figure 3 Configuration of test cases. Example with 3 PTCs.

Note that the MTC does not communicate directly with the SUT. The whole communication with the SUT is done by the PTCs via the points of control and observation (PCOs). On the other side the PTCs do not have the possibility to communicate with each other directly. All coordination is done by the MTC via the coordination points (CPs).

The simplicity of the configurations allow a uniform coordination of the PTCs in all test cases, no matter how many PTCs are involved. The coordination is separated from the communication with the SUT. Further the communication with the SUT is divided into smaller parts, which are easier to handle. Each PTC only has to know it's own part of the communication.

3.3 A user model and the test step library

The verbal descriptions of the test cases only use a small number of actions and reactions. Actions and reactions are subsumed as events. Examples are *X takes handset off hook*, *Y dials some numbers* or *Z is ringing*. These events reflect the perception of the human user. With TTCN in contrast, the message flow on the protocol interfaces has to be described.

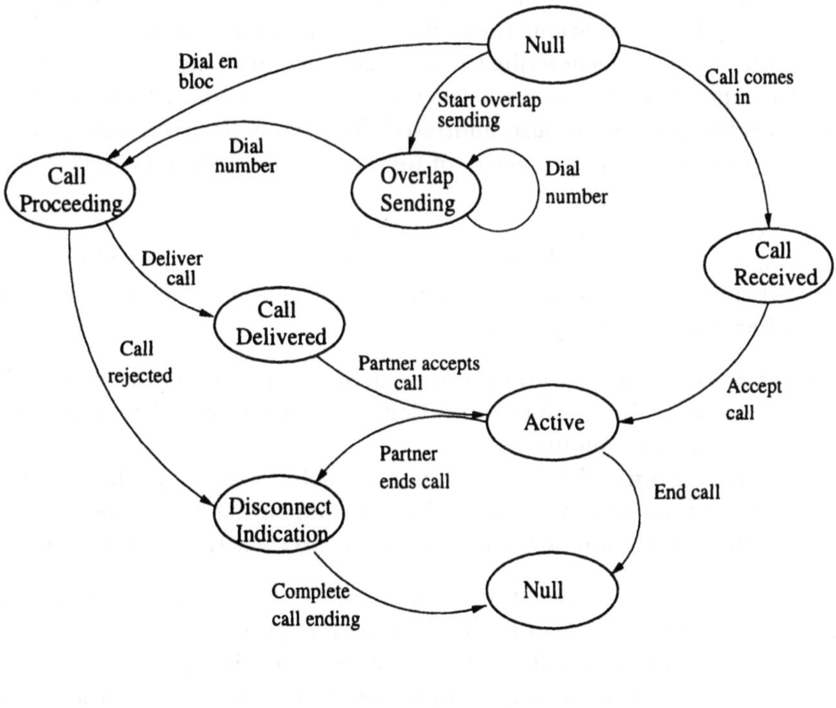

Figure 4 A model of the user's behaviour

The idea now is to model the behaviour of the user in order to fix all possible events. When it is known which flow of protocol messages corresponds to which event, then the events can be coded as test steps. When such a test step is parameterized in a suitable way, then it can be used by different PTCs.

In this style a test step library is built which hides details of the protocol and which reflects the perception of the human user. The test step library is meant to constitute a set of components from which test cases can easily be constructed.

The model of the user's behaviour is sketched in Figure 4 in the style of a state transition diagram. The diagram is not complete and should not be interpreted in a formal way. It simply did serve as a starting point to design the test step library. Each transition is realized as a test step.

The events in our example concerning subscriber A, i.e. *Take handset off hook*, *dial all digits*, *A and B have connection* etc. correspond in the diagram to the transitions *Start overlap sending*, *Dial number*, *Partner accepts call* etc. The events concerning subscriber B, i.e. *Terminal is ringing*, *take handset off hook* etc. correspond to the transitions *Call comes in*, *Accept call* etc.

3.4 Conventions

In order to maintain a uniform style in the test suite, a set of conventions was applied. The conventions range from a conceptual to a syntactical level. Conceptual conventions describe the test case configurations, the coordination of test components and the error treatment. The issues of default behaviour and of verdict passing are also addressed. The syntactical conventions cover aspects like names and parameters of test steps and constraints. TTCN style rules are also defined.

Since concurrency is a crucial point of the project and the number of participating testing parties should not be restricted a priori, special effort has been put into the conventions to coordinate the test components. Two cases of coordination are distinguished.

- **Synchronization** Synchronization is necessary in the case when one component is allowed to perform an action only when some other components have fulfiled some conditions.
- **Error treatment** When one component detects an error which leads to the termination of the test case, all other components have to be informed about the forthcoming termination in order to clear their connections.

The test suite writer knows the points in the test case where components have to be synchronized. So the coordination for synchronization is described explicitly in the dynamic tables of the test case. On the other hand it is not predictable at which point errors will be detected. So the coordination in the error case is described and "hidden" in defaults.

The approach in the project is to fix a small number of coordination messages and to restrict their use, in order to get a uniform coordination style.

The descriptions of the behaviour of the PTCs of a test is divided into 3 classical parts: preamble, body and postamble. The preamble may contain test steps to configure the SUT as required and to bring it into a suited starting state. The body contains the test steps of the library mentioned before. The postamble may contain steps to check internal values, to reconfigure the SUT and to bring it into a suited final state.

3.5 Message flow diagrams

When concurrency is to be described, the most important thing is to keep the overview of the expected behaviour or communication. Here a description of the expected flow of messages on several intefaces is needed. There may not be determined a total ordering on the messages a priori.

For example when two messages are expected at two different interfaces, one message at each interface, the ordering of arrival may not be predictable. So there is only a partial ordering defined on the message flow.

While the presented verbal descriptions of the test cases do not even mention protocol messages, the TTCN code is very detailed and difficult to read. So the need for an intermediate notation arises, which concentrates on the message flow. Following the terminology of the ISO (ISO 94) such descriptions can be considered as test purposes.

Message flow diagrams are an appropriate notation for this purpose. Although the standardized notation of message sequence charts (MSCs) fulfils all the requirements (ITU 97), in the project an informal variant of message flow diagrams is used. An example for the (ab-)use of notation in the project is that message boxes are used to indicate hierarchy.

The diagrams indicate the expected message flows of test cases. They also sketch the software structure of a test case. The diagrams are used at two levels. An overview diagram contains all components participating at a test case, i.e. the PTCs and the MTC. In the overview diagram only the used test steps are sketched, not the protocol messages. Then there is a specific diagram for each PTC involved which indicates the test steps used by this PTC as well as the protocol messages to send or to receive.

To illustrate this use, 3 diagrams which refine our example are presented in the appendix.

4 EXPERIENCES

This section presents experiences from the project. It covers the subjects tools test case validation, parallel editing and aspects of the notation TTCN.

4.1 Tools

The used tools were suitable and fulfiled their purpose. It has turned out that user-friendly tools are crucial for the work with TTCN. Features like syntax and semantic checks are valuable, as well as support for the analysis of test runs.

Since concurrency constructs have been especially essential in the project it did cause some delay that not all the constructs were supported by the tools from the beginning. The critical tool has been the TTCN compiler.

4.2 Validation of test cases

Validation is important to get trustworthy test cases. In the project there was the lucky situation that a reference implementation of the SUT was available. Validation consisted mainly in running the test cases against the reference implementation and in analysing the test runs. Mutual reviewing of test suites is also recommendable.

The software package TTCN Professional contains a tool called Animator which turned out to be very helpful for the analysis of test runs. It offers a debug functionality which allows the user to step through the executed commands of a TTCN test case. Equally important are traces of the runs, which record the exchanged protocol messages and coordination messages.

4.3 Parallel editing

It is not easy for several developers to work on the same suite because of the many references. It is important to fix the personal rights to modify the different parts of the suite. Parallel editing is eased by test suite fragmentation. The editor Concerto did support this possibility. For the future, the standardized modular features of TTCN are to be welcomed.

4.4 The language TTCN

TTCN is stated to be an "informal notation", but in RESTATE it was used just like any other programming language. A preliminary version of 2nd edition TTCN (ISO 97) was the basis of the project, but not all new features have been used. Concurrency has been used heavily. Defaults have been very helpful, also the Return statement. It has already been mentioned that modularity has not been used. Proceduraly defined test suite operations have not been used neither.

The conclusion is that the TTCN is suited to describe test cases not only in the restricted field of conformance test, but also to describe service test cases such as those above. However, some shortcomings have been experienced and the language could still be improved.

- **Overview** TTCN is not an easy language to read. It is difficult to keep the overview. One reason is that the principle of reference is used extensively. Space is valuable inside the narrow TTCN columns, which leads to a heavy use of references. Another reason is that the tabular style is not always well suited for the dynamic part. A big dynamic table is hard to read, which leads to the use of many small tables. In this way the information about dynamics is often spread over many tables.

- **Ordering of events** It would be nice to have the possibility to relax the strictness of the ordering of events in time. In a situation where n events are being waited for, the exact ordering of these events might not be known. It is cumbersome to write down all $n!$ possible orderings. It would be nicer to have the possibility to "collect" the events without considering the exact order.

5 CONCLUSIONS

This section presents some conclusions. They concern the expectations, a side effect, MSCs and the problem of acceptance. Finally, a summary is given.

5.1 Expectations

The main expectations of the project have been fulfiled. Namely, it has been possible to code the considered test cases in TTCN so that they are executable automatically. They are also robust with regard to protocol conformant changes of the behaviour of the SUT.

5.2 A side effect

A side effect of RESTATE has been to deepen our understanding of the test process for private communication switches and to discern a point of possible improvement. The standard of DSS1 often contains options. An enforcement of the documentation of implementation specific details will ease testing as well as development.

5.3 Message flow diagrams

The use of message flow diagrams has been proved worthy. For similar projects, the use of the standardized notation MSC '96 is recommended (ITU 97).

5.4 Acceptance

ı RESTATE, the coding of the test cases has been done by a team from ›utside the group in whose development process TTCN is thought to be integrated. The main expectations have been fulfiled, but it also has been shown that the effort to code test cases in TTCN is big and that the complexity of the language is high. Therefore it is still not clear how TTCN will be integrated in the development process.

5.5 Summary

Concluding it is stated that it is possible to manually code service test cases in TTCN, but that the effort is high. The introduction of TTCN is considered to be an investment, and it is must be estimated carefully, for which purposes the application of TTCN is justified.

It is an open point wether if the protocol interface is the right one to specify service test cases. An alternative would be to look for an interface on a higher level, where the details of the protocol do not have to be described. The choice of the right interface should depend on the goals of the test phase under consideration.

However, if TTCN is employed and coded manually, the use of an application specific test step library has been proved worthy. The use of MSCs to describe test purposes is recommended.

It would be a tempting alternative to evaluate the possibilities of automatic test case generation out of MSCs as it is described in recent papers, cf. (Ek *et al.* 97) and (Grabowski *et al.* 97) in the context of service testing .

6 ACKNOWLEDGEMENTS

The author would like to thank all people that supported the RESTATE project, especially to Jörn Münzel, to Bertram Weber and to Helmut Meyer. Further thanks to Andreas Zamperoni for the explanation of diverse tools, to Mark Andrew for reviewing and to the group around Jens Grabowski for fruitful discussions.

7 GLOSSARY

CP	Coordination Point
DSS1	Digital Subscriber Signalling No. One
ESSI	European Systems and Software Initiative
GQM	Goal Question Metrics
ISDN	Integrated Services Digital Network
MSC	Message Sequence Chart
MTC	Master Test Component
PCO	Point of Control and Observation
PIE	Process Improvement Experiment
PTC	Parallel Test Component
SUT	System Under Test
TTCN	Tree and Tabular Combined Notation

REFERENCES

Ek, Grabowski, Hogrefe, Jerome, Koch and Schmitt 97 *Towards the industrial use of validation techniques and automatic test generation methods for SDL Specifications* Proc. of the 8th. SDL Forum, Evry, France, ed. Cavalli and Sarma, Elsevier Science,1997.

ETS 95 *Integrated Services Digital Network (ISDN); Digital Subscriber Signalling System No. one (Dss1) protocol; Signalling network layer for circuit-mode basic call control.* ETS 300 403-1 & 2. European Telecommunications Standards Institute, Sofia Antipolis, 1995 (ITU-T Recommendation Q.931 (1993), modified).

Grabowski, Scheurer, Dai and Hogrefe 97 *Applying SAMSTAG to the B-ISDN protocol SSCOP* Proc. of IWTCS 10th., Cheju Island, Korea, ed. Kim, Kang and Hong, Chapman & Hall, 1997

ISO 94 *Information Technology, Open Systems Interconnection, Conformance Testing Methodology and Framework.* International Standard IS-9646-1 & 2. ISO, Geneve, 1994.

ISO 97 *Information Technology, Open Systems Interconnection, Conformance Testing Methodology and Framework.* International Standard IS-9646-3, Delivery 9.6. ISO, Geneve, 1997.

ITU 97 *ITU-TS Recommendaion Z.120: Message Sequence Charts (MSC 96).* ITU-TS, Geneve, 1997.

Münzel 97 *Quality Improvement Through Formalised Testing* Proc. of CONQUEST 1, Arbeitskreis Software Qualität Franken e.V., Erlangen, Germany, pp. 184-187, 1997.

Prevot, D'Hooge and Labbe 96 *Experiences with concurrent TTCN for GSM/SS7 testing* Proc. of IWTCS 9th., ed. Baumgarten, Burkhardt and Giessler, Chapman & Hall, 1996

8 BIOGRAPHY

Martin Anlauf graduated in 1994 from the University of Kaiserslautern, where he worked from 1995 to 1996 as assistant. In 1996 he started to work at Bosch Telecom in the software technology department. His interests include testing and formal description techniques.

9 APPENDIX

The following diagrams illustrate the flow of messages of the example test case. The use of test steps is indicated with comment boxes. The overview diagram containes all participating parties. The specific diagrams concentrate on only one testing party.

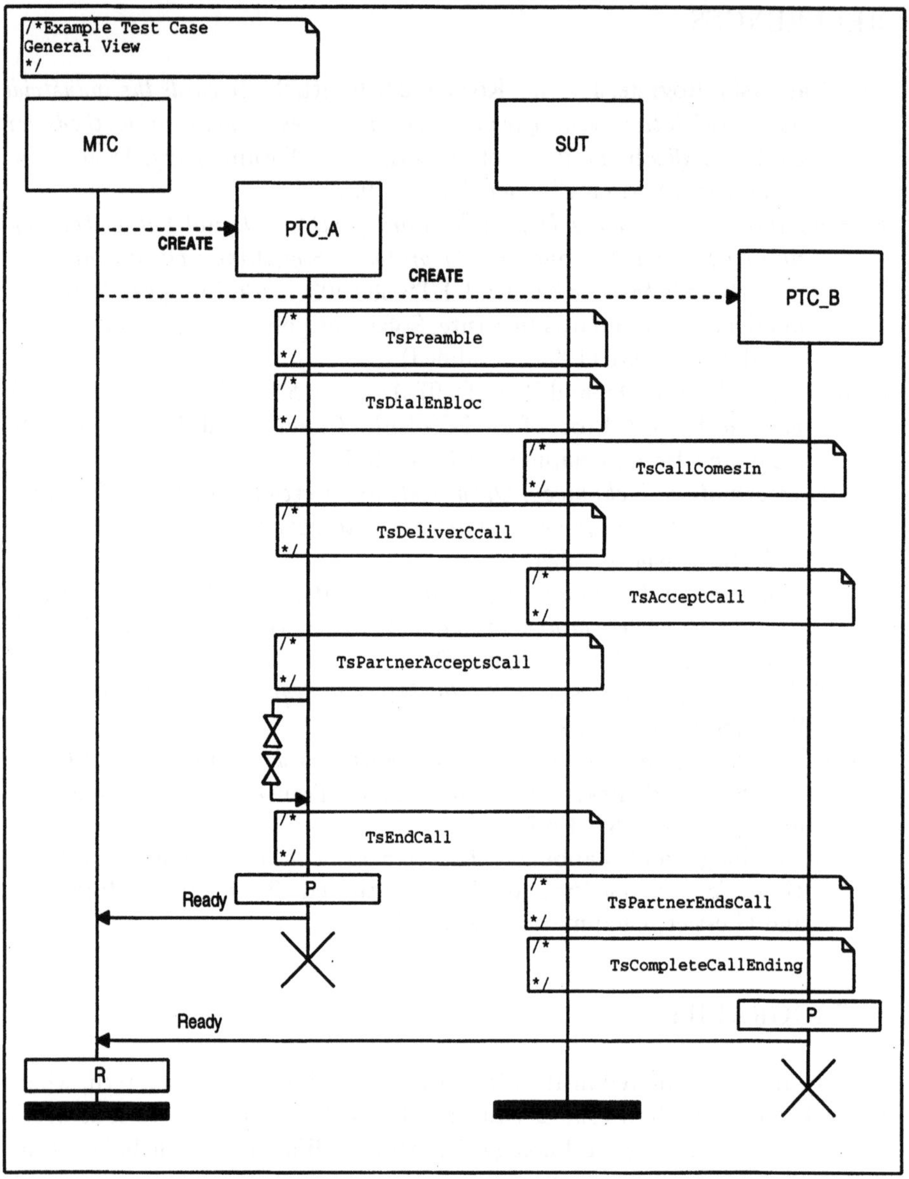

Figure 5 Overview diagram

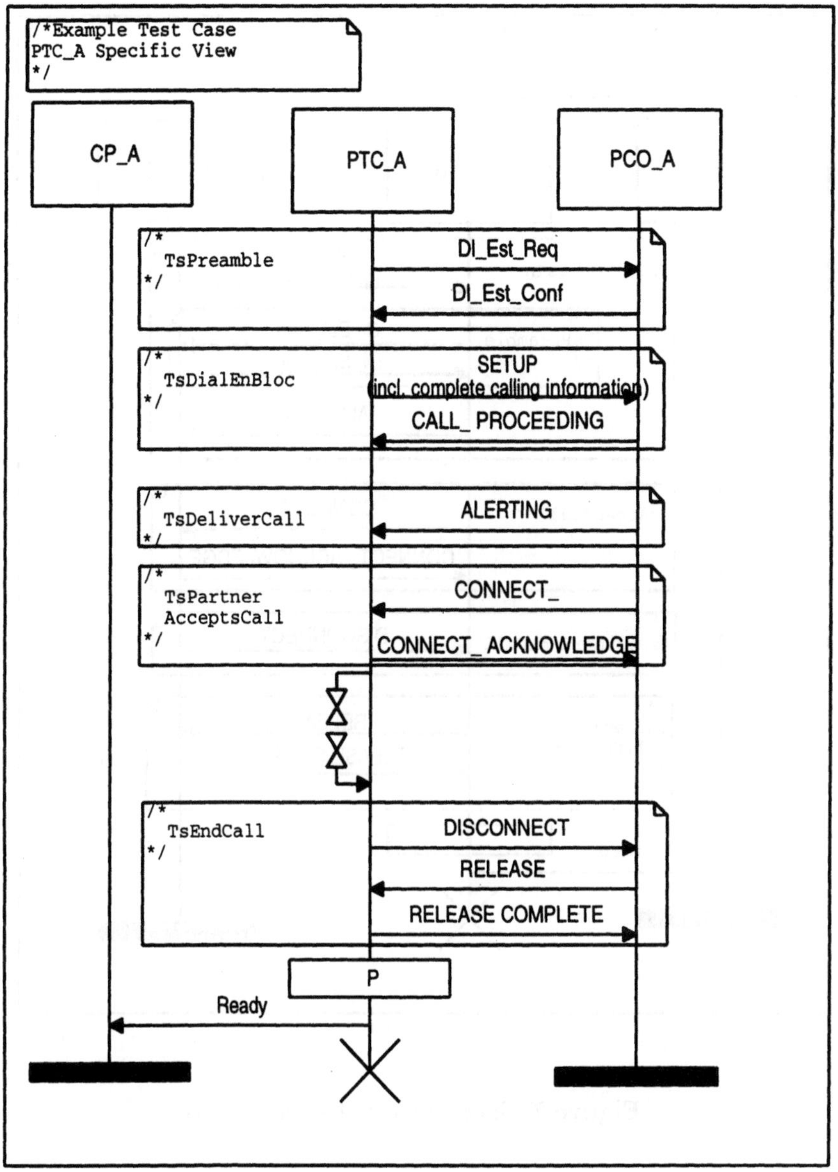

Figure 6 Specific view of subscriber A

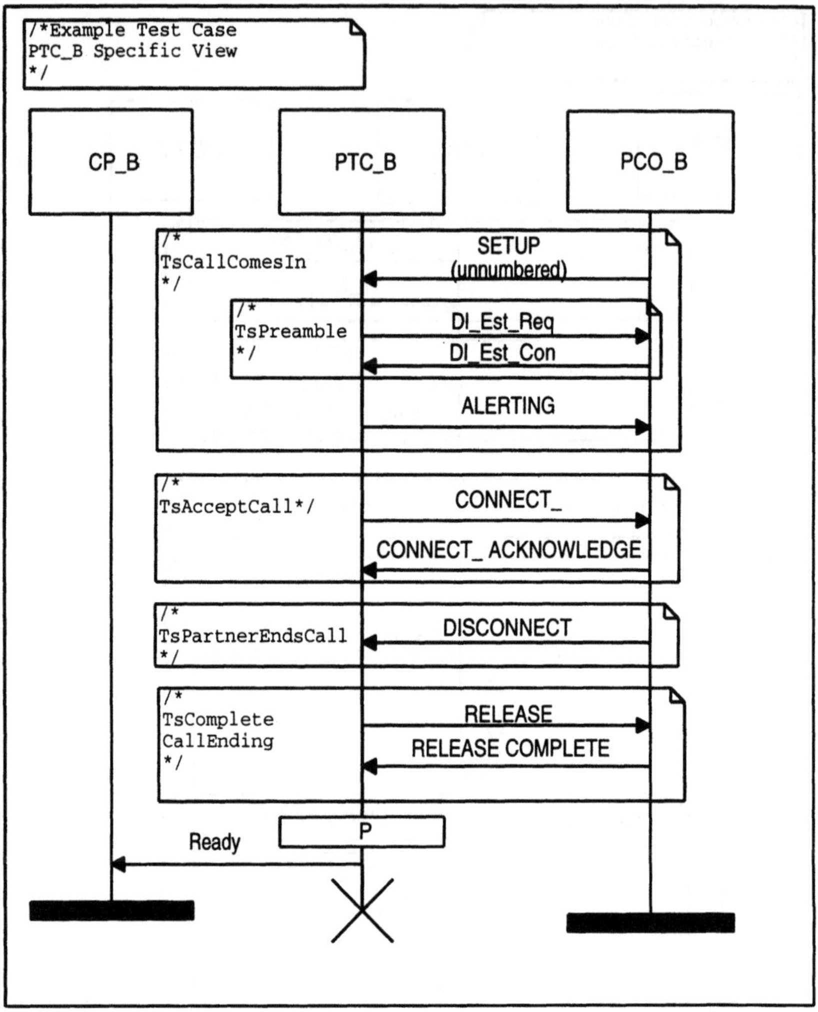

Figure 7 Specific view of subscriber B

A guided incremental test case generation procedure for conformance testing for CEFSM specified protocols[1]

C. Bourhfir[2], R. Dssouli[2], E. Aboulhamid[2], N. Rico[3]

[2] *Département d'Informatique et de Recherche Opérationnelle, Pavillon André Aisenstadt, C.P. 6128, succursale Centre-Ville, Montréal, Québec, H3C-3J7, Canada. E-mail: {bourhfir, dssouli, aboulham}@iro.umontreal.ca.*

[3] *Nortel, 16 Place du commerce, Verdun, H3E-1H6*

Abstract

This paper presents an incremental method for automatic executable test case and test sequence generation for a protocol modeled as communicating extended finite state machines (CEFSMs) with asynchronous communication. Instead of testing the protocol by computing the product of all CEFSMs, we test it by incrementally computing a partial product for each CEFSM C, taking into account only transitions which influence (or are influenced by) C, and generating test cases for it. The partial product for C represents the behavior of C when composed with parts of the other CEFSMs. Experimental results show that this method can be applied to systems of practical size. We also propose a method which reduces the size of the product machine for certain systems.

Keywords

CEFSM, Partial product, Reachability analysis, Control and data flow testing, Executability

---

1. This work is supported by a Nortel grant.

1 INTRODUCTION

To ensure that the entities of a protocol communicate reliably, the protocol implementation must be tested for conformance to its specification. In principle, finite state machines (FSMs) model appropriately the control portions of communication protocols. However, in practice most protocol specifications include variables and conditional statements based on variable values. Therefore the Extended Finite State Machine (EFSM) like model should be used. Quite a number of methods have been proposed in the literature for test case generation from EFSM specifications using data flow testing techniques and/or control flow testing techniques [Ural 1991, Huang 1995, Chanson 1993, Bourhfir 97]. However, these methods are applicable when the protocol consists only of one EFSM. For CEFSM specified protocols, other methods should be used. To our knowledge, very few work has been done in this area, and most existing methods deal only with communicating finite state machines (CFSMs) where the data part of the protocol is not considered. An easy approach to testing CFSMs is to compose them all-at-once into one machine, using reachability analysis, and then generate test cases for the product machine. But we would run into the well known state explosion problem. Also, applying this approach to generate test cases for CEFSMs is unpractical due to the presence of variables and conditional statements. To cope with the complexity, methods for reduced reachability analysis have been proposed for CFSMs [Rubin 1982, Gouda 1984, Lee 1996]. The basic idea consists in constructing a smaller graph representing a partial behavior of the system and allowing one to study properties of communication. In this paper, we present a method which can be used to test a CEFSM based system or a part of it after a correction or enhancement. Our method does not compose all machines but decomposes the problem into computing a partial product (defined later) for each CEFSM and generating test cases for it. To compute the partial product for a certain CEFSM, we compute the product of this CEFSM with parts of the other CEFSMs which influence (or are influenced by) it, i.e., we consider all transitions in the CEFSM and a subset of the transitions of each CEFSM which are involved in the communication with it when computing the partial product. We call it partial product since not all transitions in all machines are considered. By doing so, the complexity of test generation for a communicating system is significantly reduced which enables to test real communicating systems. In the rest of the paper, we use the term "CEFSM in context" to describe the partial product for a CEFSM. Our objective is not to cover all global transitions in the product of all CEFSMs but to cover all transitions in all CEFSMs and all global transitions as well as all data-flow paths in each partial product.

The organization of this paper is as follows. Section 2 describes the EFSM and CEFSM models. Section 3 presents the EFTG tool for automatic test generation for EFSM based systems. In Section 4, an algorithm for computing the partial product for a CEFSM is presented. Section 5 presents an example. In section 6, the algorithm for generating executable test cases for a system modeled by a set of CEFSMs is presented. Section 7 analyzes the results. Finally, section 8 concludes the paper.

2 THE EFSM AND CEFSM MODELS

Definition1. An EFSM is formally represented as a 8-tuple <S, s_0, I, O, T, A, δ, V> where

1. S is a non empty set of states,
2. s_0 is the initial state,
3. I is a nonempty set of input interactions,
4. O is a nonempty set of output interactions,
5. T is a nonempty set of transitions,
6. A is a set of actions,
7. δ is a transition relation $\delta : S \times A \rightarrow S$,
8. V is the set variables.

Each element of A is a 5-tuple t = (initial_state, final_state, input, predicate, block). Here "*initial_state*" and "*final_state*" are the states in S representing the starting state and the tail state of t, respectively. "*input*" is either an input interaction from I or empty. "*predicate*" is a predicate expressed in terms of the variables in V, the parameters of the input interaction and some constants. "*block*" is a set of assignment and output statements.

Definition2. A communicating system is a 2n-tuple (C_1, C_2,...,C_k, F_1, F_2,..., F_k) where

- C_i = <S, s_0, I, O, T, A, δ, V>
- F_i is a First In First Out (FIFO) for C_i, i=1..n.

Suppose a protocol Π consists of k communicating CEFSMs: C_1, C_2,...,C_k. Then its state is a k-tuple <$s^{(1)}$, $s^{(2)}$,..., $s^{(k)}$, m_1, m_2,...,m_k> where $s^{(j)}$ is a state of C_j and m_j, j=1..k are set of messages contained in F_1, F_2,...,F_k respectively. The CEFSMs exchange messages through bounded storage input FIFO channels. We suppose that a FIFO exists for each CEFSM and that all messages to a CEFSM go through its FIFO. We suppose in that case that an internal message identifies its sender and its receiver. An input interaction for a transition may be internal (if it is sent by another CEFSM) or external (if it comes from the environment). The model obtained from a communicating system via reachability analysis is called a global model. This model is a directed graph G = (V, E) where V is a set of global states and E corresponds to the set of global transitions.

Definition 3. A global state of G is a 2n-tuple <$s^{(1)}$, $s^{(2)}$,..., $s^{(k)}$, m_1, m_2,...,m_k> where m_j, j=1..k are set of messages contained in F_1, F_2,...,F_k respectively.

Definition 4. A global transition in G is a pair t = (i,α) where $\alpha \in A_i$ (set of actions). t is firable in $s = <s^{(1)}, s^{(2)},..., s^{(k)}, m_1, m_2,...,m_k>$ if and only if the following two conditions are satisfied where α = (input, predicate, output, compute-block).

- A transition relation $\delta_i(s,\alpha)$ is defined
- $input = null$ and $predicate = True$ or $input = a$ and $m_i = aW$, where W is a set of messages to C_i, and $predicate = True$.

After t is fired, the system goes to $s' = <s'^{(1)}, s'^{(2)},..., s'^{(k)}, m'_1, m'_2,...,m'_k>$ and messages contained in the channels are m'_j where

- $s'^{(i)} = \delta(s^{(i)},\alpha)$ and $s'^{(j)} = s^{(j)} \quad \forall(j \neq i)$
- if $input = \varnothing$ and $output = \varnothing$, then $m'_j = m_j$
- if $input = \varnothing$ and $output = b$ then $m'_k = m_k b$ (C_k is the machine which receives b)
- if $input \neq \varnothing$ and $output = \varnothing$ then $m_i = W$ and $m'_j = m_j \quad \forall(j \neq i)$
- if $input \neq \varnothing$ and $output = b$ then $m_i = W$ and $m'_k = m_k b$

The next section summarizes the approach for automatic test generation for EFSM based systems presented in [Bourhfir 1997]. This approach uses control flow testing techniques, data flow testing techniques and symbolic evaluation.

3 AN EFSM TEST CASE GENERATION ALGORITHM

The algorithm EFTG (Extended Finite state machine Test Generator) presented in [Bourhfir 1997] generates executable test cases for EFSM specified protocols which cover both control and data flow. The control flow criterion used is the UIO (Unique Input Output) sequence [Sabnani 1986] and the data flow criterion is the "all-definition-uses" criterion [Weyuker 1985] where all the paths in the specification containing a definition of a variable and its uses are generated. A variable is defined in a transition if it appears at the left hand side of an assignment statement or if it appears in an input interaction. It is used if it appears in the predicate of the transition, at the right hand side of an assignment statement or in an output interaction. After reading the specification, EFTG generates a dataflow graph. For each state S in the graph, EFTG generates all its executable preambles (a preamble is a path such that its first transition's initial state is the initial state of the system and its last transition's tail state is S) and all its postambles (a postamble is a path such that its first transition's start state is S and its last transition's tail state is the initial state). To generate the "all-definition-uses" paths, EFTG generates all paths between each definition of a variable and each of its uses and verifies if these paths are executable, i.e., if all the predicates in the paths are true. To evaluate the predicates along each transition in a definition-use path, EFTG interprets symbolically all the variables in the predicate backward until these variables are represented by constants and input parameters only. If the predicate is false, EFTG applies a heuristic in order to make the path executable. The heuristic uses "Cycle Analysis" in order to find cycles (a sequence of transitions such that the start state of the first transition and the ending state of the last transition are the same) which can be inserted in the path so that it becomes executable. After this step, EFTG removes the paths which are included in others, completes the remaining paths (by adding postambles) and adds paths to cover the transitions which are not covered by the generated test cases. EFTG discovers more executable test cases over the other methods [Ural 1991, Chanson 1993, Huang 1995] and enables to generate test cases for specifications with unbounded loops.

In the next section, we present a method for testing a CEFSM in context. This method uses the information computed by EFTG, when used to generate test cases for the CEFSM in isolation.

4 GUIDED TEST CASE GENERATION FOR A CEFSM

Testing a CEFSM C in isolation reduces the cost of testing. However, the key problem in testing C resides in its interaction with the other CEFSMs because the test cases generated for C do not consider its interaction with the other CEFSMs; also, some executable test cases for C may no longer be executable when it is combined with the others. Therefore, since testing each CEFSM separately is not sufficient, and since computing the complete product is too costly, we use a middle approach. We test a CEFSM based system by testing each CEFSM in context. We are not aware of any similar work for CEFSMs, but the problem of testing in context is studied for CFSMs in [Petrenko 1996] where a basic framework for testing in context has been given. The method computes a so-called approximation of the specification in context and the idea consists in reducing testing in context to testing in isolation. The method tries to give the most general solution to this problem.

In this paper, we generate a partial product for a CEFSM and then use the EFTG tool to generate test cases for it. In general, in order to test a system modeled by several CEFSMs, we first test each CEFSM in isolation to make sure that it is correct, then the global system is tested. The next algorithm computes the partial product for a CEFSM and generates executable test cases for it using EFTG. Moreover, the algorithm uses the test cases generated by EFTG for that CEFSM in isolation, as a guide, to compute its partial product. In that case, the generated test cases of the partial product cover the data and control flow criteria used in EFTG.

We call our method a *guided* procedure because it uses the already generated test cases as a guide to choose the transitions which will participate in the partial product generation. The process of generating the partial product for CEFSM C_n is divided in the following four steps:

Step1: Test case generation for all CEFSMs in isolation. This step consists in calling EFTG to generate test cases for all CEFSMs in isolation. The test cases generated for each CEFSM are executable test cases which cover both the control and data flow criteria used in EFTG. This step is done once.

Step 2: The marking process. Suppose that the system to be tested is modeled by C_1, C_2,...,C_k and suppose that we want to test the machine C_n in context. We use the test cases generated by EFTG in Step 1 to mark the transitions in all the paths which trigger C_n (or are triggered by C_n). We shall call the first set of transitions Pr(C_n) and the latter Po(C_n). Determining Pr(C_n) and Po(C_n) can be very costly if exhaustive reachability analysis is considered. For this purpose, our method uses the test cases generated by EFTG in Step 1 as a guide. If a transition in C_n receives (sends) a message from (to) a CEFSM C_i and since this message is sent (received) by some transitions in C_i which belong necessarily to test cases generated by EFTG for C_i in isolation, we mark all the transitions in these test cases. By marking all the transitions in each test case, we insure that transitions preceding (following) the transition which sends (receives) the message participate in the partial product generation. When the test case which contains the transition sending (receiving) a message to (from) C_n is marked, we verify if it contains transitions receiving (sending) messages from (to) other CEFSMs. If this is the case, for each such transition T, the same procedure is repeated in order to mark the paths in the machine sending (receiving) the message received (sent) by T. The next algorithm marks all the transitions in the machine C_n, then marks all the transitions in the other machines which can be part of the preambles and postambles of transitions in C_n. If there are cycles between 2 machines, the algorithm Marking detects them enabling procedures Marking-Backward and Marking-Forward to end.

```
Algorithm Marking (machine under test Cn)
Begin
   For each transition T in Cn Do
      Mark T;
      If T receives an internal input M from another machine
         Marking-Backward(Sender of M,Cn,M)
      For each internal output statement O in T Do
         Marking-Forward(Cn,Receiver of O,O)
End;
```

At this stage, we suppose that the test cases for each CEFSM in isolation are already generated by EFTG. The goal of procedure Marking-Backward (Sender,Receiver,M) is to mark all the paths in Sender which contain a transition sending message M to Receiver (i.e., determine the set Pr(C_n)).

In the same way, procedure Marking-Forward(Sender,Receiver,M) marks the paths in Receiver which have transitions receiving message M from Sender (i.e., determine the set Po(C_n)).

```
Procedure Marking-Backward(Sender,Receiver,M)
Begin
 For each test case TC in Sender
  If one or more transitions in TC send M to Receiver Then
    For each unmarked transition T in TC Do
      If T receives an internal message M' and a call to the recursive
      procedure was not made with (Sender of M',Sender, M') Then
         Marking-Bakward(Sender of M', Sender, M')
    Mark all the transitions in TC
End;
```

```
Procedure Marking-Forward(Sender,Receiver,M)
Begin
 For each test case TC in Sender
  If one or more transitions in TC receive M from Sender Then
    For each unmarked transition T in TC Do
      For each internal output statement O in T such that a call to the recursive
      procedure was not made with (Receiver,Receiver of O, O) Do
          Marking-Forward(Receiver,Receiver of O, O)
    Mark all the transitions in TC
End;
```

At the end of this step, all marked transitions will participate in the generation of the partial product.

Step 3: Partial product generation.

Definition 5 . A partial product for C_n is defined as:

$PP(C_n) = C_n \times \{marked\text{–}transitions\}$. $PP(C_n)$ is also called C_n in context.

After the marking process, the partial product for C_n is performed. At each time, among the possible transitions which can be picked, only the marked ones are chosen.

The unmarked transitions do not participate in the partial product computation because they do not influence the machine under test. Procedure ComputePP computes a partial product for a certain CEFSM and is similar to an ordinary algorithm for reachability analysis. The main difference is that among the possible transitions which can be picked, only the marked ones are chosen (according to definition 4 given in Section 2).

Procedures AddGlobalState and AddGlobalTransition add a new global state and a new global transition to the graph of the partial product respectively. Functions Push (resp. Pop) are functions which add (remove) a global state to (from) the stack.

```
Procedure ComputePP(CEFSM M)
Begin
 Create a new global state NewState <s0^(1), s0^(2),..., s0^(k), m_1, m_2,...,m_k>
 AddGlobalState(NewState)
 Push(NewState,Stack)
 While Stack is not empty Do
    CurrentState ← Head(Stack)
    If a marked firable transition T exists in some machine Ci then
      Create a new Global State NewState
      If transition T has an internal Input Then Remove the Input from its FIFO
      If T has internal outputs Then Add the outputs to the corresponding FIFOs
      AddGlobalState(NewState) if it does not exist   , AddGlobalTransition(T)
      Push(NewState)
    Else  Pop(Stack)
 End;
```

Step4: Test case generation for the partial product. After the partial product for C_n is computed, its test cases are generated automatically with EFTG since the result of the product of C_n with the marked transitions is also an EFSM (according to definition 1 presented in Section 2). In that case, we do not have the problem which is inherent to the composition of FSMs: is the product of two FSMs still an FSM? These test cases cover both control and data flow in the partial product and we guarantee the coverage of the "all-definition-uses + all transitions" criterion in the partial product. We should also note that our method is more adapted to test a CEFSM in context when the global system is made of more than two CEFSMs because if the system is made of only 2 CEFSMs, then the partial product for one CEFSM is in general the complete product. In the next section, the process of testing in context each CEFSM in a communicating system is presented.

5 AN EXAMPLE

Consider the example in Figure 1. We have 3 processes B1, B2 and B3 which communicate with each other and with the environment. B1 reads an external input with parameter c and assigns c*c to R in t1. In state s2, if B1 receives input Int1 from B3, it responds with Int2 containing the value of R. R can be modified in t3 before being sent to B3. In t4, if R is even, a message Int3 is sent to B2 with the computed value of L ("Ext" means a message to/ from the environment while "Int" means an internal message; also, "?msg" means an input interaction while "!msg" means an output interaction).

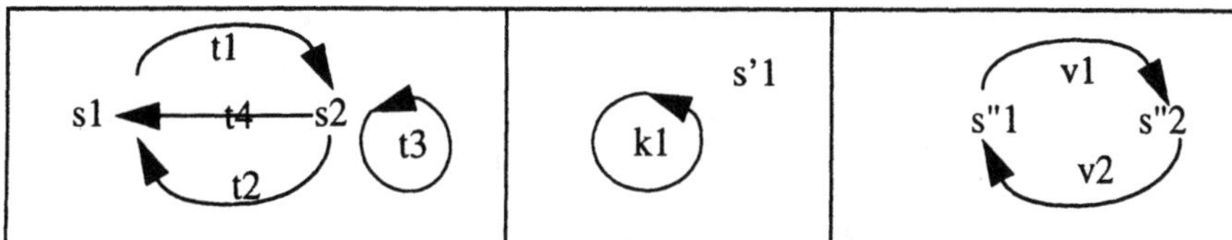

(a) B1	(b) B2	(c) B3
t1: ?Ext1(c) R:=c*c; t3:(R is odd) R:=R+1; t2: ?int1 !Int2(R); t4: (R is even) L:=R/2; !Int3(L);	k1: ?Int3(L) !Ext4(L);	v1: ? Ext2 !Int1 v2: ?Int2(R) !Ext3(R)

FIGURE 1. A system of CEFSMs: B1, B2 and B3.

The next section illustrate the process described in Section 4.

5.1 Test case generation for the partial product for B3

Step 1: Test case generation for each machine in isolation. Table 1 presents the executable test cases generated by EFTG for each CEFSM in isolation (for more detail on this phase, see [Bourhfir 1997]).

Test cases for B1	Test cases for B2	Test cases for B3
• t1, t2 • t1, t3, t2 • t1, t3, t4 • t1, t4	• k1	• v1, v2

TABLE 1. Executable test cases for B1, B2 and B3 in isolation.

Step 2: The marking process. Suppose that we want to generate the partial product for B3. Then, all transitions in B3 are marked. Since v1 sends a message to B1, which can be received by t2 then all transitions in all test cases of B1 which contain t2 are marked (i.e., t1, t2 and t3). Before marking these transitions, we first take a look at the test cases containing t2. t1 receives an external transition and does not receive nor send any internal message. t2 sends Int2 to B3, but since all transitions in B3 are marked, nothing happens. t3 does not contain any internal input or output message. Transition k1 in B2 is not marked because it is not influenced by any transition in B3, nor does it influence any transition in B1.

Step 3: Generation of the partial product for B3. After the Marking process, the partial product for B3 is computed. This latter is shown in Figure 2. It contains 11 transitions and 6 states. For clarity purposes, we kept the old names of the transitions. The detail about each transition can be found in Figure 1.

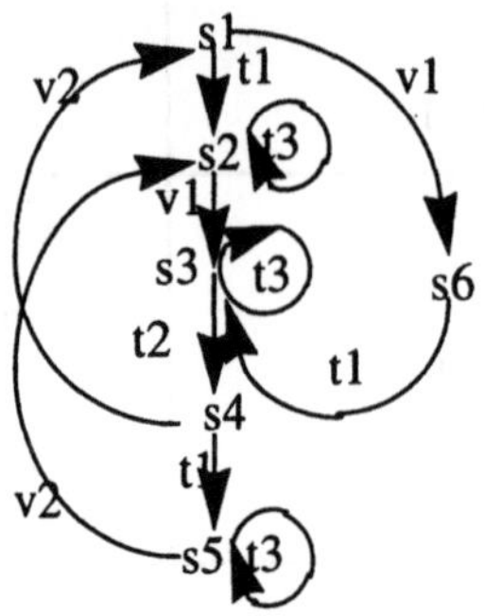

FIGURE 2. partial product for B3.

Step 4: Test cases of the partial product for B3. After the partial product for B3 is computed, its test cases are generated by EFTG. These test cases cover all the definition-uses criterion in the partial product for B3. In term of transition coverage, they cover all transitions in B3, 75% of transitions of B1 and no transition in B2. These test cases are as follow:

1) t1, t3, v1, t3, t2, v2
2) t1, t3, v1, t2, v2
3) t1, v1, t3, t2, v2
4) t1, v1, t2, t1, t3, v2, v1, t2, v2
5) t1, v1, t2, t1, v2, t3, v1, t2, v2
6) t1, v1, t2, t1, v2, v1, t3, t2, v2
7) t1, v1, t2, t1, v2, v1, t2, v2
8) t1, v1, t2, v2
9) v1, t1, t3, t2, v2
10) v1, t1, t2, v2

In the same manner, to compute the partial product for B2, t1, t3, t4 and k1 are marked. This partial product has 7 transitions and 4 states and 6 test cases. In terms of transition coverage, the test cases cover 75% of the transitions in B1, 100% of the transitions in B2 and 0% of the transitions in B3. If we consider both partial products of B2 and B3, we find that they cover 100% of the transitions in each machine as well as the "all-definition-uses" criterion for the partial products for B2 and B3.

5.2 Test case generation for the complete product of B1, B2 and B3

We computed the complete product machine (see Figure 3). This latter has 30 transitions and 12 states and 1483 executable definition-uses paths. This number corresponds to the product of 3 small CEFSMs. For big communicating systems, testing the product machine becomes unpractical. Therefore testing each CEFSM in context may be the only solution for large systems.

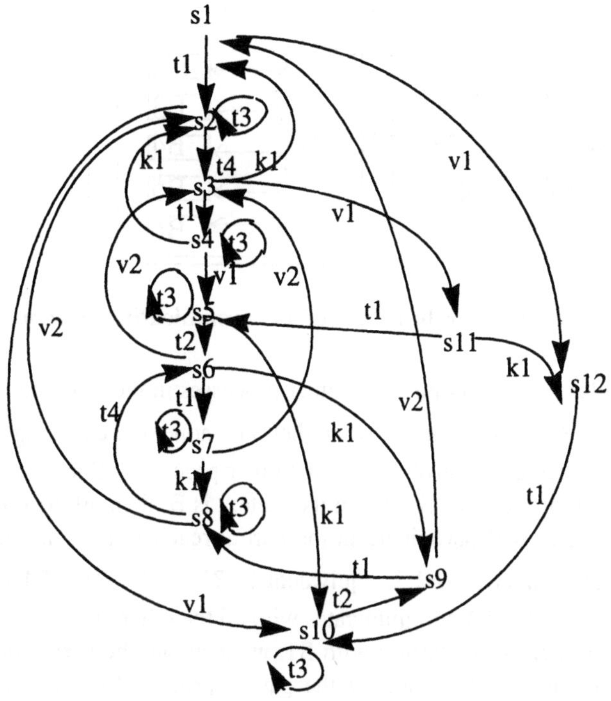

FIGURE 3. The complete product of CEFSMs in Figure 1.

6 A TEST CASE GENERATION ALGORITHM FOR CEFSM SPECIFIED PROTOCOLS

In this section, we will present the general algorithm CEFTG for automatic test generation for CEFSM based systems. This algorithm uses EFTG as well as the algorithm described in Section 4. CEFTG generates executable test cases incrementally by generating test cases for the partial product for each CEFSM until the desired coverage is achieved. The process of generating test cases for the global system may stop after generating the partial product for some CEFSMs (not all of them). In that case, these test cases cover the all definition-uses criterion for the partial product machines (this include all transitions in the CEFSM as well as transitions in other CEFSMs). In term of transition coverage, these test cases cover 100% of the transitions for each CEFSM for which a partial product was computed and p% of the transitions of each other CEFSMs, where $p = (marked-transitions)/(all-transitions)$ for some CEFSM.

First, lets take a look at the example in Figure 4. The example is inspired from an existing communicating system. The global system is made of two blocks A and B. Block A communicates with block B and with the environment, while block B communicates only with block A.

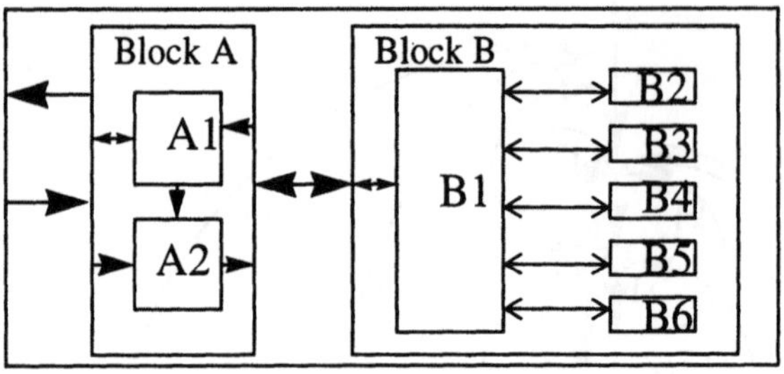

FIGURE 4. Architecture of a communicating system.

In this particular case, B1 controls all other processes in block B and none of the processes Bi $2 \le i \le 6$ is influenced by another, i.e., none of the messages sent by a process does influence the control flow of the other processes. Suppose we want to compute the partial product for B1, when its context is B; then all the transitions in B will be marked and will participate in the partial reachability analysis. In other words, the partial product for B1 is equivalent to $B1 \times B2 \times B3 \times B4 \times B5 \times B6$. This is due to the fact that B1 communicates with all the other processes in block B. To test block B, if instead of starting with B1, we generate the partial products for B2,..,B6 then we can avoid generating the partial product for B1, because after generating the partial product for B6, the test cases for all the partial products cover all the transitions in all processes (even B1). In that case, if our goal is to simply cover each transition in each process without generating a partial product for each machine, then the order in which CEFSMs are chosen is very important. In the next section, we present some metrics which help choosing, among several CEFSMs, the CEFSM which has less interaction with the others.

6.1 Metrics for measuring complexity of communication

In Section 6.2, we present an algorithm which builds progressively the test cases for a communicating system by generating test cases for the partial product for each component machine. Normally, a partial product is generated for each CEFSM. But in some cases, we may not need to do so because the test cases of the partial product for certain machines may cover all transitions in all CEFSMs (in term of transition coverage and not all-definition-uses coverage). For systems having an architecture similar to that of block B in Figure 4, it is more interesting not to generate a partial product for B1 and a method should be used to avoid computing the partial product for the primary block which in this case is the complete product of all machines.

Let M be a set of CEFSMs. We define the following metrics for each CEFSM C.

- Input/Output Count (IOC): The number of internal input and output interactions in C is considered as a metric for measuring the complexity of communication that may take place during the composition of the CEFSMs in M. IOC of C is defined as, IOC(C) = # of input/output messages involving CEFSMs in M (number of internal input and output messages).
- CEFSM Count (CC): The CC for C is the number of CEFSMs which communicate with C.

To prevent the generation of the complete product when the goal is to cover each transition in each CEFSM, we first compute an IOC for each CEFSM. In each iteration, the algorithm will compute the partial product for the machine with the smallest IOC and eventually the smallest CC. After the partial product for a CEFSM is computed, its test cases are generated and the user can stop the execution of the algorithm when the desired coverage is achieved. In that case, for Block B, the execution of the algorithm may stop after the partial products of B2, B3,.., B6 are computed since, by that time, all transitions in B1 are already covered by the generated test cases of the partial products of B2,..,B6. This is particularly interesting if B1 is huge since it communicates with all machines. Also, the architecture of the system presented in Figure 4 is very common, and the method we are presenting in this paper can be very useful for such systems.

In the next section, we present an algorithm for automatic test generation for a CEFSM based system which uses the metrics described above.

6.2 An incremental algorithm for test generation for CEFSM based systems

The next algorithm presents our method to generate test cases for a communicating system. The user can either let the algorithm generate a partial product for each CEFSM and its test cases or stop the execution of the algorithm when the coverage achieved by the test cases for some partial products is satisfactory ("all definition-uses + all transitions" coverage for each partial product and "transition coverage" only of each CEFSM for which a partial product was not computed. After computing the partial product for one or many machines, Compute Coverage determines how many transitions in each machine are covered by the generated test cases (in term of transition coverage). After computing the partial products of all CEFSMs and generating the test cases for them, we verify if some machines are still not completely covered. If this is the case, this means that some transitions in these machines may not be accessible when the communication between the machines is considered (possible specification errors or an observability and controllability problem).

```
Algorithm CEFTG(Nb: the number of CEFSMs)

Begin
  For i = 1 to Nb Do
     Read each CEFSM specification
     Generate Executable test cases for each CEFSM in isolation using EFTG
  i ← 1
  While i ≤ Nb Do
     Choose the CEFSMi with the lowest IOC (and CC) which has not been
     chosen yet
     Marking(CEFSM_i)
     Ai ← ComputePP(CEFSM_i)
     Generate executable test cases for A_i using EFTG
     ComputeCoverage
     If the user is satisfied with the coverage then Break
     Else   i ← i + 1
End;
```

7 RESULT ANALYSIS

First, we would like to mention that all algorithms presented in this paper were implemented in C++ and integrated to EFTG. In the example in Figure 1, we can clearly see that k1 in B2 and t4 in B1 have no influence on B3. In this case, the marking algorithm marked only the transition t1, t2, t3, v1 and v2. Now lets compare the partial product for B3 with the complete product of B1, B2 and B3. Removing k1 from the complete product leads to remove states s8, s9, s10 and s12 and their transitions. Also, removing t4 from the resulting machine leads to remove states s2 and s3 and their transitions. The new resulting machine is nothing else but the partial product for B3 (Figure 2). In other words, the partial product for B3 is the ***projection*** of the complete product on {t1, t2, t3, v1, v2.}. Also, if we consider the test cases of the partial product and the test cases of the complete product, then removing all test cases containing k1 and t4 leads to remove 1475 test cases. To test B3 in context, we only need 10 and not 1483 test cases. The test cases of the partial products for B2 and B3 can be used to test the global system, and we may not need to generate the partial product for B1 which is the product of B1, B2 and B3. It is important to note that the test cases generated by our method cover all the global states and global transitions in each partial product (not global transitions in the complete product) since not all transitions participate in the partial product generation. They also cover all the "definition-uses" paths in each partial product since test case generation is performed by EFTG. Compared to other methods [Lee 1996], our coverage criterion is very strong. In that case, for huge systems, our incremental test generation strategy can be combined with a reduced reachability analysis technique. In Procedure ComputePP, at each step, we can choose a marked transition which has not been tested yet to reduce the size of the partial product. But, the coverage criterion induced by this method is weaker.

Since we divide the problem of generating test cases for a CEFSM based system by incrementally generating test cases for each partial product, we divide also the complexity of the problem by not considering all transitions in all CEFSMs at once. As we use the test cases generated by EFTG for the machines in isolation to determine the preambles and postambles for the transitions in the machine under test (MUT), we may not generate all preambles and postambles for the MUT (because, in the test cases generated by EFTG, we attach the shortest executable preamble to each definition-use path and not all preambles). In that case, the test cases for the partial product for a CEFSM may not be exhaustive. For this reason, we implemented another marking algorithm which uses all the preambles and all the postambles of each state, instead of the test cases, to guide the marking process of the transitions which will participate in the partial reachability analysis.The results obtained on the examples presented in this paper were similar. For systems where transitions may have many preambles and postambles, the second marking algorithm will achieve better coverage. Also, in order to find the shortest executable preamble to one transition, EFTG generates all its executable preambles and chooses the shortest executable one.

For the product in Figure 3, EFTG generates more than 20 000 preambles and postambles. For this reason, we implemented and instrumented the Dijkstra Algorithm [Dijkstra 1959] to find the shortest preambles and the shortest postambles for each state. As the shortest preamble to a certain state may not be executable, we intend to explore this part further in the near future by looking at algorithms which find all shortest paths between two nodes or paths of length k [Monien 1985, Watanabe 1981].

8 CONCLUSION

In this paper, we presented the algorithm CEFTG which generates automatically and incrementally executable test cases for a CEFSM specified protocol by generating test cases for each CEFSM in context. It is important to note that our method computes a "partial product" for each CEFSM with parts of the other CEFSMs. Also, since the method is guided by the test cases generated by EFTG when used to test each CEFSM in isolation, and since in [Bourhfir 1997] we demonstrated on some examples that EFTG generates more executable test cases than other existing methods, we are quite confident that the partial products computed by our method are representative of the behavior of each machine in its context. Our method has many advantages. First, it may take less time and space than the all-at-once reachability analysis. Second, incremental test generation can significantly reduce the effort for re-testing of a large protocol due to a correction or enhancement so that we won't have to re-test the entire system. To find the paths containing transitions triggering (or triggered by) the machine under test, our method uses some of the information already generated by EFTG when used to test the machines in isolation. This method can also detect deadlocks in the CEFSMs and can be used to detect inaccessible transitions, by detecting the transitions which are not covered by the generated test cases. As we mentioned earlier, our method can easily incorporate a reduced reachability analysis technique such as [Lee 1996] which will decrease the size of the partial product for each CEFSM. Finally, since many published methods can only be applied to simple examples or to CFSM based systems, we consider this work as a step towards data-flow testing of real communicating systems.

9 REFERENCES

C. Bourhfir, R. Dssouli, E. Aboulhamid, N. Rico (1997) , "Automatic executable test case generation for EFSM specified protocols", IWTCS, Chapman & Hall pp. 75-90.

Chanson, S. T. and Zhu, J.(1993) A Unified Approach to Protocol Test Sequence Generation. In Proc. IEEE INFOCOM.

E. W. Dijkstra (1959) , "A note on two problems in connection with graphs", Numer. Math. 1 pp. 269-271.

M. G. Gouda, Y. T. Yu (1984), "Protocol Validation by Maximal Progressive Exploration", IEEE Trans. on Comm. Vol. 32, No. 1, January.

Huang, C.M. Lin, Y.C. and Jang, M.Y. (1995) An Executable Protocol Test Sequence Generation Method for EFSM-Specified Protocols (IWPTS), Evry, 4-6 September.

David Lee, Krishan K. Sabnani, David M. Kristol, Sanjoy Paul (1996), "Conformance Testing of Protocols Specified as Communicating Finite State Machines- A Guided Random Walk Based Approach", IEEE Trans. on Comm. Vol. 44, No. 5, May.

Monien, B. (1983) "The complexity of determining paths of length k", Proc. Int. Workshop on Graph Theoretical Concepts in Computer Science, Trauner B Verlag, Linz, 241-251.

A. Petrenko, N. Yevtushenko, G. Bochmann, R. Dssouli (1997), "Testing in context: framework and test derivation", a Special Issue on Protocol Engineering of Computer Communication.

J. Rubin, C. H. West (1982), "An Improved Protocol Validation Technique", Computer Networks, 6, April.

Ural, H. and Yang. B. (1991) A Test Sequence Selection Method for Protocol Testing. IEEE Transactions on Communication, Vol 39, No4, April.

K.Sabnani, A.Dahbura (1985), "A new Technique for Generating Protocol Tests", ACM Comput. Commun. Rev. Vol 15, No 4, September.

Watanabe (1981), "A fast algorithm for finding all shortest paths", Inform. Process. Lett. 13, 1-3.

Weyuker, E.J. and Rapps, S. (1985) Selecting Software Test Data using Data Flow Information. IEEE Transactions on Software Engineering, April.

PART EIGHT

Testing from the Industrial Perspective
Invited Talk

Testing and Test Generation: State of the Art and Future Expectations

Anders Ek, Telelogic AB, Box 4148, S-203 12 Malmö, Sweden, tel. +46 40 17 47 12, fax +46 40 17 47 47, email anders.ek@telelogic.com

Abstract

Formal methods, testing and test generation are in this paper discussed from a pragmatical industrial perspective and in particular as seen from a CASE tool vendors point of view. Since a CASE tool vendor survives by convincing potential customers that they will make money by buying (sometimes expensive!) tools, he needs good sales arguments. So, how do formal methods, testing and test generation fit into this? Essentially the idea is to show that a development process supported by tools based on these concepts is more efficient, giving higher quality to a lower cost, than the currently used process .

The SOMT method provides such a process based on object oriented analysis and formal methods, and the requirements and testing track of this method is the main subject of this paper. As a complement to the method also the necessary tool support is discussed and exemplified with features from the Telelogic Tau tool set.

Keywords

MSC verification, test generation, use case, MSC, SDL, TTCN, UML

1 INTRODUCTION

Testing is an important part of any development project and this is true also for projects using object oriented techniques and an incremental, use case centred development method. In this paper I will discuss a development method with these characteristics. The focus of the discussion will be on the requirements and testing track that is a part of the method. The method is an elaboration of a tool specific method called the SOMT method (Telelogic, 1996) that is aimed at giving an efficient development process for certain classes of applications. The SOMT method is mainly intended for the development of reactive, distributed, real time, embedded and communicating systems. Characteristic for this type of applications is that they are difficult and expensive to test and that they very often are appropriate to design using object oriented methods.

The SOMT method is based on combining the strength of object oriented analysis and use case centred design with formal methods, specification level testing and test case generation. The notations used in this method are UML, SDL, MSC and TTCN but if needed it is easy to generalise the method to other notations provided they have the necessary level of formalism.

In this paper I will discuss both the method itself and the tool support required to use the method. When relevant I will give examples from the Telelogic Tau tool set (Telelogic 1998a). In section 2 I will give an overview of the method, in section 3 discuss various aspects of the requirements and testing track and finally in section 4 comment on the current usage in industry and the expectations we may have on the future development in this area.

2. THE SOMT METHOD

When discussing a development method it is convenient to discuss it in terms of activities and models, where the activities are different tasks that have to be performed and the models are the products (usually documents, source code etc.) that are produced by the activities. Even if the activities are presented here in one specific order this is just for the convenience of description. An actual project, in particular if it is using an incremental and iterative approach, will perform several activities in parallel and also iterate and alternate between different activities. Anyway, the following activities will have to be considered in a development project according to the SOMT method:

- requirement analysis,
- system analysis,
- system design,
- object design, and
- targeting/implementation.

Each of the activities has two different aspects, one architecture aspect that defines the structure and behaviour of the application and one requirements and testing aspect that defines how initial requirements are transformed into tests that are used to verify the correct functionality of the system. An overview of the

SOMT method is illustrated in Figure 1. In this figure the notation has also been assigned to the different. This is however only to give an indication of the most common usage, sometimes other variants are more convenient like using SDL in the system analysis or UML in the object design and the method does not exclude these possibilities.

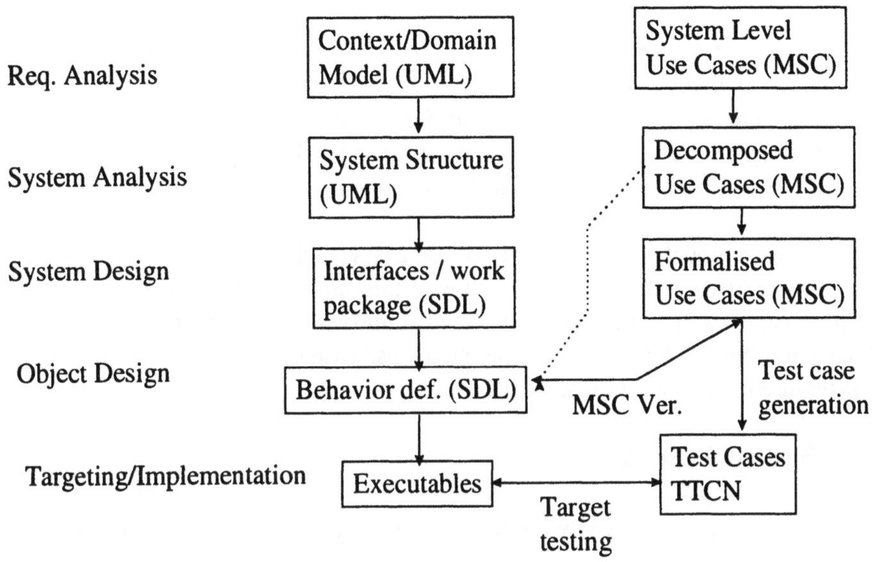

Figure 1. Summary of the SOMT method.

The rest of this section briefly describes the different activities.

2.1 Requirements Analysis

The requirements analysis is the first activity in the SOMT method. It is focused on the external aspect of the application to be built. The purpose of the activity is to capture and analyse the problem domain and the user requirements. For this purpose the system is viewed as a black-box and only the objects and concepts visible on the system boundary and outside the system are modelled.

For our purposes there are two essential models produced:

- a requirements object model including context diagrams and a problem domain model, and

- a use case model.

The requirements object model may be a conventional object model, i.e. one or more diagrams illustrating a number of objects and their relations (including inheritance and aggregation relations) but it may also be a fully described model, including behaviour, of the external view of the system. The purpose of the object model is two-fold:

- to use context diagrams to describe the system and the external actors that interact with the system, and
- to document all the concepts found during the requirements analysis and the relations between them in order to ensure that developers and users have a common understanding of the problem domain.

The use case model consists of a set of use cases, each described using MSCs and sometimes structured text. The purpose of this model is to capture and validate requirements from a user's point of view to make sure that the system will solve the right problem. As we will see below the use case model is also the first part of the requirements/testing track of SOMT.

2.2 System Analysis

In the system analysis activity the focus is on analysing the internal structure of the application. The purpose of the activity is to describe the architecture of the system and identify the objects that are needed to implement the required functionality. The models produced in this activity are:

- an analysis object model, to describe the architecture of the system in terms of objects and subsystems
- an analysis use case model which shows the interaction between the objects and subsystems in each use case.

The analysis object model is a conventional object model that forms the input to the object design phase. The analysts should in this activity be concerned with identifying the objects that must be present in the system, the responsibilities of the different objects and how they interact to fulfil the requirements posed on the system.

The analysis use case model is an elaboration of the use cases from the requirements analysis. In the system analysis the use cases are expressed in terms of the internal structure of the system and defines how the subsystems and objects in the system interact to provide the desired functionality.

2.3 System Design

The focus of the system design is on interfaces, reuse and work packages. The goal is to get an implementation structure that makes it possible for different teams to work on different parts of the system in parallel and to have precise definitions of their interfaces. To accomplish this the architecture track of the system design should product models like

- a design module structure, describing the source modules of the design,
- an architecture definition, containing SDL system/block diagrams, that defines the structure of the resulting application and that gives precise definitions of the static interfaces.

The requirements/testing track that is of more interest for us in this paper is in this activity focused on formalising the use cases to make them precise enough to be possible to verify against the application in a simulated environment. In other words, to prepare for the actual testing task.

2.4 Object Design

The purpose of the object design is to create a complete and formally verified definition of the system. The main work here is of two kinds:

- to do the coding of the behaviour of the objects in the system, and
- to verify that the system fulfils its requirements.

The models used in this activity are mainly SDL diagrams, in particular SDL process graphs that are used to define the behaviour of active objects.

The testing/verification task is done in a simulated environment using SDL simulators for interactive debugging and MSC verification for checking that the system fulfils its requirements as expressed in the formalised use cases.

2.5 Targeting/Implementation

The implementation and testing activities are aimed at producing the final application, i.e. executable software and hardware. The activities in this phase are very much depending on the execution environment of the application but usually include:

- either using an automatic code generation tool to produce the code from the SDL design or manually implementing the SDL design,

- integrating the code to the hardware requirements by means of using real-time operating systems and cross-compilers to generate the executable for the hardware,
- implementing and executing test cases in the target environment based on the design use cases from the system design activity.

From our perspective the most interesting part is the implementation and execution of the test cases. In the SOMT method this is done by generating TTCN test cases from the MSC use cases and executing the test cases against the application to verify that the targeting/implementation was successful.

3 THE TESTING TRACK IN SOMT

3.1 Use Cases

The testing track of the SOMT method starts with a use case oriented requirements capture. The term 'use case' has its origins in object oriented analysis methods and was established in the classical book by Jacobson (1992) but is actually an old idea that in different shapes has been practised for many years. The idea is to focus on the users of the system, called 'actors' in use case terminology. For each actor we analyse in what ways he would like to use the system. These different ways of using the system form the use cases.

So, each use case describes essentially a set of possible sequences of events that take place when one or more actors interact with the system in order to fulfil the purpose of the use case. A use case is thus simply a description, in one format or another, of a certain way to use the system. It has been found to be a very efficient way to capture a user's view of the system and the concept of use cases is now used in a number of object-oriented methods. A difference compared to most other approaches is that SOMT puts some more effort in the formalisation of the use cases to be able to use them for verification purposes during the object design. The formalisation is done using the MSC notation, in particular using the extensions provided in the 1996 version of the MSC standard as will be described in the next section.

3.2 Message Sequence Charts

A message sequence chart (MSC) is a high-level description of the message interaction between system components and their environment. A major advantage of the MSC language is its clear and unambiguous graphical layout which immediately gives an intuitive understanding of the described system behaviour. The syntax and semantics of MSCs are standardised by the ITU-T as recommendation Z.120.

There are various application areas for MSCs but for our purposes in the SOMT method the most important are:

- to define the requirements of a system,
- to check the consistency of SDL specifications using MSC verification, and
- to form as a basis for specification of TTCN test cases .

There have been two versions of the Z.120 recommendation published, one in 1992 and one in 1996. The 1992 version included the conventional sequence diagram notation that made it possible to essentially describe simple sequences of events. This recommendation was quickly implemented in a number of commercial tools but two problems were reported from people using MSCs in industrial applications. Both problems concerned how to handle complexity in the requirements specifications.

The first issue was that when people started using MSCs they quickly found out that they wanted to decompose the MSCs into smaller MSC that could be reused in different ways to cover different cases. In MSC'92 it was suggested to use the condition symbols for this. If one MSC ended with a condition 'Idle' and another started with the same condition it was informally meant that the second could follow the first one. The problem was that it was difficult to get a high level view of the actual use cases and the overview was lost.

The second problem was that the number of MSC that was needed to describe the requirements was found to be very large. To a large extent the reason for this was that all different combinations of exceptional cases and alternative possibilities had to be described in separate MSCs. This lead to either an explosion in the number of MSCs needed or the introduction of informal comments into the MSCs. Most actual users chose the informal comment solution. Unfortunately this made all verification tools useless.

However, both of these problems were solved in the revised Z.120 recommendation issued in 1996. This was mainly accomplished by introducing two new concepts:

- high-level MSC diagrams and
- inline expressions.

The high-level MSC diagrams (HMSCs) solves the overview problem. The HMSCs simply describe how other MSCs (either MSC'92 style MSCs or HMSCs) can be combined to define a more complex behaviour. For example, the HMSC in Figure 2 shows a situation where we first have the MSC 'ConnectionEst' and then one or more 'DataTrans' followed by one 'Disconnect' MSC. Note that when the flow lines between the MSC references branch, this indicates an alternative between the subsequent MSCs.

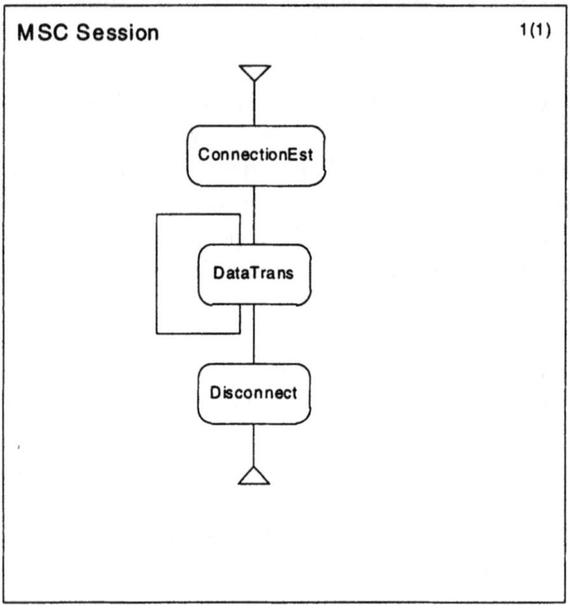

Figure 2. A high-level message sequence chart.

The combinatorial explosion of the number of MSC caused by similar, but not identical MSCs was solved by the addition of inline expressions in MSC'96. The idea is that a number of the messages and other events in an MSC can be framed and defined to be e.g. optional or an alternative to other events. Essentially this gives us a possibility to describe several similar, but slightly different, MSCs in one MSC. The inline expression possibility reduces the number of MSCs that need to be written substantially while they still preserve the intuitive syntax of MSC'92 and the formality of the requirement specification

From the SOMT point of view the new 1996 version of MSC meant that the developers in industry could start capturing the full set of requirements and still remain in the formal MSC notation.

3.3 Decomposing the MSCs

In the requirements analysis the MSCs are usually on an application level of abstraction with one instance axis for each actor and one instance axis for the system to show the external view of the use cases. In the system analysis the MSCs are elaborated and the system instance axis is replaced with instance axis for the different subsystem and/or objects in the system. the purpose is to see how the functionality of the use case is distributed among the subsystems.

3.4 Keeping the MSCs Consistent and Formal

In the system design activity the designers of the application formalises the interfaces between different subsystems. Some typical tasks done is to precisely specify signal lists, signal names, parameters of signals and the structure of data.

In the testing track the main tasks are to keep the use case MSCs consistent with the precisely defined static interfaces and to do a thorough review of the use cases to make sure that they are precise enough both to act as an correct requirement for the implementers of the system and as a basis for testing of the requirements.

This may sound like a trivial activity but it involves quite a lot of work and it is very important if we want to get a smooth development and testing in the rest of the project.

3.5 Design Level Testing and MSC Verification

In the object design activity there are two major tasks to be performed. First of all the details of the application is defined in SDL. This means in practise that the complete dynamic behaviour of the objects in the system ('processes' with SDL terminology) is coded using the state machines, abstract data types and the other concepts of SDL. The result is a complete definition of the application described as a set of communicating SDL processes whose combined behaviour is defined by the SDL run time model. Since SDL includes a precise definition of the run time semantics it is fairly straight forward to build tools that can simulate an SDL system using the predefined run time semantics. The Tau tool set contains both SDT Simulator for debugging SDL systems and tools like the SDT Validator that can automatically explore the state space of the SDL system.

This executable property of SDL is used in the SOMT method for the requirements/testing task to be performed in object design; the verification of the use case MSCs against the SDL specification. Basically this is a kind of testing performed in a simulated environment with the purpose of verifying that the different execution paths described by the MSCs indeed are implemented in the application. This is most efficiently done by simply letting a state space exploration tool perform a search through the state space and in parallel evaluate the MSC seen as a restriction on the possible execution sequences. In the Tau tool set the SDT Validator is designed to do this (See Ek, 1993).

This design/specification level debugging is very convenient in particular for distributed systems where a test in a real target environment would include downloading the application in a distributed testing hardware which is a fairly complex process. Also for real time systems the possibility to test in a simulated environment is very efficient since real time application often involve specific hardware that may not be available until late in the project. By testing in a simulated environment the testing process can start earlier.

3.6 Test Case Generation

However, even if the design is verified in a simulated environment there still is a need to do a real testing in the target environment. To facilitate this the SOMT method uses the TTCN language as test notation for target testing.

The generation of TTCN test cases from the MSC use cases is a problem that has some intricacies that may not be immediately noticed. One is that an MSC use case may, in particular if HMSCs are used, actually be a description of many test cases. The reason is that the MSC may branch and the decision of which branch to choose depends on what aspect of the use case we want to test. This is a very convenient feature in the MSC since it gives a compact definition of the use case and a state space exploration based MSC verifier can deal with this. However it is a problem when trying to test in a target environment and thus it is a complication when mapping to TTCN.

Another problem is that often the MSC use cases doesn't contain a complete description of e.g. all data values to use when sending signals to the application. The MSC verification can handle this but the data values are needed when doing real testing on target.

To overcome these problems a flattening and further formalisation of the MSC use cases are needed before moving to TTCN. In the Tau environment this can be accomplished by using the MSC verification to produce test purpose MSCs as a side effect of verifying the use case MSC. The test purpose MSCs are simple MSCs that contain no branching but sufficient details to act as input to the TTCN generation. This is illustrated in Figure 4.

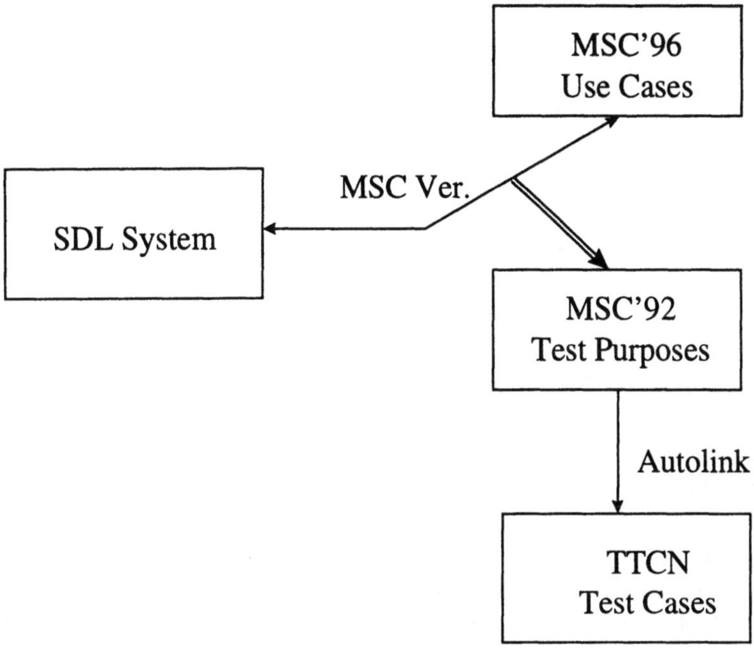

Figure 4. Generation of test purpose MSCs and TTCN test cases based on use case MSCs.

Another complexity arises from the circumstances under which the generated TTCN test cases should be used. If they are only to be used for an in-house automatic verification step, where the generated tests are executed on the run time platform in order to check that the adaptation of generated code works as expected, then the quality in terms of readability of the test suite etc. is not too important. However, if parts of the SDL system are implemented by hand or maybe even by a different organisation than the one that develops the SDL and TTCN the situation is different. Then the TTCN specification will be official and thoroughly inspected by people. In this situation it is essential that the TTCN test suite is nicely structured and readable. To a large extent the readability of a test suite depends on minor details like the names of constraints and parameters. Unfortunately this structuring and naming is very difficult or impossible to generate automatically. Fortunately, quite a lot of effort has been put into solving this by allowing the user to guide and control the automatic generation of test cases including e.g. controlling the naming and parametrisation of constraints.

The Autolink feature in the SDT Validator (see Ek *et al* 1997, Schmitt *et al* 1997 and Telelogic 1998b) is the result of a joint project between Telelogic and the Institute for Telematics in Luebeck. This project has to a large extent been focused on these practical aspects of the test generation problem. The basis of the Autolink test generation features is the state space exploration possibilities in the SDT Validator but a substantial amount of the actual work has been to develop methods that solves the more practical problems. The result is a product that now is used to produce TTCN test cases to be published as official ETSI recommendations (see Schmitt *et al* 1998).

4 STATUS AND EXPECTATIONS IN INDUSTRY

The SOMT method has been the recommended methodology guideline for people using the Tau tools since it first was published in Telelogic (1996) and the combined usage of UML, SDL, MSC and TTCN is now spreading in industry.

SDL is now an accepted development notation in the telecom part of industry where many applications have been developed using automatic code generation from SDL designs. It is also used in standardisation where SDL is used as a specification notation, e.g. the INAP specification done by ETSI is specified in SDL. Nowadays, the tool support for SDL based development and code generation from SDL are very good and both our Telelogic Tau tools and other tools like Verilogs ObjectGeode are now successfully competing on the CASE tools market.

The combination of UML for analysis and SDL for design is starting to get used but is not really wide spread yet, even though predecessors of UML and object oriented analysis has been used for many years as a complement to SDL in the early stages of development. However, since UML now is standardised and is replacing older techniques like OMT and Booch I would expect the usage of UML to increase very rapidly in the future.

MSCs and similar notations have been used for many years in industry to capture requirements and is an established method. Traditionally however MSC has been used in a fairly informal way making it difficult to use it directly in tools for verifying consistency with the design. To a large extent this is the case because the first formal version of MSC as standardised in 1992 did suffer from a number of weaknesses compared to the informal versions used in industry. As a consequence MSC is heavily used but today mainly to informally specify the requirements and also often extended with informal comments to overcome the limitations in MSC'92. The typical SDL developer of today would use the MSCs as input to a manual simulation task. He would take the MSC and manually run the SDL simulator according to the MSC requirements, creating simulation scripts that can be reused for regression testing. This is not a bad strategy but involves a substantial amount of manual work and, which may be a more severe problem, it makes it necessary to maintain two levels of manually created and maintained descriptions. Both the requirement MSCs and the simulation scripts need to be maintained. As the new 1996 version of MSC is starting to get tool support and acceptance in industry this situation has started to change and MSC verification is beginning to be used as an industrial testing technique (see e.g. Ek *et al*, 1997).

On the target testing market most major telecom suppliers today have had their own proprietary test notation and test environment. However, as more and more commercially available test platforms based on TTCN are released and new test specifications using TTCN are published we have at Telelogic seen a clear trend towards TTCN as a platform for testing tools. This has raised the interest for automatic generation of TTCN code considerably since there is a very strong request from our customers to try to reduce the manual work needed for developing tests.

To summarise the industrial perspective it is, as always, driven by commercial considerations. To make money from new products they have to be delivered on time, at reasonable price and quality. All of these aspects lead to a need to streamline the development process and to use tool support to automate as much as possible of the necessary task in the development process. It pays off to let the engineers do engineering and use computers for routine tasks.

As a consequence all new CASE tools features that promise to cut down the manual development costs are very interesting for the development industry. This is good news for example for test generation tool developers, since the test generation tools definitely match this description. However, there are two challenges facing the introduction of the new tools:

- They must be able to handle industrial size problems.
- They must fit into the development processes used in industry.

Both of these problems are difficult, but with new commercial tools like the Telelogic Tau tools that are able to handle real applications, but still are based on formal techniques, the first problem is beginning to get a solution, even if there of course still is some potential for improvement.

The second problem, i.e. to change the way of working for a development department with maybe several hundred developers, a large design base of existing code and several already released product versions is often an even more challenging problem. Fortunately, from a CASE tool vendors point of view, the development departments are also faced with tough requirements from their respective customers. They need to shorten their development time and increase their productivity with a maintained high quality. This is usually not possible without changing the work process and in fact the entire CMM and quality certification movement is targeted on making the development processes visible so they can be enhanced and made more efficient. This gives CASE tools that try to automate e.g. the testing, a very good opportunity to be introduced as part of a more streamlined development process.

So, the need and the market opportunity for design level testing, test generation and efficient test environments are here and it is up the tool vendors and testing community to face the challenges and provide the solutions.

6 REFERENCES

Ek, A. (1993), Verifying Message Sequence Charts with the SDT Validator. In *Proc. 6th SDL Forum*, Darmstadt, 1993, North-Holland.

Ek, A., Grabowski, J., Hogrefe, D., Jerome, R., Koch, B. and Schmitt, M. (1997) Towards the industrial use of validation techniques and automatic test generation methods for SDL specifications, In *Proc. 8th SDL Forum*, Evry, France, Elsevier.

Jacobson, I. et al (1992), Object-Oriented Software Engineering, Addison-Wesley.

OMG 1997, *The Unified Modelling Language Version 1.1*, Object Management Group.

Schmitt, M., Koch, B., Grabowski, J., Hogrefe, D. (1997) Autolink - a tool for automatic and semi-automatic test generation. *In Proceeding of the Seventh GI/ITG Technical Meeting on Formal Description Techniques for Distributed Systems*, Berlin, June 1997.

Schmitt, M., Koch, B., Grabowski, J., Ek, A., Hogrefe, D. (1998) Autolink - Putting test generation into practise, In *Proc IWTCS 98*, Tomsk, Russia, Chapman & Hall

Telelogic (1996), Tau methodology guidelines part 1 - The SOMT method, *Telelogic Tau 3.1 Documentation*, Telelogic.

Telelogic (1998a), *Telelogic Tau 3.3 Documentation,* Telelogic.

Telelogic (1998b), Validating an SDL System, *Telelogic Tau 3.3 Documentation,* Telelogic.

7 BIOGRAPHY

Anders Ek has studied computer science and engineering at Lund Institute of Technology in Sweden. After his graduation in 1986 he started working at Telia, the national Swedish telecom operator. At Telia he did research on formal methods and associated development methodology. Since 1992 he has been working with development of tools and methods at Telelogic, a commercial CASE tool vendor providing tools based on formal description techniques.

Interoperability Test Suite Derivation for the ATM/B-ISDN Signaling Protocol

Jaehwi Shin and Sungwon Kang
Korea Telecom Research & Development Group
Sochogu Umyundong 17, Seoul 137-792, Korea
e-mail: {jhshin, kangsw}@sava.kotel.co.kr

Abstract

In this paper, we propose a test derivation method suitable for testing interoperability for the class of communication protocols like the ATM/B-ISDN signaling protocol. For this, we begin by defining the notion of interoperability test case. Next we select an effective interoperability test architecture by carefully comparing advantages and disadvantages of three candidate test architectures. Then we present an interoperability test suite derivation method based on the definition of interoperability test case. For that, we give an algorithm for deriving from a given pair of FSMs a composite FSM and an interoperability test suite in parallel. In addition, occurrence orderings of output messages for all interoperability test cases are analyzed and their regularity is exploited to reduce the length of interoperability test cases.

Keywords

Interoperability testing, Test architecture, Signaling protocol, Test optimization

1 INTRODUCTION

Conformance testing that checks whether an implementation is correct with respect to the relevant standards has limitation in ensuring interoperability. Thus, even a pair of two conforming implementations can fail to interoperate [RafC 90][APRS 93]. Two main causes of non-interoperation of conforming implementations are ambiguity of protocol standards and incompleteness of conformance testing. Thorough validation of standards is necessary to prevent the former cause of non-interoperation whereas some sort of direct testing of interoperation is necessary to overcome the latter cause of non-interoperation.

There are some research work on interoperability testing done in the past. [RafC 90] deals with interoperability test suite generation based on reachability analysis. However, it is not based on a rigorous definition of interoperability and considers only the case where lower testers exist between two Implementations Under Test (IUTs). [VerB 93] expounds experiences with interoperability testing of FTAM protocol that uses only a single tester between two IUTs and thus has limited capability. [APRS 93] derives conformance test suite and interoperability test suite separately and later manually combines them to reduce the number of conformance test cases. [KanK 97] develops methods for systematically dealing with symmetric communication protocols but uses an interoperability test architecture that does not observe the interface behavior between two IUTs.

In this paper, we propose a test derivation method suitable for interoperability testing of the class of communication protocols like ATM/B-ISDN signaling protocol. Although there is no consistent agreement on definition of interoperability and methodology of interoperability test derivation, for systematic derivation of interoperability test suite it is essential to first state what precisely is meant by interoperability testing. Thus, we first discuss the meaning of interoperability testing and the conditions for a test case to be an interoperability test case.

Next, we investigate interoperability test architectures. By carefully comparing advantages and disadvantages for three candidate interoperability test architectures, we single out among them the most effective test architecture. The selected test architecture has testers for the sides of IUTs facing the external environment and a monitor between two IUTs so that it can check the internal interface between the two IUTs.

Next, we present an interoperability test suite derivation method based on the definition of interoperability testing. For that, we give an algorithm for deriving from a given pair of Finite State Machines(FSMs) a composite FSM and an interoperability test suite in parallel. By applying the algorithm to the ATM signaling protocol as defined by ATM Forum UNI 3.1 specification[AF UNI] and ATM Forum PNNI specification[AF PNNI1][AF PNNI2], 26 interoperability test cases are derived.

Since communication protocols have multiple interfaces in general, when messages of a test case are output concurrently at interfaces, it is difficult to capture correct orderings efficiently in test case description. Thus, we analyze the occurrence orderings of output messages for all derived interoperability test cases and use the regularity to reduce the length of interoperability test cases.

This paper is organized as follows: Section 2 introduces ATM signaling protocol defined by UNI 3.1 and PNNI and describes FSMs for them. In Section 3, we state the conditions for a test case to constitute interoperability test case and classify interoperability test cases according to their message interaction patterns. In Section 4, we consider three test architectures for interoperability testing of ATM signaling protocol and select the most appropriate test architecture by comparing their merits and demerits. In Section 5, based on the definition of interoperability test case and the selected test architecture, we describe a method for deriving interoperability test suite from specifications given in FSM model. In Section 6, we show the application results. Section 7 concludes the paper.

2 OVERVIEW OF THE ATM SIGNALING PROTOCOL

Switch as equipment implementing ATM signaling protocol can be classified into local switch that handles subscriber call and transit switch that performs call transfer function. Thus, there are three interconnection combinations for a pair of switches: i.e., (local switch, local switch), (local switch, transit switch), and (transit switch, transit switch).

Though a respective interoperability test suite can be obtained for each combination using the same method to be shown later in this paper, we only consider interoperation of two local switches.

In local switches, ATM call progresses through both User-Network Interface(UNI) and Network-Network Interface(NNI). We consider ATM Forum UNI 3.1 Specification[AF UNI] and ATM Forum PNNI Specification[AF PNNI1][AF PNNI2] for UNI and NNI specifications for the local switch. The UNI 3.1 specification provides functions and procedures for the ATM user and the ATM network to access each other. The PNNI specification provides functions and procedures to establish and clear calls through ATM networks.

Among the messages for UNI 3.1 signaling layer, SETUP, CALL PROCEEDING, CONNECT, and CONNECT ACKNOWLEDGEMENT messages are used to establish an ATM call and RELEASE and RELEASE COMPLETE messages are used to clear the ATM call. Figure 1 is the FSM model of the Signaling layer of UNI 3.1 specification.

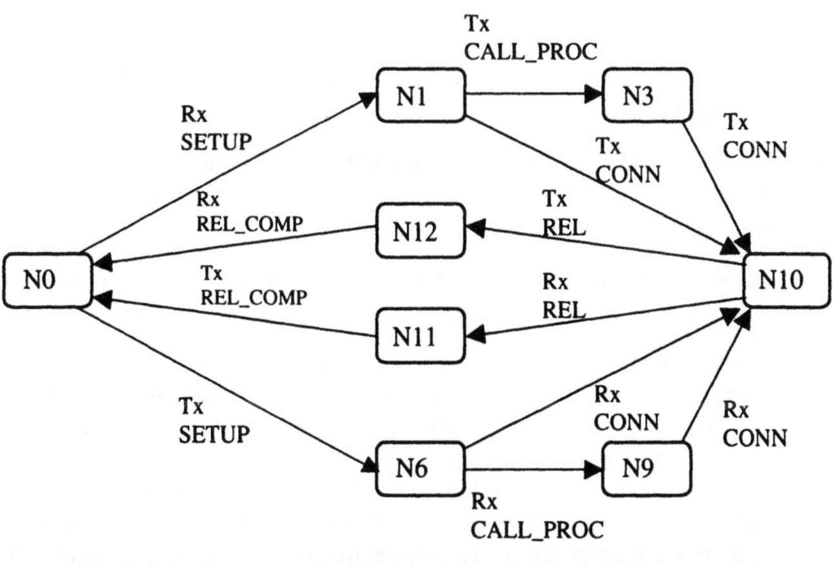

Figure 1 FSM model for UNI 3.1 network-side signaling protocol.

Among the messages for PNNI signaling layer, SETUP, CALL PROCEEDING, ALERTING and CONNECT messages are used to establish ATM calls and RELEASE and RELEASE COMPLETE messages are used to clear ATM calls. Figure 2 is the FSM model of the Signaling layer of PNNI specification.

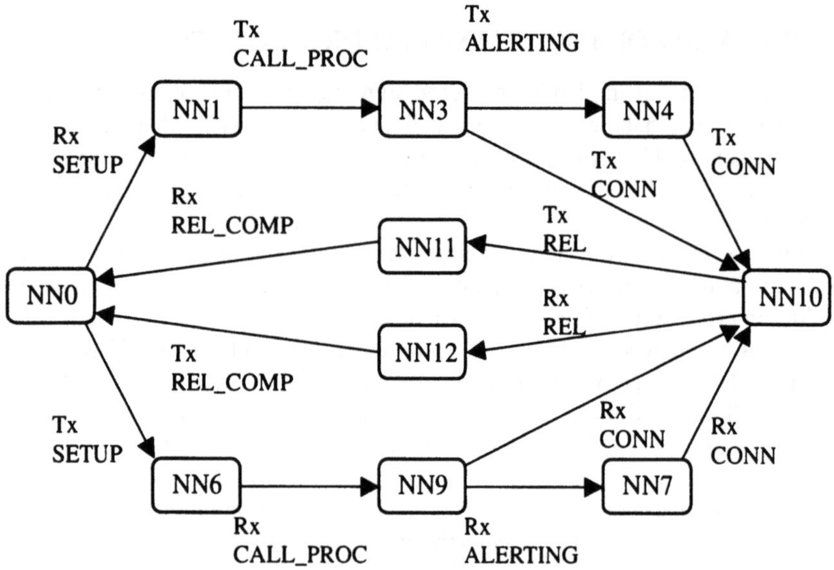

Figure 2 FSM model for PNNI signaling protocol.

3 DEFINITION OF INTEROPERABILITY TESTING AND TYPES OF INTEROPERABILITY TEST CASES

As already mentioned in Introduction, there is no established consistent framework for interoperability testing either in academy or in industry. ISO and ETSI define interoperability briefly as follows:

·ISO/IEC JTC1 DTR-10000[ISO DTR10000]: The ability of two or more systems to exchange information and to mutually use the information that has been exchanged.

·ETSI ETR 130[ETSI ETR130]: Protocol Interoperability: the ability of a *Distributed System* to interchange PDUs via the *Communication Platform.*

A common characteristic of interoperability testing that the above definitions point to is as follows: Compared to the conformance testing that checks whether an implementation comforms to its specification, *interoperability testing* checks two interconnected equipment for correct operation. That is, interoperability testing checks correct response of two IUTs when an external input is applied to the interconnected IUTs whereas conformance testing checks for correct response of IUT when an input is applied to it. This distinction can be clearly seen with Figure 3 and Figure 4. The viewpoint of regarding the response of implementations for a single input as constituting a test case is called 'Single Stimulus Principle'[AraS 92]. It lets us have the most fine-grained test cases and has been adopted in [LuBP 94] and [KanK 97].

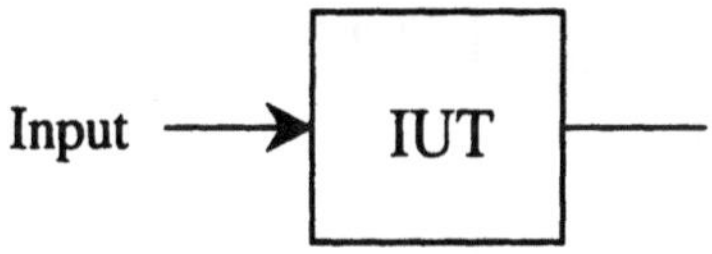

Figure 3 Conceptual diagram of conformance testing.

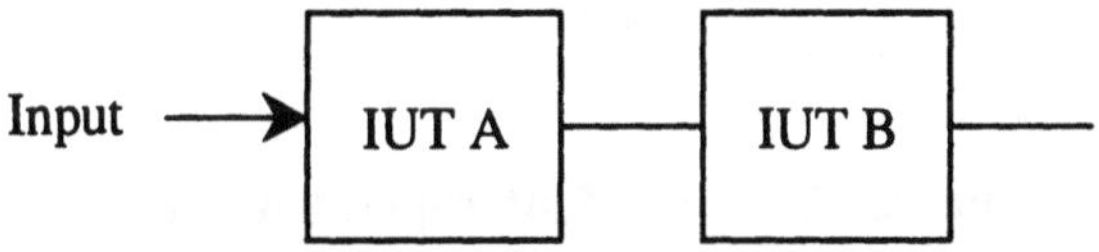

(a) Interconnection of two IUTs and application of an input

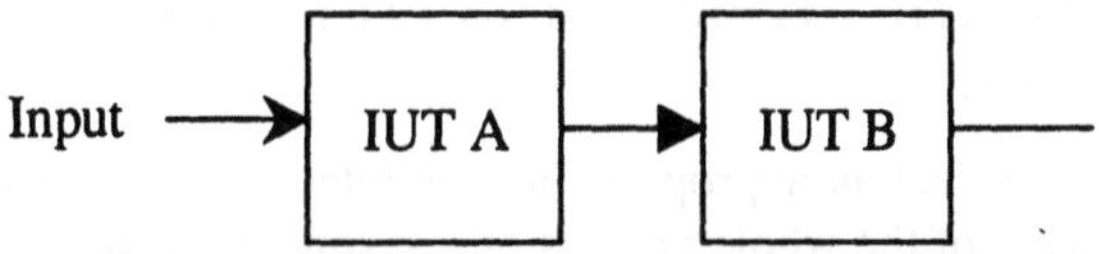

(b) Minimal requirement for an interoperability test case

Figure 4 Conceptual diagram of interoperability testing.

On the other hand, not all applications of an external input message lead to an interoperability test case. According to the above definition of interoperability, an *interoperability test case* should (1)check whether two IUTs exchange information and (2)use the information that has been exchanged. When an external input is applied to the IUT A as depicted in Figure 4(a), at least one message must be transferred from IUT A to IUT B in order to meet the requirements (1) and (2).[1] Therefore, in this paper, we consider a sequence of interactions as one interoperability test case if the interactions begin at a stable state and end at a stable state and satisfies Figure 4(b) when an external input is applied to a system composed of two IUTs.

There can be various patterns of message exchanges that satisfy Figure 4(b). Analyzing ATM signaling protocol composed of UNI 3.1 and PNNI, we can classify all possible patterns of such interoperability test cases as in Figure 5. This classification is used in Section 6 to minimize the length of test cases.

In Figures 3, 4 and 5, quadrangle arrows indicate external input messages and triangle arrows indicate output messages.

1 In this section, to simplify discussion, we consider only the cases where external inputs are applied to IUT A.

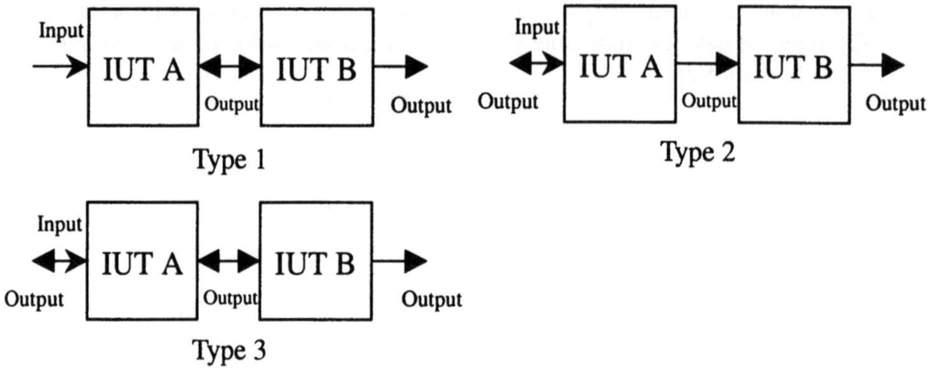

Figure 5 Type classification of interoperability test cases.

4 SELECTION OF THE INTEROPERABILITY TEST ARCHITECTURE

It is very important to select an appropriate test architecture since test architecture affects the effectiveness of the whole process of test suite derivation and efficiency of the resulting test suite. For this, we consider as candidates the three test architectures in Figure 6. The arrowhead in the figure indicates functionality. So a double-headed arrow represents a *Point of Control and Observation*(PCO) and indicates that the tester has both observability and controllability. A single-headed arrow as in Figure 6(b) represents a *Point of Observation* (PO) and indicates that the tester has only observability and hence is a monitor.

Now, we consider interoperability test case execution under each of the three test architectures. We select a Type 1 test case for the purpose of illustration.

- Test Architecture I: Test starts by Tester A sending a message to IUT A. Then, IUT A sends a message to IUT B and Tester C only reads the message. As this is a Type 1 test case, IUT B sends messages to IUT A and Tester B, and these messages are read by Tester C and Tester B, respectively. After the message exchange finishes, Tester A, Tester B, and Tester C send STATUS ENQUIRY messages to verify the reached states of IUT A and IUT B. In particular, Tester C sends STATUS ENQUIRY in two directions, one to IUT A and the other to IUT B.
- Test Architecture II: Test starts by Tester A sending a message to IUT A. Then, IUT A sends a message to IUT B and Monitor C only reads the message. As this is a Type 1 test case, IUT B sends messages to IUT A and Tester B, and these messages are read by Monitor C and Tester B, respectively. After the message exchange finishes, Tester A and Tester B send STATUS ENQUIRY messages to verify the reached states of IUT A and IUT B. Unlike Test Architecture I, Monitor C does not send a STATUS ENQUIRY.

- Test Architecture III: Test starts by Tester A sending a message to IUT A. Then, IUT A sends a message to IUT B but this message is not read. IUT B sends messages to IUT A and Tester B, and only the message sent to Tester B is read. After the message exchange finishes, Tester A and Tester B send STATUS ENQUIRY messages to verify the reached states of IUT A and IUT B. In this test architecture, interface C is not examined at all.

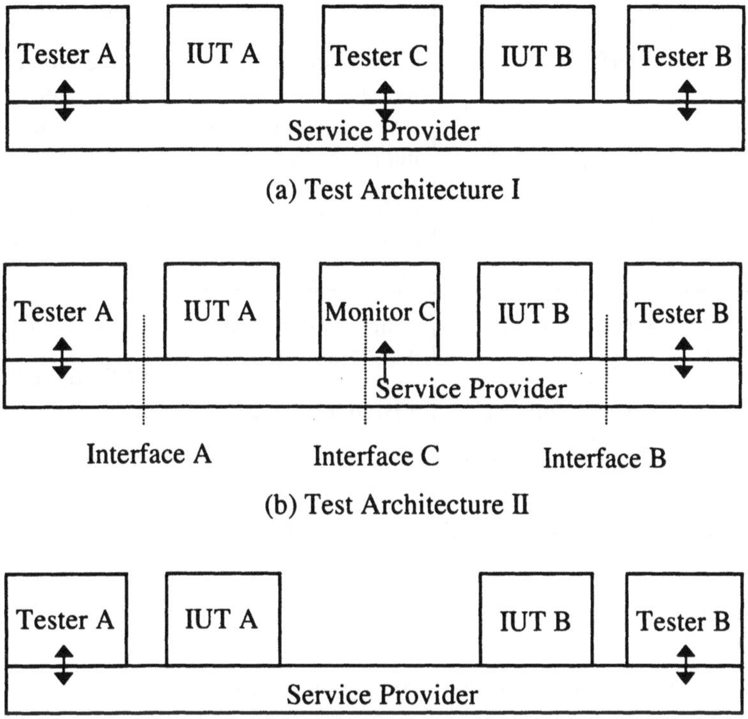

(a) Test Architecture I

(b) Test Architecture II

(c) Test Architecture III

Figure 6 Interoperability test architectures.

Based on the above test execution outlines, comparison results for the three test architectures can be summarized as in Table 1. We consider which test architecture is most appropriate for interoperability testing of local switch ATM signaling protocol. First, Test Architecture III has the lowest analytic power and cannot directly analyze behavior at PNNI. Although Test Architecture I is most powerful, the structure of derived test cases are also most complex, which makes it difficult to adopt it for practical use. Therefore, we select Test Architecture II for interoperability test suite derivation of the ATM signaling protocol.[2]

2 Although Test Architecture III has limited error detection capability, Architecture III is very efficient when test resource(such as the number of links, the number of processors, etc) is limited or handling of the link between IUT A and IUT B is difficult.

Table 1 Comparison of interoperability test architectures

	Test Architecture I	Test Architecture II	Test Architecture III
Characteristics	• Generic test architecture • Use Tester at interface C • Detection of both message error and state transfer error at interface A, B and C	• Use Monitor instead of Tester at interface C. • Detection of both message error and state transfer error at interface A and B • Detection of message error at interface C	• Use neither Tester nor Monitor at interface C • Detection of both message error and state transfer error at interface A and B • Errors at interface C are not detected
Merits and Demerits	• Most difficult to realize • Maximum test coverage	• Simpler to realize than Test Architecture I • High test coverage	• Simplest to realize • Minimum test coverage

5 DERIVATION OF INTEROPERABILITY TEST SUITE

When two IUTs (IUT A and IUT B) implementing specifications A and B are tested with Test Architecture II, an interoperability test suite can be derived as follows.

Interoperability Test Suite Derivation Procedure

Step 1. Derivation of FSM M_A and M_B for the control flow of the specifications A and B.

Step 2. Composition of FSM M_A and M_B.

Step 3. Derivation of interoperability test case skeletons by selecting among all the transitions of the composite FSM only the transitions that satisfy the definition of interoperability test.

Step 4. Derivation of interoperability test cases by parameterization of interoperability test case skeletons. (One or more interoperability test cases can be obtained by parameterization of one skeleton.)

We define FSM model in Section 5.1 and present an Algorithm for Step 2 in Section 5.2.

5.1. FSM Model

We mentioned the necessity of representing a protocol specification with an FSM to formally describe the derivation process of composite FSM. Let M be the FSM for the protocol specification of an IUT. Then M can be represented as follows:

Definition 1 A FSM is a 5-tuple $<\mathbf{S}, S_0, \mathbf{L_{in}}, \mathbf{L_{out}}, \mathbf{Tr}>$ where:

(1) $\mathbf{S}=\{S_0, S_1, ..., S_{n-1}\}$ is a set of states of M,

(2) $S_0 \in \mathbf{S}$ is the initial state,

(3) $\mathbf{L_{in}}=\{v_1, v_2, ..., v_m\}$ is a set of input symbols,

(4) $\mathbf{L_{out}}=\{u_1, u_2, ..., u_k\}$ is a set of output symbols, and

(5) $\mathbf{Tr} \subseteq \{S\text{—}v/U\rightarrow S' \mid S,S' \in \mathbf{S} \wedge v \in \mathbf{L_{in}} \wedge U \subseteq \mathbf{L_{out}}\}$ is a set of transitions.

Bold letters in Definition 1 represent sets. " S—v/U→S' " describes a transition where S, S', v, and U represent, respectively, starting stable state, the final stable state, an input symbol and a set of output symbols.

Let Π be the composite FSM describing the combined behavior of two FSMs in Test Architecture II of Figure 6(b). Let M_A and M_B be the FSMs describing specification A and B, respectively. In communication protocols, it is usually the case that at most two messages are output in response to a stimuli from the external environment. Thus we assume that, for M_A and M_B, U in Definition 1 consists of at most two output symbols, each for different interfaces. Then Π can be defined as follows:

Definition 2 The composite FSM Π is a 5-tuple $< \mathbf{S_\Pi}, S_{\Pi,0}, \mathbf{L_{\Pi,in}}, \mathbf{L_{\Pi,out}}, \mathbf{Tr_\Pi}>$ where:

(1) $\mathbf{S_\Pi}=\{S_{\Pi,0}, S_{\Pi,1}, ..., S_{\Pi,n-1}\}$ is a set of global state, and $S_{\Pi,i}=(S_A, S_B)$, where S_A and S_B are states of M_A and M_B respectively.

(2) $S_{\Pi,0} \in \mathbf{S_\Pi}$ is the initial state.

(3) $\mathbf{L_{\Pi,in}}=\{v_{\Pi1}, v_{\Pi2}, ..., v_{\Pi m}\}$ is a set of input symbols. It consists of input messages for the interface A and B.

(4) $\mathbf{L_{\Pi,out}}=\{U_{\Pi1}, U_{\Pi2}, ..., U_{\Pi k}\}$ is a set of output symbols. Each element of $\mathbf{L_{\Pi,out}}$ is of the form [u1, u2, u3, u4], where u1, u2$\in \mathbf{L_{out,A}}$ and u3, u4$\in \mathbf{L_{out,B}}$. In more detail, u1, u2, u3, and u4 represent messages transmitted by M_A via interface A, M_A via interface C, M_B via interface C, and by M_B via interface B.

(5) $\mathbf{Tr_\Pi} \subseteq \{S_\Pi\text{—}v_\Pi/U_\Pi\rightarrow S_\Pi' \mid S_\Pi,S_\Pi' \in \mathbf{S_\Pi} \wedge v_\Pi \in \mathbf{L_{\Pi,in}} \wedge U_\Pi \in \mathbf{L_{\Pi,out}}\}$ is a set of transitions.

5.2. Interoperability Test Suite Derivation Algorithm

Given FSMs M_A and M_B describing specifications A and B, respectively, the composite FSM Π and the interoperability test suite IOPTS can be derived in parallel by the following Algorithm:

Algorithm: (Interoperability Test Suite Derivation)

Input: FSMs $M_A=<\mathbf{S_A}, S_{A,0}, \mathbf{L_{in,A}}, \mathbf{L_{out,A}}, \mathbf{Tr_A}>$ and $M_B=<\mathbf{S_B}, S_{B,0}, \mathbf{L_{in,B}}, \mathbf{L_{out,B}}, \mathbf{Tr_B}>$. Let $\mathbf{L_{in,A,E}}$ be the subset of $\mathbf{L_{in,A}}$ consisting of messages on interface A, and $\mathbf{L_{in,B,E}}$ be the subset of $\mathbf{L}_{in,B}$ consisting of messages of interface B.

Output: FSM Π=<$\mathbf{S}_\Pi$, $S_{\Pi,0}$, $\mathbf{L}_{\Pi,in}$, $\mathbf{L}_{\Pi,out}$, $\mathbf{Tr}_\Pi$> and Interoperability Test Suite **IOPTS** $\subseteq$ $\mathbf{Tr}_\Pi$.

```
L_Π,in:=L_in,A,E ∪ L_in,B,E
S_Π,0:=(S_A,0,S_B,0)
S_Π:={S_Π,0}
Tr_Π:={}
IOPTS:={}
NEW:=S_Π
while NEW ≠ ∅ do begin
        gs_i:=choose-any(NEW)
        NEW:= NEW − {gs_i}
        Input := L_Π,in
        while Input ≠ ∅ do begin
                v:=choose-any(Input)
                Input :=Input − {v}
                U_out:=[ , , , ]
                gs_f:=state(gs_i|v)
                U_out:=output(gs_i|v)
                L_Π,out:= L_Π,out ∪ {U_out}
                Tr_Π:=Tr_Π ∪ {gs_i — v/U_out → gs_f}
                if (gs_f ∉ S_Π)
                then begin
                        S_Π:=S_Π ∪ {gs_f}
                        NEW:=NEW ∪ {gs_f}
                end
                if (U_out[2] ≠ NULL ∨ U_out[3] ≠ NULL)
                then
                    IOPTS:=IOPTS ∪ {gs_i — v/U_out → gs_f}
        end while
end while
```

In the algorithm, choose-any(A) is a function that chooses an arbitrary element of A, and state(gs_i|v) and output(gs_i|v) represent, respectively, the final stable state and the output messages when an input v is applied to a stable state gs_i. U_out consisting of output messages is a four dimensional vector such that the meaning of each element is the same as that of the element of $\mathbf{L}_{\Pi,out}$ in Definition 2.

The transition set $\mathbf{Tr}_\Pi$ obtained by the Algorithm includes the initial stable state before a transition occurs, input at that state, output by the input and the final stable state reached by the transition. However, $\mathbf{Tr}_\Pi$ contains conformance test cases in addition to interoperability test cases. **IOPTS** is the pure interoperability test suite without conformance test cases which have been filtered out by applying the second if-clause.

6 APPLICATION TO THE ATM SIGNALING PROTOCOL

For Step 1 of the interoperability test suite derivation in Section 5, we derived the FSM in Figure 7 from the two FSMs in Figures 1 and 2 that describes the entire behavior of ATM signaling protocol.[3] [4]

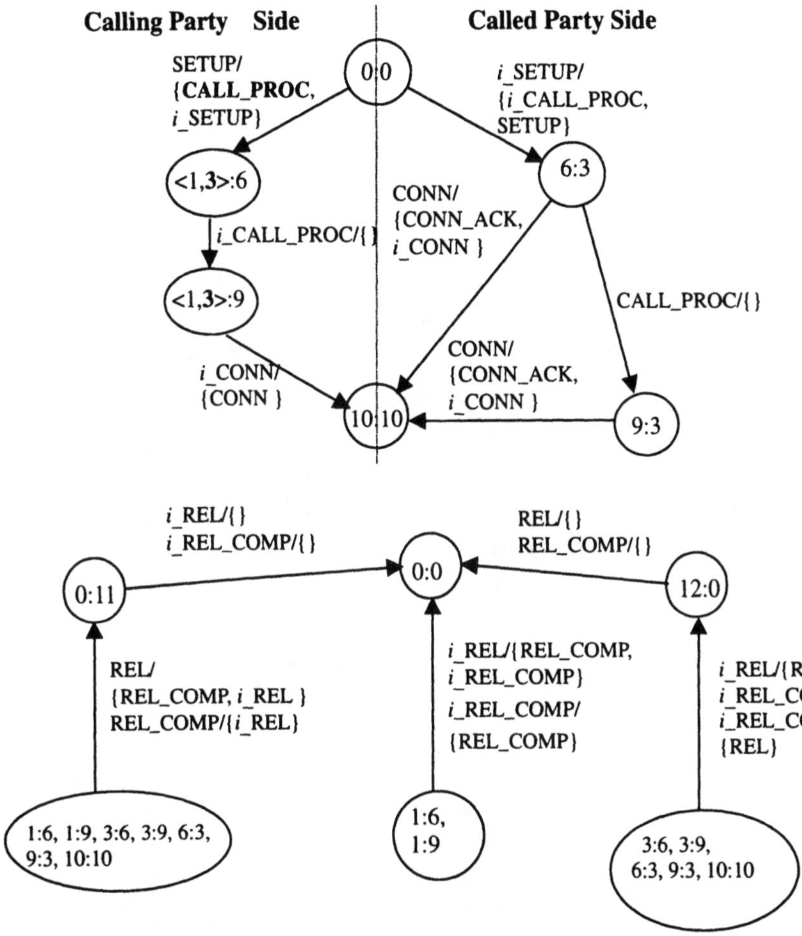

Figure 7 FSM model for the ATM signaling protocol.

Figure 8 depicts composite FSM obtained by applying the Algorithm to the FSM in Figure 7 where $M_A = M_B$. In Figure 8, a global state, for example, (3;9,3;6)

---

[3] Specification for the entire behavior of the signaling protocol for the ATM local switches has not been published. Thus Figure 7 need be inferred from UNI 3.1 specification and PNNI specification.

[4] Sending of CALL_PROC in response to SETUP by the UNI 3.1 network-side is an optional feature. It is assumed here that IUT A and IUT B both send CALL_PROC.

indicates that UNI state and NNI state of IUT A are N3 and NN9, respectively, and UNI state and NNI state of IUT B are N6 and NN3, respectively. Transitions in bold line in Figure 8 are interoperability test cases for the ATM signaling protocol.

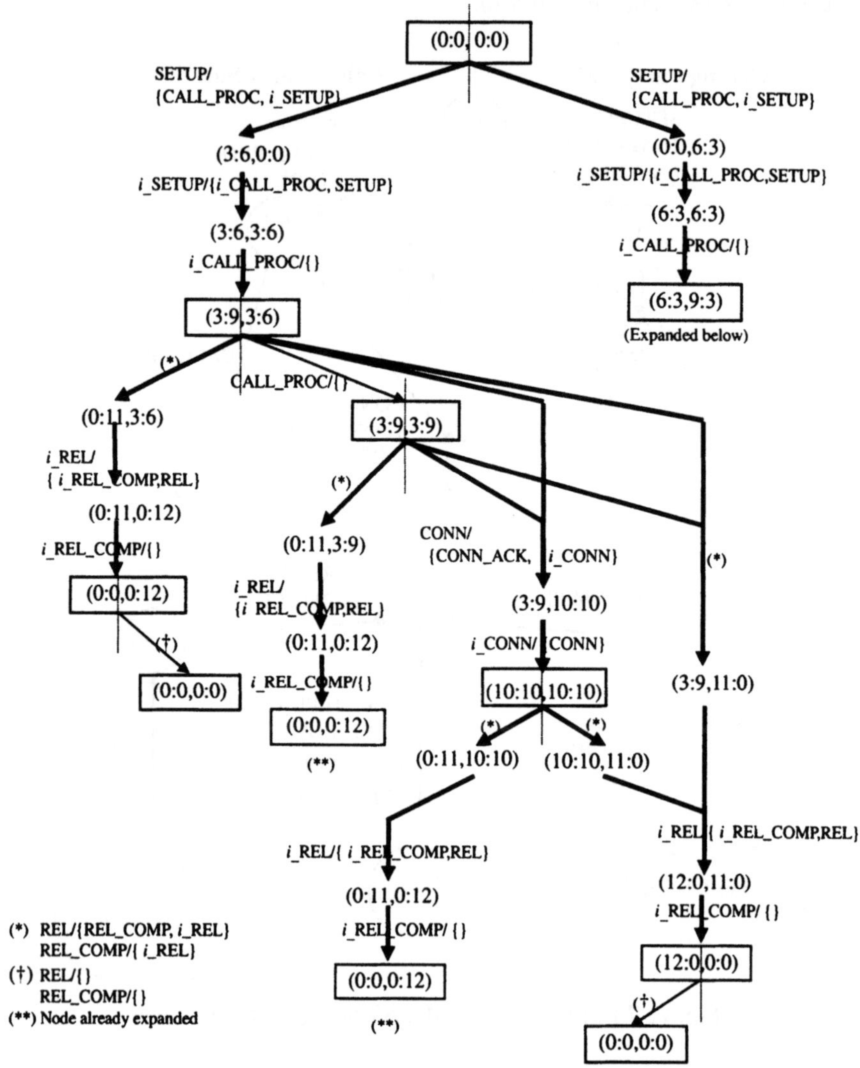

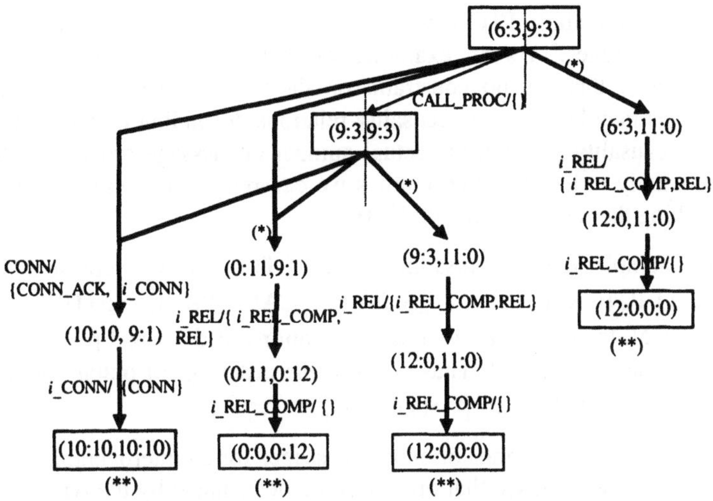

Figure 8 Composite FSM Π for the ATM signaling protocol.

6.1. Test Suite Structure

Derived test cases are grouped according to the call processing function in Figure 9. As seen from Appendix 1, both of /ESTABLISH/A_TO_B/ and /ESTABLISH/B_TO_A/ groups contain 3 test cases, and both of /CLEARING/A_TO_B/ and /CLEARING/B_TO_A/ groups contain 8 test cases. In the figure, A_TO_B and B_TO_A mean that ATM call direction is, respectively, from Tester A to Tester B and the reverse direction. Test group /CLEARING/COMMON/ has 4 test cases and is to test call clearing after a call has been established regardless of the call direction. In Appendix 1, a suffix attached to the end of an input message indicates the direction of the message. For example, REL_A represents RELEASE is transferred from Tester A to IUT A. The rightmost column of the table is the type of each test case assigned according to Figure 5.

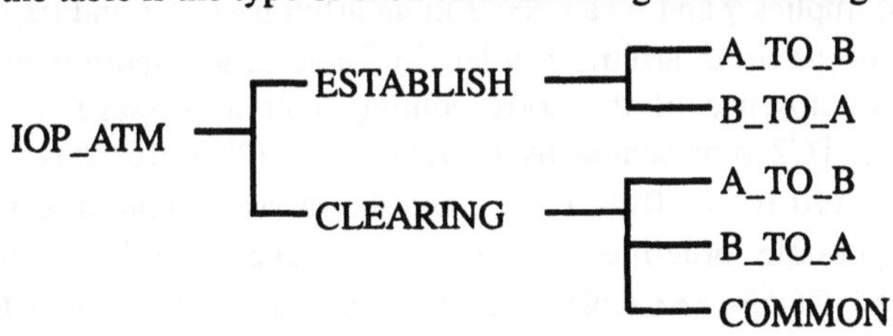

Figure 9 The interoperability test suite structure for ATM signaling protocol.

6.2. Ordering of Output Messages

The output from each test case of the derived interoperability test suite consists of 3 or 4 messages and is expressed as [α, β, γ, δ] using a vector notation. To describe it using (sequential) TTCN, we must consider all possible orderings of expected output messages.[5] As an example, we must allow 4!=24 cases for 4 output messages if there is no constraint on the message sequence. However, output messages may have certain causality relations depending on the protocols at hand. By the causality relation, the actual number of message orderings allowed can be significantly reduced, which implies we can reduce the length of test cases when describing them using sequential TTCN.

For example, if we express the output of TC1 which is represented by (0:0,0:0) — $SETUP_A$/[CALL_PROC, *i*_SETUP, *i*_CALL_PROC, SETUP] → (3:9,3:6) by [α, β, γ, δ], then there are two causal relations of β→γ and β→δ according to the ATM signaling protocol. Thus, all possible orderings of output messages for TC1 can be represented by α→β→(γ,δ) and β→(α,γ,δ).

If we calculate orderings of output messages for all interoperability test cases in a similar way, we can see that orderings are determined by the type of an individual test case and the direction of the input message as shown in Table 2.

Table 2 Output message ordering patterns

Types	Sender of the input message	Output messages	Possible orderings of output messages
1	Tester A	[, β, γ, δ]	β→(γ,δ)
	Tester B	[δ, γ, β,]	
2	Tester A	[α, β, , δ]	α→β→δ and β→(α,δ)
	Tester B	[δ, , β, α]	
3	Tester A	[α, β, γ, δ]	α→β→(γ,δ) and
	Tester B	[δ, γ, β, α]	β→(α,γ,δ)

In Table 2, (γ,δ) implies γ and δ can occur in an arbitrary order and (α,γ,δ) implies α, γ and δ can occur in an arbitrary order. In Table 2, an empty place in vector notation indicates absence of the corresponding output message. According to Table 2, because TC2 represented by (3:9,3:6) — $CONN_B$/[CONN, , *i*_CONN, CONN_ACK] → (10:10,10:10) is Type 2 and the message input direction is from Tester B, all possible orderings of output messages can be represented as CONN_ACK → i_CONN → CONN and i_CONN → (CONN_ACK, CONN).

When two IUTs interwork, we cannot know in advance in which sequence messages will be produced. Table 2 shows compressed ordering patterns from which all the possible message orderings can be extracted when we are to realize

[5] When we use concurrent TTCN, we do not need to calculate the occurrence orderings of the output messages because each PCO (or PO) checks messages independently. But, the use of concurrent TTCN is yet in its early stage.

test cases using TTCN. Thus, we can obtain minimal length test cases by exploiting these patterns. Figure 10 shows the TTCN description for TC2 using this principle.

```
LTb! CONN
  LTb? CONN_ACK
    LTc2? i_CONN
      LTa? CONN
  LTc2? i_CONN
    LTb? CONN_ACK
      LTa? CONN
    LTa? CONN
      LTb? CONN_ACK
```

Figure 10 TTCN description for TC2.

LTa and LTb represent PCOs, respectively, at interface A and at interface B. LTc2 represents a PO at interface C that reads messages sent by IUT B. The number of orderings considered in Figure 10 is 3.

7 CONCLUSION

To ensure interoperation of communication system, interoperability testing is esential in addition to conformance testing. In this paper, we presented systematic interoperability test suite derivation method for the class of communication protocols like ATM signaling protocol.

We analyzed in detail the conditions for a test case to constitute an interoperability test case and showed that interoperability test cases for ATM signaling protocol can be classified into three types based on the patterns of message interactions. After a careful comparison of three candidate interoperability test architectures for interoperability testing of two equipment implementing ATM signaling protocol, we selected Test Architecture II as it provides a good test coverage at a low cost.

Next, we presented a method to derive interoperability test cases that satisfy interoperability test case conditions under the selected test architecture. It includes an algorithm that performs composition of FSMs and derivation of interoperability test suites in parallel. The algorithm produced 26 interoperability test cases for the ATM signaling protocol.

Furthermore, we showed that possible orderings of output messages are determined by the types of test cases and the direction of the initial external input message. We used this regularity to reduce the lengths of interoperability test cases.

In this paper, we considered only control flow and message types. If we further take into account information elements of messages, more than one test cases may be obtained from a single test case (skeleton). Currently we are investigating how such aspect can be incorporated into our method.

As an alternative method of deriving interoperability test cases, one can think of extracting a test suite that includes both control flow and data flow at once, by deriving first an extended-FSM(EFSM) for each entity and composing the two EFSMs. Although this method has one less step than one in this paper, further study is needed to see which method is superior to the other since using the extended-FSM increases the complexity of test derivation.

8 REFERENCES

[AF UNI] ATM Forum af-uni-0010.002, "ATM UNI interface specification version 3.1"

[AF PNNI1] ATM Forum af-pnni-0055.00, "Private Network-Network Interface Specification Version 1.0(PNNI v1.0)," Letter Ballot, March 1996.

[AF PNNI2] ATM Forum af-pnni-81.00, "Private Network-Network Interface (PNNI) Version 1.0 Errata and Protocol Implementation Conformance Statement (PICS) Proforma," March 1997

[APRS 93] Arakawa, N., Phalippou, M., Risser, N. and Soneoka, T., "Combination of conformance and interoperability testing," Formal Description Techniques, V (C-10) M. Diaz and R. Groz (Eds.), Elsevier Science Publishers B. V. (North-Holland), 1993.

[AraS 92] Arakawa, N. and Soneoka, T., A Test Case Generation Method for Concurrent Programs, *Protocol Test Systems, IV*, J. Kroon, R.J. Heijink and E. Brinksma (Eds.), Elsevier Science Publishers B. V. (North-Holland), 1992.

[ISO DTR10000] ISO/IEC JTC1 DTR-10000, "Information Technology - Framework and classification of international standard protocol," 1994

[ETSI ETR130] ETSI ETR(ETSI Technical Report) 130, "Methods for Testing and Specification(MTS); Interoperability and conformance testing/A classification scheme," April 1994

[KanK 97] Kang, S., and Kim, M., Interoperability Test Suite Derivation for Symmetric Communication Protocols, Forte/PSTV'97, Osaka, Japan, November 1997.

[LuBP 94] Luo, G., Bochmann, G. and Petrenko, A., Test Selection Based on Communicating Nondeterministic Finite-State Machines Using a Generalized Wp-Method, *IEEE Transactions on S.E., Vol. 20, No. 2*, Feb. 1994.

[RafC 90] Rafiq, O. and Castanet, R., "From Conformance Testing to Interoperability Testing," The 3rd International Workshop on Protocol Test Systems, 1990

[VerB 94] Vermeer, G. S., and Blik, H., Interoperability Testing: Basis for the Acceptance of Communicating Systems, Protocol Test Systems, VI(C-19), Elsevier Science Publishers B. V. (North-Holland), 1994.

9 BIOGRAPHY

Jaehwi Shin received the B.S. and M.E. degrees in electronic engineering from Hanyang University and Seoul National University in 1991 and 1993, respectively. Currently he is with the Korea Telecom Research and Development Group as a member of Protocol Engineering Team. His research interests include intelligent networks, communication protocol testing, digital signal processing and circuit theory.

Sungwon Kang received B.A. from Seoul National University in Korea in 1982 and received M.S. and Ph.D. in computer science from the University of Iowa in U.S.A in 1989 and 1992, respectively. Since 1993, he has been a senior researcher at Korea Telecom. In 1995-1996, he was a guest researcher at National Institute of Standards and Technology of U.S.A. In 1997, he was the co-chair of the 10th International Workshop on Testing of Communicating Systems. Currently he is the head of the Protocol Engineering Team at Korea Telecom R&D Group. His research interests include communication protocol testing, program optimization and programming languages.

Appendix 1. Interoperability test suite for ATM signaling protocol (Test Architecture II)

Number	Initial Global State	Input message / Output messages	Final Global State	Type
		Test Group: ESTABLISH/A_TO_B/		
TC1	(0:0,0:0)	SETUP$_A$	(3:9,3:6)	3
		[CALL_PROC, i_SETUP, i_CALL_PROC, SETUP]		
TC2	(3:9,3:6)	CONN$_B$	(10:10,10:10)	2
		[CONN, , i_CONN, CONN_ACK]		
TC3	(3:9,3:9)	CONN$_B$	(10:10,10:10)	2
		[CONN, , i_CONN, CONN_ACK]		
		Test Group: ESTABLISH/B_TO_A/		
TC4	(0:0,0:0)	SETUP$_B$	(6:3,9:3)	3
		[SETUP, i_CALL_PROC, i_SETUP, CALL_PROC]		
TC5	(6:3,9:3)	CONN$_A$	(10:10,10:10)	2
		[CONN_ACK, i_CONN, , CONN]		
TC6	(9:3,9:3)	CONN$_A$	(10:10,10:10)	2
		[CONN_ACK, i_CONN, , CONN]		
		Test Group: CLEARING/A_TO_B/		
TC7	(3:9,3:6)	REL$_A$	(0:0,0:12)	3
		[REL_COMP, i_REL, i_REL_COMP, REL]		
TC8	(3:9,3:6)	REL_COMP$_A$	(0:0,0:12)	1
		[, i_REL, i_REL_COMP, REL]		

Number	Initial Global State	Input message	Final Global State	Type
		Output messages		
TC9	(3:9,3:9)	REL$_{A}$	(0:0,0:12)	3
		[REL_COMP, i_REL, i_REL_COMP, REL]		
TC10	(3:9,3:9)	REL_COMP$_{A}$	(0:0,0:12)	1
		[, i_REL, i_REL_COMP, REL]		
TC11	(3:9,3:6)	REL$_{B}$	(12:0,0:0)	3
		[REL, i_REL_COMP, i_REL, REL_COMP]		
TC12	(3:9,3:6)	REL_COMP$_{B}$	(12:0,0:0)	1
		[REL, i_REL_COMP, i_REL,]		
TC13	(3:9,3:9)	REL$_{B}$	(12:0,0:0)	3
		[REL, i_REL_COMP, i_REL, REL_COMP]		
TC14	(3:9,3:9)	REL_COMP$_{B}$	(12:0,0:0)	1
		[REL, i_REL_COMP, i_REL,]		
Test Group: CLEARING/B_TO_A/				
TC15	(6:3,9:3)	REL$_{B}$	(12:0,0:0)	3
		[REL, i_REL_COMP, i_REL, REL_COMP]		
TC16	(6:3,9:3)	REL_COMP$_{B}$	(12:0,0:0)	1
		[REL, i_REL_COMP, i_REL,]		
TC17	(9:3,9:3)	REL$_{B}$	(12:0,0:0)	3
		[REL, i_REL_COMP, i_REL, REL_COMP]		
TC18	(9:3,9:3)	REL_COMP$_{B}$	(12:0,0:0)	1
		[REL, i_REL_COMP, i_REL,]		
TC19	(6:3,9:3)	REL$_{A}$	(0:0,0:12)	3
		[REL_COMP, i_REL, i_REL_COMP, REL]		
TC20	(6:3,9:3)	REL_COMP$_{A}$	(0:0,0:12)	1
		[, i_REL, i_REL_COMP, REL]		
TC21	(9:3,9:3)	REL$_{A}$	(0:0,0:12)	3
		[REL_COMP, i_REL, i_REL_COMP, REL]		
TC22	(9:3,9:3)	REL_COMP$_{A}$	(0:0,0:12)	1
		[, i_REL, i_REL_COMP, REL]		
Test Group: CLEARING/COMMON/				
TC23	(10:10,10:10)	REL$_{A}$	(0:0,0:12)	3
		[REL_COMP, i_REL, i_REL_COMP, REL]		
TC24	(10:10,10:10)	REL_COMP$_{A}$	(0:0,0:12)	1
		[, i_REL, i_REL_COMP, REL]		
TC25	(10:10,10:10)	REL$_{B}$	(12:0,0:0)	3
		[REL, i_REL_COMP, i_REL, REL_COMP]		
TC26	(10:10,10:10)	REL_COMP$_{B}$	(12:0,0:0)	1
		[REL, i_REL_COMP, i_REL,]		

Test Suite Migration Experiences for B-ISUP and B-MTP

C. Desroches, M. Li, S. Pietsch, A. Rennoch,
I. Schieferdecker, T. Vassiliou-Gioles
GMD FOKUS
Kaiserin-Ausgusta-Allee 31, D-10589 Berlin, Germany
phone: (+49 30) 3463 7000, fax: (+49 30) 3463 8000
http://www.fokus.gmd.de/tip

Abstract

It is well known that the development of ATSs and ETSs is time and cost intensive. This contribution provides practical experiences and results gained from a test suite development project in the context of today's ATM protocols. In particular we discuss the problems and effort to develop and execute the BISUP and MTP test suites within B-ISDN based on corresponding N-ISDN tests. The consideration of our approach and observations should help during the production and execution of related protocol conformance tests. We have outlined basic ideas on how to underpin methodologically the ATS migration process.

Keywords

Protocol Conformance Testing, ATM, B-ISDN, BISUP, MTP, TTCN

1 INTRODUCTION

The conformance assessment process defined in the Conformance Testing Methodology Framework (CTMF) standard (X.294) gives a summary of the necessary working steps for the preparation, execution and analysis w.r.t. classical protocol conformance testing. During the preparation phase questions related to testability, SUT configuration, Abstract Test Method and Abstract Test Suite (ATS) must be addressed. According to ITU-T Recommendation X.290 a protocol defining group is responsible for the Test Suite Structure and Test Purposes (TSS&TP) documentation. The final ATS is to be produced together by protocol experts and the test realizer (see Figure 1).

Many times TSS&TP documents are not available from the protocol defining group and the ATS is developed by testing experts with good knowledge of test cam-

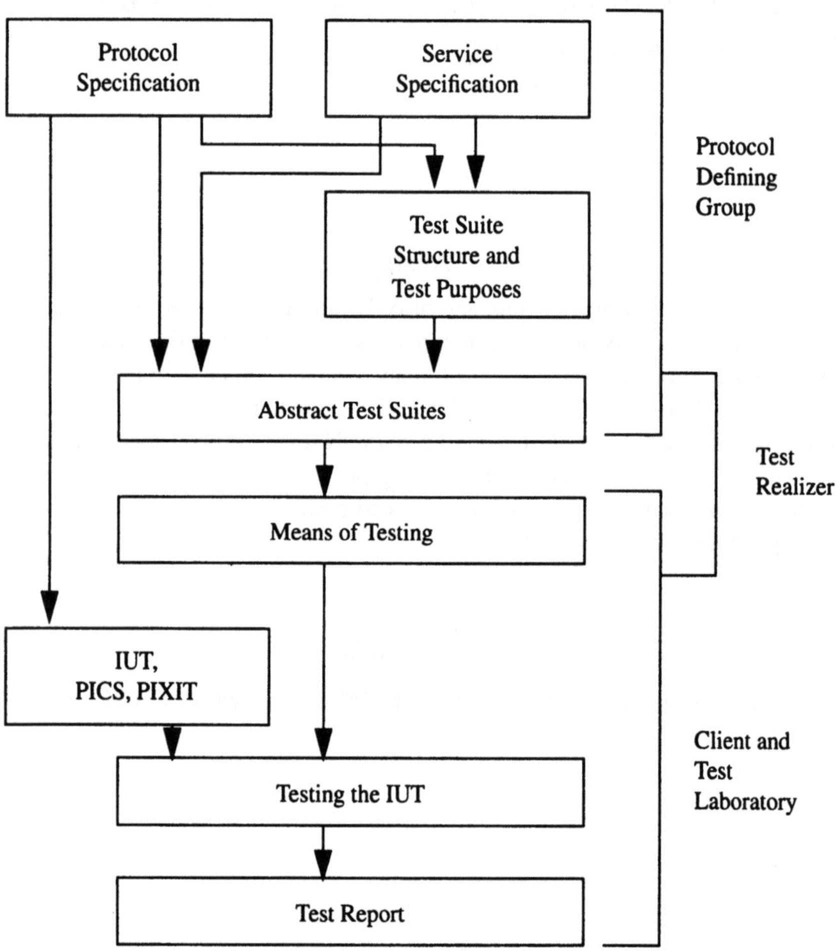

Figure 1 Relationships between parts, concepts and activities (X.290/Fig. 10).

paign, TTCN/ASN.1 and the test equipment but having less expertise of the protocols under test.

An ideal approach to develop an ATS is the tool supported derivation of tests from a formal protocol specification. There exist a lot of experiences with the generation of TTCN tests from SDL specifications (e.g. Pérez 1997, Grabowski 1997). In this context test purposes are required to be written in Message Sequence Charts (MSC). Test case generation from SDL specification is only effective if an SDL model is available. However, to develop a SDL model might be as time consuming and labour intensive as writing a test from scratch. Time pressure will enforce a manual test suite development with the disadvantage of some incompleteness and restricted means to reason about correctness and test suite coverage.

Other approaches for test case derivation propose e.g. a step-wise knowledge-base development of test cases also starting with some test purpose formalisation (Guerrouat 1996) or even request an enhanced protocol model (König 1997). The aim in the latter proposal is to consider testability during protocol engineering to decrease the efforts in protocol testing (e.g. extra testing points should be added to observe inter-module communication).

Many of these proposals are of scientific value but do not solve the telecommunication industry work of today, i.e. development of usable ATSs in very short time frames. Unfortunately (semi-) automatic test suite derivation tools are still prototypical in nature and formal protocol specifications are often missing. Manual ATS development thus becomes a challenge in balancing sound protocol knowledge with adequate testing expertise.

In this paper we report on the approaches used and results obtained in a joint research and industry project in which abstract and executable test suites were developed within the B-ISDN / ATM framework. Several ATM Executable Test Suites (ETS) are scheduled for development. These include ETSs for the User-Network Interface (UNI) and Network-to-Network (NNI) related protocols. The scope of the project includes ATM/AAL layer as well as several ATM signalling related protocols SSCOP (Q.2110), SSCF (Q.2130, Q.2140), MTP (Q.2210), BISUP (Q.2761-64) and UNI Signalling (Q.2931 / UNI 3.1 - user and network side, Q.2961, Q.2963, Q.2971). Figure 2 gives an overview of the signalling protocol stacks for UNI and NNI nodes.

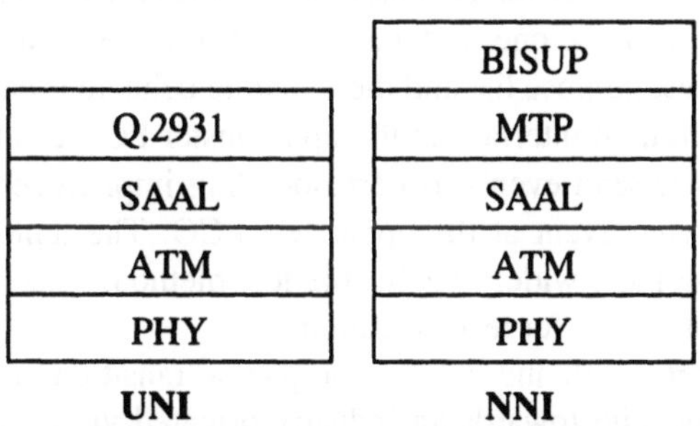

Figure 2 ATM network Signalling protocol stacks (Q.2010/Fig. 7).

Preliminary versions of some ATSs, UNI 3.1 and AAL5, were available from the ATM-Forum. These still needed to be analysed, corrected, enhanced and/or completed and made executable. Others needed to be developed directly from the protocol specification. In this paper we focus on two examples namely BISUP and MTP.

We chose these two protocols as both have pre-existing narrowband (N-ISDN) predecessors upon which the Broadband ISDN (B-ISDN) variants are based.

Further test purpose lists and even TTCN ATSs were available for the Narrowband protocol variants. Test purpose lists and the narrowband ISUP (BISUP predecessor) ATS have been standardized by the ITU-T. The narrowband MTP ATS has been specified by ETSI. Due to this prerequisites, we will report on our approaches to consider this test purposes lists and to migrate the narrowband ATS to an enhanced B-ISDN version.

It is our aim to demonstrate how existing test suites can be migrated towards a newer technology and to show different approaches during the test suite design phase. We also provide some insight on problems and dependencies to be considered during the adaptation of the ATS to a specific System under Test and a Means of Testing (MOT).

The Tektronix K1297 TTCN development tool kit (based on Forth / VxWorks) has been selected as the protocol conformance testing platform for Executable Test Suite (ETS) development.

2 TEST SUITE DESIGN

The test suite development that is described below was constraint by the following side conditions:

- For each test suite, three approaches used to derive Test Suite Structures (TSS) were considered. The state based approach which defines test groups for each stable/testable protocol state. The procedure based approach which defines test groups for each protocol procedure and lastly the PDU format based approach, which defines test groups based on PDU formats.
  It has been investigated which test suite structure fits best to the protocol description in the standard and to what extent the test suite structure of the narrowband test suite can be adapted to the broadband version.
- The basic principle for the development of test cases is that one test purpose is respected by exactly one test case, so that there is a one-to-one mapping between the test purpose list and the dynamic behaviour part of the test suite.
- Since in general, no interface at the upper tester PCO can be assumed, we treat the upper tester send events as corresponding implicit send events and do not include receiving event at the upper test PCO. The remote testing method is practically the most widely applicable test method. Therefore, we assume test operator involvement for test execution.
- Being a joint research and industry project, a transition in our thinking from an academic viewpoint towards an industry oriented viewpoint was in order. The ATSs could only use control structures and language features currently sup-

ported by the TTCN compiler. Further, the runtime environment and hardware used by Tektronix needed to be taken into consideration.

3 BROADBAND ISDN USER PART

3.1 Prerequisites

The B-ISDN user part (BISUP) protocol is part of the B-ISDN Signalling protocol stack at NNI nodes (Q.2010). It realises the signalling capabilities and functions required to support basic bearer services and supplementary services for capability set 1 B-ISDN applications. The BISUP is applicable to international B-ISDN networks and is defined by ITU-T Recommendations Q.2761 - Q.2764. Since at transit nodes BISUP has to support the ISDN user part of Signalling System No. 7 ((N)-ISUP) services there exists a strong relationship to the corresponding narrowband ITU-T Recommendations Q.761 - Q.764.

Basic functionality of BISUP refers to the initiation and release of signalling calls due to incoming messages carrying partly complete or complete call address information. B-ISUP protocol entities must accept / confirm or reject calls based on the delivered parameters and further message exchanges with adjacent network nodes. Additional functions address the blocking or reset of signalling calls.

The data type definitions of BISUP messages and parameters are available in ASN.1 [Q.2763]. The procedural specification of BISUP given in Q.2764 is subdivided into five process functions to allow for a separation of different concerns. Consequently a large number of internal coordination signals have been defined requiring an in-depth study to understand the model specification. The related SDL diagrams are incomplete (in particular due to the data part) and are therefore not suitable for use with any (semi-)automated test case generation tool. As a consequence, the entire dynamic part, including all constraints of the test suite, were implemented manually. In protocols where complex message structures exist, development of the constraints is sometimes more labour intensive as developing the test sequences.

A starting point of the ATS development is given by the existence of an incomplete test suite of the N-ISUP protocol: In Q.784 the ITU-T gives an (N)-ISUP basic call test specification which provides a test suite structure and test purpose list. For each test purpose pre-test conditions and expected message sequences have been included. Annex A of Q.784 contains a TTCN version of Q.784. The absence of test constraints and the very intensive usage of implementation dependent upper tester primitives gives reason to believe that this TTCN test suite has never been successfully executed.

3.2 ATS development

The close relationship between BISUP and NISUP led us to adopt the NISUP test purposes for the BISUP test suite. First we have distinguished between:

- test cases which are not applicable to the BISUP since NISUP messages do not have any equivalent in BISUP, and

- test cases which are still of interest in the BISUP context.

In the latter group, we updated the PDU message types and added required message exchange sequences which do not appear in NISUP (here e.g. IAA/IAR, RAM messages). A major part of the work consists in the specification of the message constraints since none of them were available. The ASN.1 message data type definition used from Q.2763 are highly substructured and therefore requires complex data constraint definitions. Sometimes it appears that the TTCN data definitions would be preferred if we would have a decision between both specifications. But in our case the ASN.1 specification was the only one available. During the analysis of message types a number of simple changes were necessary:

- The ordering of data type definitions had to be changed since the TTCN compiler did not support forward referencing and requires the parameter type definition before they could be referenced first (Q.2763 starts with the message type list).
- Many constants (e.g. "digit0") had been defined multiple times (for different parameter types). Since this causes compiler errors we have made them unique by introducing prefix identifiers (e.g. "x_digit0").

Writing a test suite always requires an early knowledge of the potential test execution configuration. We decided to keep an upper tester PCO beside the lower tester PCO to allow test cases which include IUT initiated protocol procedures. Since no interface at the upper tester PCO could be assumed we treated the upper tester send events as corresponding implicit send events and removed the receiving event at the upper tester PCO, i.e. we assume a test operator involvement. On the other side an additional lower tester PCO for the supervision of N-ISDN circuit events (e.g. check of ringing tone) become superfluous.The final basic abstract test configuration is illustrated in Figure 3.

The nature of BISUP do not require concurrent TTCN for the implementation of its test cases.

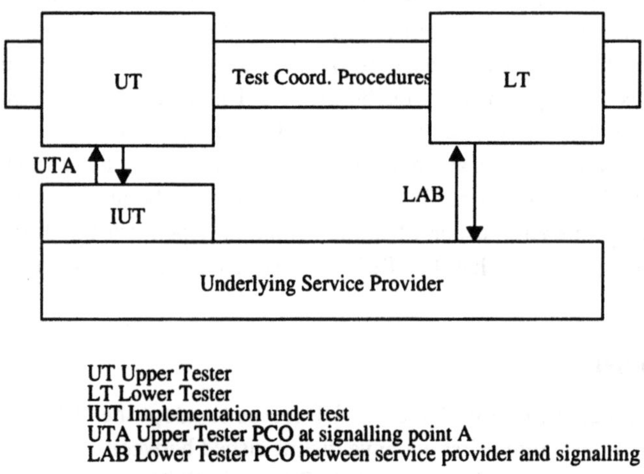

Figure 3 BISUP ATS configuration.

3.3 Test Execution

We derived 62 test cases (out of 75 tests) from the NISUP test suite. Although all of them use the same test configuration, special preparations for the IUT and tester configuration were required for those test cases, which include connection establishment / release initiated by the IUT. Since our IUT allows no manual message sending we installed a test configuration involving a network user at an UNI interface of the IUT (see Figure 4). The user behaviour at the UNI is driven by the test equipment.

This configuration is compatible with the abstract test configuration of the BISUP ATS. The test operator engagement (which is specified in the ATS via implicit send events at the UT PCO) will be realised with appropriate signalling call requests at the UNI interface UTA (using messages of ITU-T recommendation Q.2931). E.g. the establishment of an UNI signalling connection (SETUP message calling the "network node LT") leads the IUT to issue an IAM (BISUP Initial Address message) at LAB. It is clear that the IUT had to be configurated with appropriate calling address numbers of the UT "user" and the LT "network node". We had to note that this configuration is not specified within the ATS and has been selected due to the available IUT properties only.

The test suite has been compiled and executed with the Tektronix K1297 protocol tester and the Siemens EWSXpress ATM switch was used as IUT. 45 test cases were executed successfully, the remaining test cases could not be performed completely due to restrictions with the IUT. In particular, a number of IUT procedures could not been started with the upper tester configuration.

After installing a suitable basic test configuration we will extend the list of test purposes to further tests with messages of minor concern (e.g. User Part test message) and B-ISDN specific protocol procedure as consistency check. Also special test on the interworking between BISDN and ISDN are envisaged.

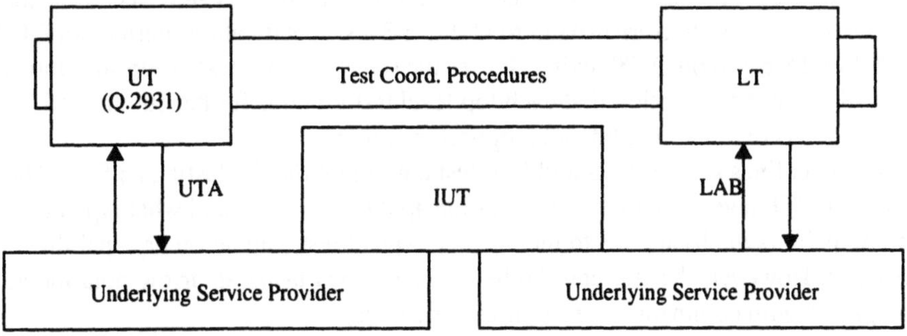

UT Upper Tester
LT Lower Tester
IUT Implementation under test
UTA Upper Tester PCO at signalling point A
LAB Lower Tester PCO between service provider and signalling point B

Figure 4 BISUP test experiment configuration.

4 BROADBAND ISDN MESSAGE TRANSFER PART

4.1 Prerequisites

The B-ISDN Message Transfer Part Layer 3 (MTP3b) protocol is part of the ITU-T Recommendation Q. series on switching and signalling network protocols. It defines the protocol for connectionless transfer of signalling messages between signalling points (SPs) which are nodes in a Common Channel Signalling network. MTP3b is sandwiched between SAAL (SSCF network side and SSCOP) below and one of various user parts (BISUP, TUP, SCCP and TCAP) above (see e.g. Figure 2).

MTP3b defines the functions and procedures required for signalling message handling and signalling network management. Signalling message handling is responsible for message discrimination where the Signalling Point (SP) determines whether a message is for itself or for another SP. If it is for itself, the message is forwarded to the message distribution function which then determines the appropriate User Part to which a message should be sent. If the message is not for itself, then it is forwarded to its message routing function which forwards messages onto the appropriate link(s) or discards the message if it cannot determine the proper receiver.

Network management is used for network reconfiguration in case of link or SP failures as well as for traffic flow control in case of network congestion. Predetermined routing information is required to achieve this.

MTP3b is specified in ITU-T Recommendation Q.2210. Q.2210 itself is derived from ITU-T Recommendations Q.704 and Q.707, which define the functions and procedures for Narrowband MTP (N-MTP). MTP3b message formats are very similar to those in N-MTP. Major differences in message formats are limited to the extension of the maximum size of a Service Data Unit due to the capabilities of the underlying protocol layer (272 octets within narrowband and 4096 octets in broadband). Further the changeover message has been extended slightly.

We re-used existing N-MTP layer 3 testing information. ITU-T Recommendation Q.782 defines the Narrowband MTP layer 3 test specification. Further an ETSI MTP3b ATS based on Q.782 exists. It uses the same top level test group structure as in Q.782 with the exception that each top level test group is further subdivided into combinations of valid, invalid and inopportune groups.

Q.782 defines a set of tests and four test configurations for testing an SS#7 Narrowband MTP layer 3. Which configuration applies to the test and what type of signalling point (switch) applies to the test, the type of test, validation or capability, is provided. However, the test case bodies are described using an arrow notation accompanied with comments (see Figure 5 for details).

We adopted Q.782 as a starting point in the development of the MTP3b ATS since it was already an established standard, and it provided a comprehensive list of test purposes, as well as detailed testing configurations for testing SPs.

4.2 Test Suite Structure and Test Purposes

Since both Q.782 and the ETSI Narrowband MTP ATS use a procedure based approach to define the ATS TSS, and since the MTP3b ATS is also based on Q.782, it was reasonable to use the same approach for defining the MTP3b TSS.

The ETSI ATS subdivides each procedure test group into valid, invalid and inopportune groups. The MTP3b ATS in comparison has no further subgrouping. Only valid test cases are provided. A separate test group containing test cases for invalid behaviour is provided.

Many of the test purposes in Q.782 were rewritten. Either the expected IUT behaviour was missing entirely from the test purpose, or was provided elsewhere, usually in the Test description, or could only be determined by analysing the message sequences themselves and referring to Q.704. For example, the test purpose for Q.782/13.9 states: "To check the actions of the system on reception of an invalid traf-

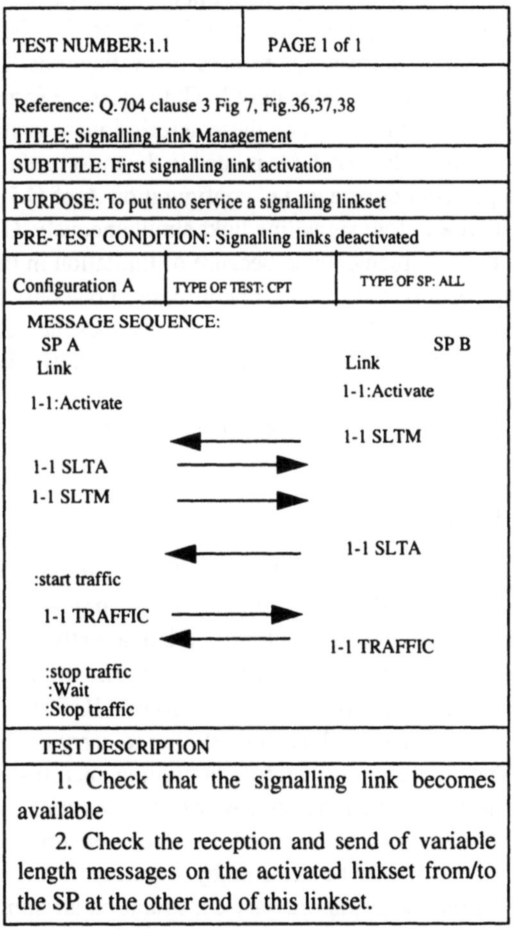

Figure 5 Sample Q.782 Test Case Description.

fic restart allowed message". In this particular case, the expected IUT behaviour is found in the test description. The test purpose was rewritten as follows: "Verify that the IUT ignores a traffic restart allowed (TRA) with an unknown OPC)". In this way, both the input is identified and the expected output or behaviour is clearly stated.

In some cases, predominantly in the Invalid Messages, where the IUTs behaviour on receipt of invalid messages is verified, test cases were split into several test cases, sometimes generating up to 16 test cases from a single test. Since a test case will stop executing on final verdict assignment, which generally occurs on observing a non-conformant behaviour, it cannot be guaranteed that all messages will have been tested, thus causing incomplete coverage. One advantage of such test cases is that it may detect compound errors, where several erroneous messages must be sent before an error appears. The principle of one test purpose one test sequence was applied.

Q.782 describes tests when initiated from the Tester side. The comments in individual test cases require that the same test be repeated, but when initiated from the IUT side. This automatically doubles the number of test cases in the ATS. Incomplete test case descriptions required further work to determine the missing behaviour and to completely define test cases. Q.782 uses arrow diagrams. This is very descriptive and simple to understand. Fortunately MTP3b message formats are simple and derivation of PDU constraints based on these descriptions was straightforward.

The MTP3b ATS includes test cases related to the most useful configuration providing maximum protocol coverage, i.e. configuration A of Q.782. It uses one signalling link set with one to four signalling links simultaneously. Other configurations were thought not to be implementable because of limitation in the number of PTCs allowed to execute concurrently. This is not the case. If one only takes the signalling links existing between the IUT and the tester, and assumes that all other SPs and signalling links can be simulated within the tester, it can be shown that at most 6 signalling links is required. The testing environment easily supports this requirement. The MTP3b ATS could therefore be extended to include test cases for all test configurations defined in Q.782.

4.3 ATS Development

The ETSI Narrowband MTP Level 3 ATS is defined using non-Concurrent TTCN. The arrow diagrams used to describe Q.782 test cases suggest that MTP3b activities are parallel in nature. Some MTP3b activities are inherently parallel in nature. E.g. MTP3b defines a load sharing procedure whereby traffic targeted for a given destination can be sent on different signalling links. How an IUT selects a signalling link and the order in which signalling links are selected to carry traffic to its destination is IUT specific. The ATS should therefore be written such that this is transparent. Further the ATS must be written in such a way that errors such as duplicate, mis-sequenced or lost messages can be detected. In a non-concurrent approach, mis-sequencing of messages can only be detected by requiring that messages be sent in a 'known' order. This places an unnecessary constraint on an implementation. Using concurrent TTCN, mis-sequencing of messages can be correctly detected without placing this same constraint on the IUT.

Using concurrent TTCN requires coordination between test components. ATS complexity thus increases. One of the major stumbling blocks in non-Concurrent TTCN when attempting to test a process which is concurrent in nature is the ordering of messages. One is forced to enumerate all possible sequences in which messages can arrive. This makes for unwieldy and unnecessarily complicated test cases. Especially this factor led us to conclude that the correct approach was use of concurrent TTCN although the ETSI ATS use no concurrence.

In some cases error indications must be communicated to the management layer. The MTP3b ATS ignores these as they are generally not observable in most cases, and tend to complicate test cases unnecessarily. Only observable, and sometimes indirectly observable behaviours are modelled in this ATS.

The MTP3b ATS went through an iterative process of defining the test configuration, test suite structure and test cases, to increase the test suite coverage and to make the test suite executable. The different versions are described below.

4.3.1 First MTB3b ATS Version

The test case structure used was intentionally very generic thus allowing for reuse with minimal change. One test case structure was designed per test configuration and depending on how many signalling links were used in the test case. This required defining highly parameterized test steps some having in excess of 6 parameters. To maintain the generic test case structure all CREATE statements were parameterized with the same test step. This test step consisted of one huge switch statement which based on an index provided as an actual parameter, would then call the appropriate test step to perform the actions associated with a given test case.

In this first approach a true master/slave relationship existed between the Main Test Component (MTC) and the Lower Tester Parallel Test Components. Intelligence was centralized in the MTC, where the MTC would send/receive PDUs to/from the Lower Testers (LTs) encapsulated within Coordination Messages (CMs) sent via CPs.

The Upper Tester (UT) was controlled by the MTC. It was responsible for generating Test Traffic and forwarding it to the IUTs' MTP3b upper interface. Its behaviour was more or less independent from the MTC, requiring no explicit PDU exchanges with the MTC and therefore was implemented as a self standing PTC.

4.3.2 Second MTP3b ATS Version

Our first attempts at viewing executed test cases, simulated in this case, was very difficult, because of the high level of parameterization, the parallel nature of execution, and the level of nesting. For these reasons, excessive test step parameterization was reduced by hard coding PCOs, CPs and other parameters directly into test steps. Further, the generic test step used in CREATE statements in test cases was replaced by direct 'calls' to the appropriate test step. This reduced the ATSs nesting depth by one. These changes improved the ATS readability tremendously, both in the GR version of ATS and during test case simulation execution on the K1297.

Further, intelligence was 'pushed down' into the LTs. As with the UT, LTs behaved independently but were nevertheless controlled by the MTC using CMs. PDU exchanges between the MTC and LTs was completely eliminated for three reasons. First, avoid the problem associated with message ordering and the sheer number of possibilities especially when several LTs executed simultaneously. Second, continuing with the same structure would make the MTC hopelessly unreadable and complicated. Lastly, the K1297 did not correctly support the use of the Meta-PDU type.

Minor changes were made to provide for more user friendly data entry. Further changes were required because of problems related to the use of forward referencing where assignment of constraints and PDUs to test case variables was not supported.

4.3.3 MTP3b ATS Version with Changeover

The MTP3b signalling network management (SNM) changeover procedure ensures that signalling traffic carried by an unavailable link is diverted to alternative link(s) while avoiding message loss, duplication and mis-sequencing. When a signalling link fails, transmission and reception of messages on the failed link is terminated, alternative link(s) are selected, the faulty link's retransmission buffer is updated and traffic is diverted to the identified alternative signalling link(s).

Buffer updating consists of transferring the concerned messages to the alternative link(s) retransmission buffer(s). Signalling links in the ATS are modelled as PCOs. Unsent messages must first be retrieved from the inactive link buffers, transferred from one PTC to another via CMs, copied field by field to another PDU sent through the PCO designated as the alternative link.

With no direct PTC to PTC communication permitted, this entailed unnecessary communication, data copying and unnecessary processing overhead. For this reason, the model was modified to permit direct PTC to PTC bidirectional communication via special CPs (LCP1_2 etc.) (see Figure 6).

The alternative link must handle traffic from the now deactivated link, its own traffic, and transmit unsent traffic from the deactivated link. Even with this new PTC to PTC communication possible, it was feared that run time performance of change-

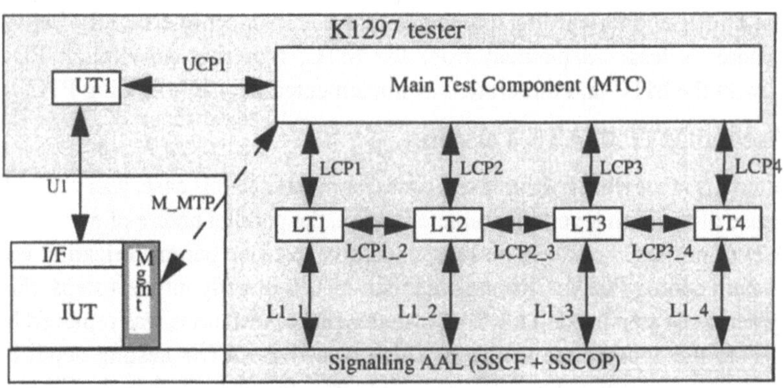

Figure 6 MTP3b Test configuration for a SEP.

over would be inadequate. The TTCN code responsible for transferring unsent data from one signalling link's buffers to the alternative signalling links buffers for re-transmission was therefore re-implemented using a test suite operation, where transparent communication between signalling links and their respective transmission buffers took place.

4.3.4 MTP3b ATS Version with Test Traffic Generation

Two testing models are defined for generating test traffic in Q.782. The first, test traffic is generated by the upper tester and delivered to the IUT via MTP Transfer request primitives. Test traffic received by the lower testers is reflected back to the IUT and should in theory be received by the upper tester to identify any loss of the test traffic.

If an IUT does not support an adequate MTP3b Upper Boundary interface, and does not provide any endpoint functionality, the ATS configuration must be changed according to Figure 7: Instead of the UT we introduce another lower tester (GEN) which takes over the test traffic generation task of the UT. Now the MTP messages are sent from a LT test component to the IUT which forwards the traffic to one of the components LT1 - LT4.

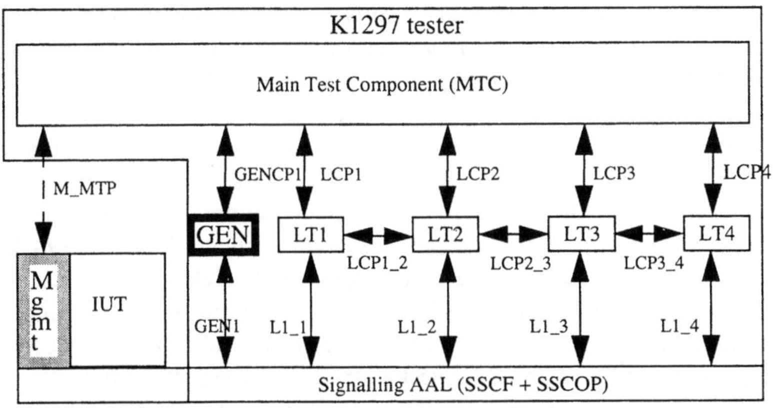

Figure 7 MTP3b Test configuration for a STP.

4.4 Test Execution

Q.782 contains 87 test case descriptions for Configuration A. The MTP3b ATS defines 204 test purposes, of which 94 are currently implemented in TTCN.

The ATS was verified using the Tektronix K1297 TTCN compiler and executable code was produced. Currently, our means of testing does not support the functionality required to perform changeover procedures and some of the ATS defined ASPs were not yet supported by the underlying SAAL implementation, so that the ETS can be used only in a restricted manner. However, the next release of the means of testing will cope with these problems.

One major stumbling block remained, namely, the unavailability of an IUT against which we could further verify our ATS. Three alternatives were considered: writing a complete set of MTP3b simulation scripts for our purposes, writing or purchasing an MTP3b emulation or lastly finding an existing MTP3b implementation, public domain if at all possible for cost reasons.

The effort required to write a complete set of simulation scripts or write our own emulation to verify our ATS was deemed to be too high. Therefore using various search engines we identified companies having MTP3b implementations. Recently, we established a joint project with a company which will provide us with a MTP3b implementation. In return they will receive a complimentary conformance test report.

5 CONCLUSIONS

The discussion on the migration of Narrowband ISDN to the Broadband ISDN technology environment has shown the various aspects of test suite development for today's ATM protocols (here BISUP and MTP). In both examples we have similar prerequisites: available TSS&TP documentation, some more or less standardized and useful ATS for N-ISDN and the same (migration) target. But due to the nature of these two protocols our experiences and efforts with this work have been very different. In particular, the introduction of parallel behaviour and the use of concurrent TTCN in the MTP ATS causes a big expense to make the final ATS design and implementation.

Due to our experiences and project results we have started research in the following directions: One aspect is that the usage of Concurrent TTCN in large and complex test suites lead us to think about some tool support during the development of ATS including concurrence. At GMD Fokus a test suite simulator (Pietsch 1998) has been implemented for verification of dynamic aspects and validation of logical correctness of concurrent TTCN test suites. The simulator has been developed using the ITEX C-Code Generator CCG for the interpretation and execution of TTCN code.

Further, from the ATS migration project arises the question of test suite generation from another viewpoint. Here we did not try to derive an ATS from a formal protocol description, but we focused on the enhancement of existing ATSs to an updated protocol standard. The question is about some test suite structure and abstract test case dynamic behaviour sequences which make an ATS suitable for periodical protocol (version) updates.

Furthermore, we see test cases related to classes of viewpoints like protocol state transitions (e.g. timeouts), procedure (e.g. data correction methods) or message format verification (PDU parameter field contents) and ask for a generic test suite structure to be adopted to different protocols and instantiated to become a helpful starting point in the ATS development process. Such documents shall serve as a constructive working platform for new ATSs and are a must to go beyond general methodology guidelines or naming conventions in the ATS production process. We think of test suite specification guidelines which are similar to specification guidelines adopted for LOTOS (Bolognesi 1995). Another area of research is to apply the framework idea known from object-oriented software development process to the area of test suite development (and migration in particular).

6 ACKNOWLEDGEMENTS

We would like to thank Tektronix Berlin GmbH, in particular Jens Kittan and Wolfgang Borgert, who supported the compilation and usage of the test suites with the K1297 protocol tester.

7 REFERENCES

Bolognesi, T., v.d.Lagemaat, J., Vissers, C. (1995): LOTOSphere: Software Development with LOTOS. Kluwer Academic Publishers.

Guerrouat, A., König, H. (1996): Automation of test case derivation in respect to test purposes. In: ICTCS'96 proceedings. Darmstadt (Germany).

Grabowski, J., Scheurer. R., Dai, Z.R., Hogrefe, D. (1997): Applying SAMSTAG to the B-ISDN protocol SSCOP. In: ICTCS'97 proceedings. Cheju Island (Korea).

ITU-T Recommendation Q.704 (1993) Messages Transfer Part. Signalling Network functions and messages.

ITU-T Recommendation Q.707 (1993) Specifications of signalling system No. 7, Testing and maintenance.

ITU-T Recommendation Q.761 - Q.764. Specification of Signalling System No. 7.

ITU-T Recommendation Q.782 (1993) Specification of signalling system No. 7. Signalling system No. 7 - MTP Level 3 test specification.

ITU-T Recommendation Q.784 (1991). ISUP Basic Call Test Specification.

ITU-T Recommendation Q.2010 (1995). Broadband Integrated Services Digital Network Overview - Signalling Capability Set 1, Release 1.

IUT-T Recommendation Q.2110. B-ISDN ATM Adaptation Layer—Service Specific Connection Oriented Protocol (SSCOP).

ITU-T Recommendation Q.2130 (1995) B-ISDN ATM Adaptation layer - service specific coordination function for signalling at the user network interface.

ITU-T Recommendation Q.2140 (1995) B-ISDN ATM Adaptation layer - service specific coordination function for signalling at the network node interface.

ITU-T Q.2210 (1995) Messages transfer part level 3 functions and messages using the services of Q.2140.

ITU-T Recommendation Q.2761 - Q.2764 (1995). B-ISDN User Part (B-ISUP) of Signalling System No. 7.

ITU-T Recommendation Q.2931 (1995). B-ISDN User-Network Interface (UNI), Layer 3 Specification for Basic Call/Connection control.

ITU-T Recommendation Q.2961 (1995). B-ISDN - Digital subscriber signalling system no.2 (DSS 2) - Additional traffic parameters.

ITU-T Recommendation Q.2963 (1996). DSS 2 - Connection modification.

ITU-T Recommendation Q.2971 (1995). UNI layer 3 specification for point-to-

multipoint call/connection control.
ITU-T Recommendation X.290 (1991). OSI Conformance Testing Methodology and Framework (CTMF) - Part 1: General Concept.
ITU-T Recommendation X.292 (1993). OSI CTMF - Part 3: The Tree and Tabular Combined Notation (TTCN).
ITU-T Recommendation X.294 (1991). OSI CTMF - Part 4: Test Realization.
König, H., Ulrich, A., Heiner, M. (1997): Design for testability: a step-wise approach to protocol testing. In: ICTCS'97 proceedings. Cheju Island (Korea).
Pérez, E., Algaba, E., Monedero, M. (1997): A pragmatic approach to test generation. In: ICTCS'97 proceedings. Cheju Island (Korea).
Pietsch, S., Schieferdecker, I. (1998): Test Suite Simulation. In: FBT'98 proceedings. Shaker Verlag, Germany.
Tektronix Protocol Test System K1297. User Manual (1996).

8 BIOGRAPHY

Claude Desroches holds an Honours BSc. in Computer Sciences from the University of Ottawa in Canada. He has worked as a software engineer in the field of telecommunications since 1987 with Bell Canada, the University of Ottawa, Hewlett-Packard/Idacom and Tekelec Inc. His research interests include Formal Description Techniques, compiler development, implementation, validation, verification, and conformance testing of protocols. He is currently a guest researcher at GMD Fokus.

Mang Li has been studying at the Institute of Telecommunication Networks, department of electrical engineering, Technical University Berlin. She has been working as a student at GMD Fokus since 1996. She is interested in conformance testing and performance testing.

Stephan Pietsch studied electrical engineering at the Technical University Berlin. He is a scientist at GMD Fokus in Berlin. His research interests are in conformance and interoperability testing of communication protocols.

Axel Rennoch studied mathematics at the Free University of Berlin. His research interests include the application of Formal Description Techniques for testing methodologies and Quality of Service considerations. Currently he is employed as a scientist at GMD Fokus in Berlin.

Ina Schieferdecker studied mathematical computer science at the Humboldt University in Berlin and received her Ph.D. from the Technical University in Berlin in 1994. She attended the postgraduate course on open communication systems at the Technical University in Berlin. Since 1993, she is a researcher at GMD Fokus and a lecturer at Technical University Berlin since 1995. She is working on testing methods for network components, and carries out research on formal methods, performance-enhanced specifications and performance analysis.

Theofanis Vassiliou-Gioles studies electrical engineering at the Technical University Berlin. His main fields of research interests include protocol validation, conformance and interoperability testing and distributed testing methodologies. Currently he is employed at GMD Fokus.

PART NINE

Testability

Moreover we are concentrating on qualitative, rather than quantitative, concepts of testability. The latter could be built upon the former, cf. Section 6.

The rationale of the chosen framework will be discussed thoroughly in separate papers, e.g. in an investigation of system properties required by specifications [3]. Technically, it draws implicitly on the mathematical properties of Galois schemes [2], based on Galois connections, which have already proved useful in Computer Science [9,11]. This paper is part of ASPEKTE, an effort to introduce a theory of specification as a basic common language for specification, verification and testing.

2 RELATED WORK ON TESTABILITY

Testability, observability and controllability have been described both formally and informally, and both as qualitative and as quantitative notions.

Informal approaches

[13] defines testability on the one hand as the degree to which a system or component facilitates the establishment of test criteria and the performance of tests to determine whether those criteria have been met, and on the other hand as the degree to which a requirement is stated in terms that permit establishment of test criteria and performance of tests to determine whether those criteria have been met.

In [20], testability is viewed as the probability that a system will fail on its next execution during testing (with a particular assumed input distribution) if the software includes a fault.

In [5, 8, 18, 21], as in many other sources, testability is circumscribed as the existence of features, properties or characteristics that facilitate the testing process of implementations. Particular aims are to reduce effort or cost and to facilitate the easy application of testing methods and the detection or isolation of existing faults. The testability measure should be a vector of measures of particular testability aspects. [10] identifies observability and controllability as two important factors of testability.

Formal approaches

Carnap's seminal paper [7] on testability, dating back to 1936, seems to have been quite ignored in Computer Science. Since Kalman's control-theoretical paper [17], controllability and reachability concepts (of states and systems) refer to the possibility of directing a system from given states to other states, while observability concepts refer to the possibility of identifying the unknown state of a system by means of the relationship between certain inputs fed to the system and the outputs obtained from it in a finite time interval.

Early definitions of controllability and observability concepts in the context of abstract machines and automata were proposed by [16] and [19]. [16] defines controllability for Mealy automata as the property that all states can be reached from one another with the appropriate inputs. The automaton is called initial-state determinable if its unknown initial state can be determined by experiments feeding inputs and observing outputs—a generic definition that depends on the kinds of

Qualitative Notions of Testability

B. Baumgarten,
GMD, TKT
Rheinstr. 75
64295 Darmstadt, Germany
baumgart@darmstadt.gmd.de

H. Wiland
Karlstr. 9a
68307 Mannheim, Germany

Abstract

Testability, and design for testability, are widely discussed practical issues in software engineering, especially in protocol engineering. Existing definitions (or circumscriptions) of testability seem to be either quite vague, or, if more or less formal, then dedicated only to very special system models. Testability is usually decomposed into aspects like observability and controllability, and these are defined either as qualitative properties or as quantitative measures. We identify a set of qualitative testability properties that we define completely independently from any special system model, only in terms of user-relevant aspects like possible, desired and undesired system observations or outcomes of experiments.

Keywords

Testability, Specification theory, Conformance, Semantics

1 INTRODUCTION

In theory and practice of testing, testability is often emphasized as a desirable property of specifications and systems. There exist various useful informal discussions of testability [5,10,18,20,21]. Related formal notions of observability and controllability were defined in mathematical systems theory for abstract machines [16, 19].

In the following we use a general semantic framework for system specification based on the specification context between systems, system properties and system observations, to discuss general qualitative concepts of testability. We deal only with testability aspects of specifications, not with test implementation matters.

experiments permitted. In [19], a non-deterministic Mealy automaton is called observable if a state and an input performed and an output observed in this state determine the next state reached.

In [8, 18], controllability and observability are quantified in the setting of finite state machines.

3 SPECIFICATION CONTEXTS

3.1 Definition

When we deal with specifications, we usually do so within some context, characterized by a population of systems under consideration, the set of their possibly relevant properties, and a set of possible observations of interest. What is practically considered as a system, a property, or an observation in a context depends e.g. on

- the chosen part of reality or field of thinking (e.g. we may be dealing with protocol entities and not with white mice),
- the chosen level of abstraction (e.g. on the level of data base access protocols we may not care about signal wave shapes), and
- the envisaged meaningful uses and abuses of the system (e.g. we may not care about the colour of the inside of the casing of our monitor, even though it certainly has some colour, and we could find it out if we desired to).

In [3], a **specification context** is defined as a quintuple (*Systs*, *Props*, *Obs*, *has_property*, *permits*), where

- *Systs* is a set of **systems**,
- *Props* is a set of **properties**,
- *Obs* is a set of **observations**,
- *has_property* is a relation between systems and properties, and
- *permits* is a relation between systems and observations, which, for the purposes of this paper, is non-empty as well as left- and right-total.

In practice, specification contexts will be described by a selected combination of natural language, technical vocabulary and mathematics. In this paper we will not be very much concerned with *Props* and *has_property*.

The last item of the above definition concerns three non-degeneracy assumptions dictated by practical considerations. Specification contexts without systems or observations are of little interest. Non-observable systems can be ignored, they are transcendental w.r.t. the given context: we could not (or would not care to) notice such a *sys*, even if we were standing right in front of it. In order to simplify later definitions, we work only with observations that can really be made on at least one system. Formally, our non-degeneracy assumptions amount to:

$$Systs \neq \emptyset \wedge Obs \neq \emptyset, \qquad \text{(i)}$$
$$\forall sys \in Systs : \exists\, obs \in Obs : sys \ \mathit{permits} \ obs, \qquad \text{(ii)}$$

$\forall obs \in Obs : \exists\, sys \in Systs : sys\ permits\ obs.$ (iii)

Relationships and transitions between different system contexts, which play an important role in questions of abstraction, refinement and modelling, and for which it sometimes pays to relax (i,ii,iii), will not be investigated here. Instead, we will mostly confine ourselves to an arbitrary single specification context *Cont* = (*Systs*, *Props*, *Obs*, *has_property*, *permits*). Definitions are then implicitly referring to *Cont*. For example, a "behaviour" means a "behaviour in *Cont*."

3.2 Observations

Observations are a central notion in this paper. We explicitly assume that *Obs* comprises all observations that are both imaginable and of potential interest, in the sense that they may be permitted, desired or undesired. It does not matter whether they are direct or indirect, active or passive, long or short, atomic or complex. As to complexity, for example, on a certain level of specifying a vending machine, both the insertion of a coin and the dispensal of a soft drink may be atomic observations, but we also need more complex observations such as "I inserted the required coin but nothing happened in the next 20 seconds," and "I inserted the required coin and received a drink within 5 seconds." A realizable combination of repeated and/or parallel observations, for example, must also be considered as one complex observation if it is relevant w.r.t. to system conformance.

Later definitions will be implicitly based on the informal assumption that each observation is made in finite time. As the saying goes, "infinity is where things happen that don't." [3] shows that system requirements based on infinite observations may turn out to be void.

3.3 Behaviours

Apart from what we know, or assume, about all systems in *Systs*, whatever we find out about a system, we find out exclusively by means of observations. Every system permits only certain observations. Thus, the (visible) **behaviour of a system** consists of the observations that can be made of it:

$$sys\_beh: \begin{cases} Systs & \rightarrow & \mathbf{P}(Obs) \\ sys & \mapsto & \{obs \in Obs \mid sys\ permits\ obs\} \end{cases}$$

A **behaviour** *Beh* in general can be defined as a set of possible observations, $Beh \subseteq Obs$. At this point, mathematical purists might suggest that a system simply is a behaviour, i.e. a subset of *Obs*. While this would not really change our results, it might be counter-intuitive for many readers; therefore, we stick to the separate set *Systs*.

Telling two systems apart takes at least one observation possible for one system and impossible for the other, even if it only consists of a name, a serial number or a position in space. Systems *sys1* and *sys2* are indistinguishable if they "behave identically," i.e. if $sys\_beh(sys1) = sys\_beh(sys2)$, the full abstractness issue of [12].

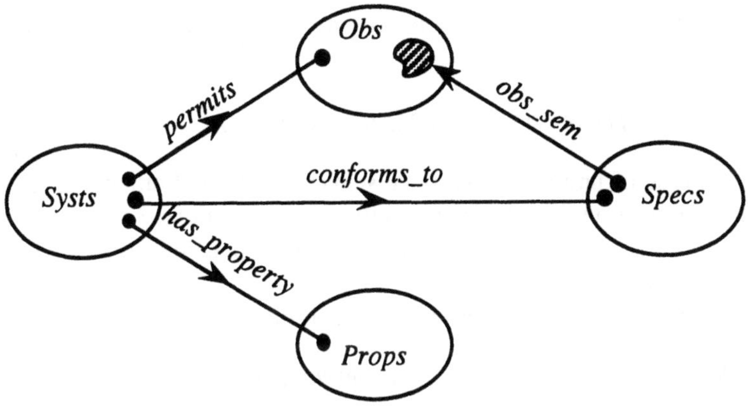

Figure 1: The semantic embedding of specifications into a specification context

3.4 Specifications and Conformance

From a user and tester standpoint, the purpose of a specification is to describe the range allowed for the behaviour of the desired systems. Such a behaviour range is—like the behaviour of a single system—also a subset of *Obs*, i.e. a behaviour. If we allow for nondeterminism in systems, the differences between a system behaviour and a specified behaviour range all but vanish. Therefore, we speak uniformly of the "behaviour" both of systems and of specifications.

Systems and specifications defined by means of a formal description technique often comprise internal aspects, like operations on internal variables in programs or invisible steps in transition systems. "Invisible behaviour" often amounts to visible behaviour in a different context permitting other observations, such as program code inspection. Within a given context, invisible aspects of a system are meaningful merely as auxiliary constructs to define (visible) behaviour.

Loosely spoken, a **system specification** is "anything that defines a behaviour." The latter, i.e. the set of permitted observations, is usually not listed literally, in particular if it is infinite. Rather, it is inferred from the chosen syntactic form of specification, taken from a set *Specs* of specification terms, via an **observational semantics** (cf. Figure 1)

$$obs\_sem\colon Specs \rightarrow \mathbf{P}(Obs).$$

Two important forms of specification terms are a (finite) list of required properties, such that $Specs \in \mathbf{P}_{fin}(Props)$, cf. [3], and a term in a formal description language used both for specifications and for abstract systems. In the latter case, $Specs = Systs$, and *obs_sem* coincides with *sys_beh*; this approach can be used nicely in process algebra [12].

Proposing specifications without commitment to *Obs* or *obs_sem* may lead to misunderstandings about valid system behaviour and jeopardizes the practical use of specifications.

A system **conforms to** a specification (term) *spec* if its behaviour stays within the range permitted by *spec* :

$$sys\ conforms\_to\ spec :\Leftrightarrow sys\_beh(sys) \subseteq obs\_sem(spec).$$

Unlike the case of untimed string trace semantics, this definition does not imply that an inactive system conforms to arbitrary specifications, or that a less active system conforms to a more active specification: in practical contexts, temporary inactivity is observable.

We call an observation *obs* **valid** for a specification *spec* if

$$\exists\ sys \in Systs\text{: } sys\ conforms\_to\ spec \wedge sys\ permits\ obs.$$

Otherwise, we call it **invalid**. Any valid observation may have come from a conforming system. Any invalid observation obtained tells us that the investigated system is non-conforming. Note that in pathological cases allowed observations may be invalid (!), as, to take an informal example, when you allow a child to play in the mud, but not to get dirty ...

We call an observation *obs* **validating** for a specification *spec* if

$$\forall\ sys \in Systs\text{: } sys\ permits\ obs \Rightarrow sys\ conforms\_to\ spec.$$

Only in very nice contexts do validating observations exist. A non-validating observation is one that may have come from a non-conforming system.

The framework described is fairly general and applies equally well to systems outside of digital information processing. Specializations typical of the latter field of application will be outlined and many of the practical notions about testing—cf. [14] for a representative list—will be formalized within the framework of specification contexts in ASPEKTE papers.

4 FLAVOURS OF REFUTABILITY AND VALIDATABILITY

The term "testability" can refer to specifications, systems, and testing environments. In this paper we deal with testability properties as possible properties of specifications. Some questions about specifications that are of obvious interest are

- Can systems be "proven to be conforming" to a specification by testing?
  Can they be "proven to be non-conforming"?
- Can all systems be proven to be conforming or non-conforming, respectively, or only some, or none?
- Can the diagnosis only be made if the—non-deterministically achieved—test result allows it, so that a decisive observation cannot be enforced?
  Or is there a method that guarantees to elicit the decisive observations?
- And if a diagnosis is possible in finite time, can this time be bounded in advance?

Thus, qualitative aspects of the **testability of a specification** are whether, through testing, conformance to this specification can be

- **validated** or **refuted**
- for arbitrary or only for some conforming resp. non-conforming systems under observation, or for none of them (**strongly** or **weakly**),
- possibly only with luck (**non-deterministically, ND**) or with a guarantee (**deterministically, D**),
- in bounded or arbitrary finite time. We will deal with this latter aspect in a separate paper [4] and will merely touch upon it in Section 5.

By "through **testing**" we mean that even though the specification context may be known, the observer does not know in advance which of the systems she is observing, i.e. which one is the system "under test". Any information about this system's behaviour is collected exclusively by means of observations.

Due to problems with non-determinism and infinite cardinality, we cannot expect to identify the full behaviour of a system by means of observations. Generally, an observation obtained may merely be one among various possible ones, and it does not tell us which other observations were possible. If we can obtain observations repeatedly—say, we can perform some experiment over again, or other experiments, as well—then we may obtain other observations. However, unless our context has nice properties, we will never be sure about the full set of possible observations of the (unknown) system we are observing—or of the full set of possible outcomes of an experiment (cf. 4.2). Even if we actually already obtained all possible observations or outcomes, we would generally not know that we did, i.e. that they make up the full set and that further repetitions will not yield anything new. Note also that we should not expect to obtain information from several observations that cannot be obtained than from any single observation, cf. Section 3.2.

However, an sufficiently "well-behaved" population *Systs* of systems may permit inferences about system behaviour on the basis of partial information. For instance, if none of the objects in the population ever changes colour, it is sufficient to determine the colour of an object just once in order to know its colour at all times. Advance knowledge about the specification context saves observation work.

We assume that a specification distinguishes between valid and invalid observations not only "in principle" but also in a practical and constructive manner: we do not deal with any "oracle-problem" here. Limits imposed by computability and complexity problems deserve a separate treatment.

A specification *spec* is called (**contextually**) **contradictory** if it does not allow a single observation (nor any system, due to (ii)). A specification *spec* is called (**contextually**) **void** if it allows all observations, $obs\_sem(spec) = Obs$, such that every system conforms. [3] treats non-trivial aspects of voidness. In practical situations, such pathological specifications should not arise.

The observational framework developed above permits us to formalize at least "non-deterministic" refutability and validatability, more precisely: without recourse to notions of (non)determinism. It is only in addressing determinism that we have to distinguish between observer and system behaviour, as we do in Section 4.2.

4.1 If it may take luck ...

Let *spec* be a non-void specification. Our definitions entail that the observation of any non-conforming system can reveal its non-conformance, though possibly only with luck, namely if we make one of the observations that the system permits and which are not valid by way of *spec*. Non-determinism lies in the possibility that the relevant invalid *obs* merely may, but need not necessarily, turn up.

THEOREM 1 [REFUTABILITY]:
Every non-void specification *spec* is **ND-refutable** in the following sense:

$$\exists Evidence \subseteq Obs\text{: } Evidence \neq \emptyset \ \wedge \ \forall sys \in Systs\text{:}$$
$$(\forall obs \in Evidence\text{: } sys \ permits \ obs \Rightarrow \neg \ sys \ conforms\_to \ spec)$$
$$\wedge \ (\neg \ sys \ conforms\_to \ spec \Rightarrow \exists obs \in Evidence\text{: } sys \ permits \ obs \ ).$$

Proof: Take $Obs \setminus obs\_sem(spec)$ as *Evidence* . ∎

While everything is quite trivial as far as refutation is concerned, things are slightly more complicated if it comes to validation. There, we can distinguish between the possibilities to validate conformance of some or of all conforming systems by means of observation. We will use "non-deterministically" not as the contrary of "deterministically" but rather as the more general notion, of which the latter is but a special case.

Let *spec* be a non-contradictory specification. In agreement with well-known limits to the power of testing, we will see that our definitions do not entail that the observation of a conforming system can generally reveal its conformance, not even with a streak of luck in the observations obtained. In some rare contexts and for some specifications, however, conformance can really be validated by means of observation. We refrain from defining validatability or refutability conditions for systems (with respect to specifications), because in the testing situation the system under test, in particular its full behaviour, is usually unknown.

A non-contradictory specification *spec* is **weakly ND-validatable** if some conforming system can actually be shown to be conforming, though possibly only by luck. By non-degeneracy arguments this amounts to saying that there exists an observation that can only be obtained from systems that conform to *spec*, i.e.

$$\exists \ obs \in Obs\text{: } \forall sys \in Systs\text{: } sys \ permits \ obs \Rightarrow sys \ conforms\_to \ spec.$$

In order to facilitate the comparison of this formula with the one in the next definition we note in passing that it is equivalent to

$$\exists \ Evidence \subseteq Obs\text{: } Evidence \neq \emptyset \ \wedge$$
$$\forall sys \in Systs, obs \in Evidence\text{: } sys \ permits \ obs \Rightarrow sys \ conforms\_to \ spec.$$

Let us discuss why these formulas reflect our intuitively formulated idea. We find out about conformance by some evidence in the form of a suitable, possibly complex, tell-tale observation. We can be sure to have found out about conformance, if we can infer from this observation that the observed (unknown) system is conforming. All we know about the system is that it permits the observation we made; apart from that it could be any member of *Systs*. Therefore, any system permitting this observation must be conforming, if we want to be certain of our diagnosis. Conversely, if the formula holds and a tell-tale *obs* turns up, then the system we have just observed must be conforming.

In Figure 2, *spec1* is not even weakly ND-validatable: every observation may have come from the non-conforming system *b*.

A non-contradictory specification *spec* is **strongly ND-validatable** if any given conforming system can actually be shown to be conforming, at least by luck. This amounts to the existence of a set of observations that can only be obtained from systems conforming to *spec* and such that each conforming system permits at least one observation in this set, i.e.

$$\begin{aligned}&\exists Evidence \subseteq Obs\colon Evidence \neq \emptyset \;\wedge \forall sys \in Systs\colon\\&\qquad (\forall obs \in Evidence\colon sys\ permits\ obs\ \Rightarrow sys\ \ conforms\_to\ spec)\\&\qquad \wedge\ (sys\ \ conforms\_to\ spec \Rightarrow \exists obs \in Evidence\colon sys\ permits\ obs\ ).\end{aligned}$$

Non-emptiness could be dropped, because it can be derived.

Again: why does this formula reflect the intuitively formulated idea?—We find out about conformance by some tell-tale observation *obs*. All tell-tale observations form a set *Evidence*. Permitting an observation in evidence must again be reserved for conforming systems, otherwise this would not indicate conformance reliably. Hence the second line. Now we want this validation to be possible for every conforming system. Thus, every one of them must permit some the tell-tale observations in *Evidence*.

Due to space limitations, we confine ourselves to the above two discussions of the agreement between intuition and formal definition. Readers are invited to analyse in a similar manner each of the notions defined in the remainder of this text.

THEOREM 2 [STRONG IMPLIES WEAK]:
Strong ND-validatability implies weak ND-validatability.

The proof is another simple exercise, if we note that *Evidence* above must be non-empty. ■

From a practical point of view, different as these two notions of validatability look, they do have a similar effect on testing: testing a conforming system against a (weakly or strongly) ND-validatable specification may, but need not, prove that the system conforms. When a conforming system is not identified as conforming by observation, then it does not matter too much whether that happened because it was one of the systems for which no telltale observation exists, or because the system happened this time not to yield a—principally possible—telltale observation.

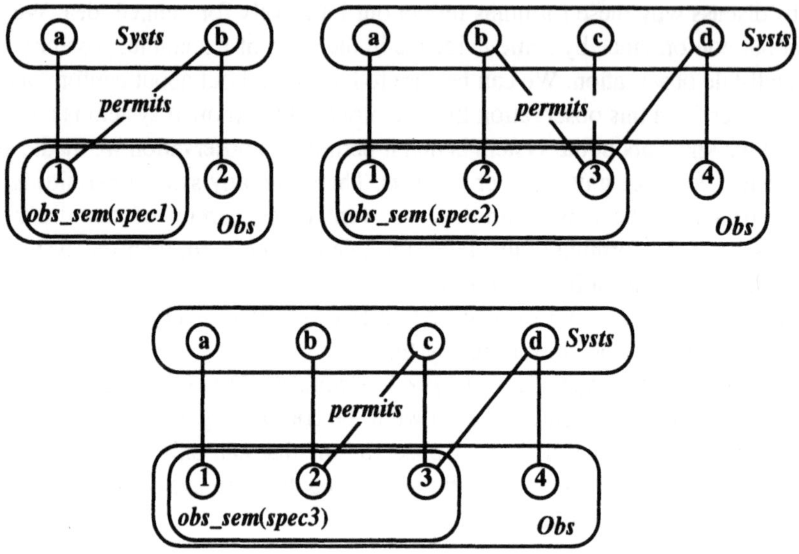

Figure 2: Three specifications in their specification contexts

In Figure 2, *spec2* is weakly but not strongly ND-validatable: system *a* can be validated (shown conforming) by observation 1, *b* can be validated by 2, but not by 3; *c* cannot be validated at all. *spec3* is strongly ND-validatable: *a* can be validated by 1, both *b* and *c* can be validated by 2, though *c* possibly only with luck.

4.2 Experimental specification contexts

Of course a tester's work is much alleviated if it does not take luck to discover possible invalid behaviour. It is desirable to have methods to arrive with certainty at certain results, be they confirmations (with which we do not deal here[1]), refutations or validations. In order to treat the enforceability of refutations or validations formally, we introduce a slight refinement in our framework. We distinguish between a behaviour part determined by the observer, called experiments, and possible outcomes of their interaction with the observed system. Non-determinism w.r.t. the outcome may creep in both by random elements in the experiment and by non-determinism in the system behaviour.

An **experimental specification context** is a specification context *ExpCont* = (*Systs*, *Props*, *Obs*, *has_property*, *permits*) (hence fulfilling (i,ii,iii)), where *Obs* is a left-total relation between a set *Exps* of **experiments** and a set *Outs* of **outcomes**, and thus a subset of *Exps* × *Outs*. This structure is related with the observation frameworks of [6] and even closer with the observation schemes of [1]. For example, both *Systs* and *Exps* may be a set of timed automata, while *Outs* may consist of the timed traces of their common transitions.

[1]) A confirmation is basically a non-refuting observation, but often linked with probabilistic aspects that are topics of ongoing research in the philosophy of science.

The possibility that the experiment *exp* performed on system *sys* yields the outcome *out*, *sys permits* (*exp*, *out*), is also written as (*sys*,*exp*) *may_yield out*, while *poss_outs*(*sys*,*exp*) := {*out* ∈ *Outs* | (*sys*,*exp*) *may_yield out*} denotes the set of possible outcomes of *exp* performed on system *sys*. By requiring *Obs* to be left-total, we postulate from an experimental specification context that every experiment performed on a system yields at least one outcome,

$$\forall\ sys \in Systs,\ exp \in Exps\text{: } poss\_outs(sys,exp) \neq \emptyset.$$

This means practically that we never wait eternally for an experiment to end (how could we, anyway?), and that we treat test log entries like "nothing happened, so we broke off after one hour" or "test equipment could not be connected to the system under test" as possible outcomes. Note also that an experiment may comprise a full-fledged hierarchy of sub-experiments, with intermediate decisions based on the part of the outcome observed so far.

Similarly as for observations, any performable combination of experiments is considered as another experiment, such that we cannot obtain from a performable combination of experiments more information than from a single experiment.

Similarly as various observations can be obtained from a system, a given experiment performed on a given system may yield various outcomes. By the same arguments as at the beginning of Section 4, even if all outcomes that may be yielded by a fixed unknown system and a known experiment have actually occurred and have been registered, this fact will generally remain unnoticed by the observer.

The definition of a specification carries over without change such that a specification defines which outcomes are allowed for which experiments. We call an outcome *out* **valid** for a specification *spec* and an experiment *exp*, if (*exp*,*out*) is a valid observation. Otherwise, we call it **invalid**. Valid and invalid outcomes, respectively, form the sets *valid*(*spec*,*exp*) and *invalid*(*spec*,*exp*). We call an outcome *out* **validating** for a specification *spec* and an experiment *exp*, if (*exp*,*out*) is a validating observation.

By simple application of our previous testability definitions we obtain …

THEOREM 3 [EXPERIMENTAL ND-REFUTABILITY AND ND-VALIDATABILITY]: In the experimental specification context *ExpCont*,

- any non-void specification *spec* is ND-refutable in the sense that
  $\exists exp \in Exps,\ out \in Outs\text{: } \forall sys \in Systs\text{:}$
  $\quad (sys,exp)\ may\_yield\ out \Rightarrow \neg sys\ \ conforms\_to\ spec;$
- a non-contradictory specification *spec* is weakly ND-validatable iff
  $\exists exp \in Exps,\ out \in Outs\text{: } \forall\ sys \in Systs\text{:}$
  $\quad (sys,exp)\ may\_yield\ out \Rightarrow sys\ \ conforms\_to\ spec;$
- a non-contradictory specification *spec* is strongly ND-validatable iff
  $\exists Evidence \subseteq Exps \times Outs : \forall\ sys \in Systs\text{:}$
  $\quad (\forall (exp,\ out) \in Evidence\text{: } (sys,exp)\ may\_yield\ out \Rightarrow sys\ \ conforms\_to\ spec)$
  $\quad \wedge\ (sys\ conforms\_to\ spec \Rightarrow \exists (exp,\ out) \in Evidence\text{: } (sys,exp)\ may\_yield\ out).$ ■

4.3 If it shall not take luck ...

A non-void specification *spec* is **weakly D-refutable** if there exists an experiment by which some non-conforming system will definitely reveal that it is non-conforming, again even if the observer does not know which particular system "under test" she is observing. This is the case iff there exist at least one non-conforming system *sys0* and one experiment *exp* such that performing *exp* on *sys0* yields nothing but invalid outcomes, or formally:

$\exists exp \in Exps, sys0 \in Systs\colon poss\_outs(sys0,exp) \subseteq invalid(spec,exp)$.

Informally spoken, membership to *invalid*(*spec*,*exp*) is assumed to be effectively decidable. Oracle problems are not considered in this paper.

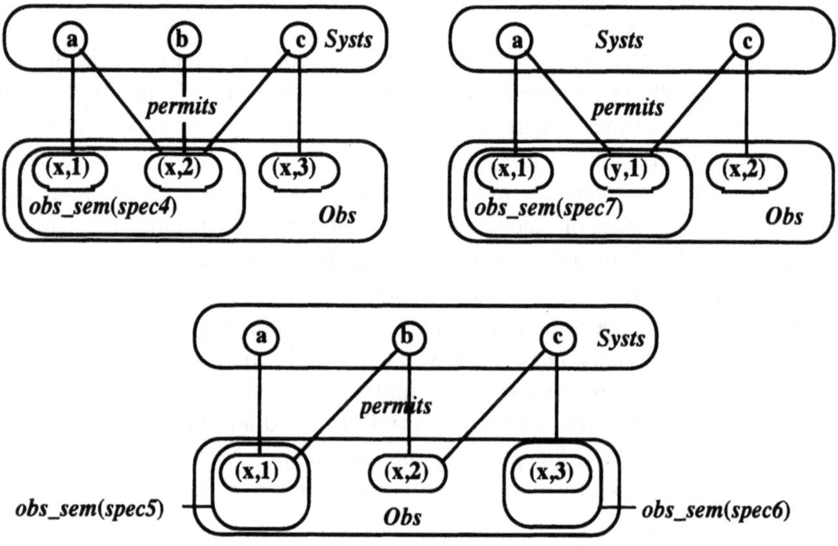

Figure 3: Four specifications in their experimental specification contexts

spec4 in Figure 3 is not even weakly D-refutable. There is only one possible experiment, *x*. No experiment, performed on an unknown system, guarantees to reveal non-conformance (if any) on the basis of each possible outcome. While outcome 3 would indeed reveal non-conformance, each experiment with each valid system may equally well produce the outcome 2, which may have come from the invalid system *c*.

A non-void specification *spec* is **strongly D-refutable** if there exists an experiment by which any non-conforming—possibly unknown—system will definitely reveal that it is non-conforming. This is the case iff there exists at least one experiment *exp* that—performed on any non-conforming system *sys*—can only yield invalid outcomes, or formally:

$\exists exp \in Exps$: $\forall sys \in Systs$:

$$\neg\ sys\ \ conforms\_to\ spec\ \Rightarrow poss\_outs(sys,exp) \subseteq invalid(spec,exp).$$

In Figure 3, *spec5* is weakly but not strongly D-refutable; we will always observe non-conformance while experimenting with *c*, but need luck to detect non-conformance if *b* is submitted to experiments. *spec7* is strongly D-refutable and demonstrates that generally the right experiment must be chosen: only *x* will reveal non-conformance.

A non-contradictory specification *spec* is **weakly D-validatable** if there exists an experiment by which some possibly unknown conforming system can be forced to reveal that it is conforming. This is the case iff there exists at least one experiment *exp* and one conforming system *sys0* such that *exp* performed on *sys0* can only yield outcomes that can only be obtained from a conforming system, or formally:

$$\exists exp \in Exps,\ sys0 \in Systs\colon\ poss\_outs(sys0,exp) \subseteq validating(spec,exp).$$

In Figure 3, *spec5* is not weakly D-validatable; *spec6* is weakly D-validatable but not weakly D-refutable.

A non-contradictory specification *spec* is **strongly D-validatable** if there exists an experiment by which any possibly unknown conforming system can be forced to reveal that it is conforming. This is the case iff there exists at least one experiment *exp* that—performed on any conforming system *sys*—can only yield validating outcomes, or formally:

$\exists exp \in Exps$: $\forall sys \in Systs$:

$$sys\ \ conforms\_to\ spec\ \Rightarrow poss\_outs(sys,exp) \subseteq validating(spec,exp).$$

In Figure 3, *spec6* is weakly D-validatable but not strongly D-validatable. *spec7* is strongly D-validatable and demonstrates that generally the right experiment must be chosen.

The fact that the specification *spec7* in Figure 3 is both strongly D-refutable and strongly D-validatable is not coincidental:

THEOREM 4 [EQUIV. OF STRONG D-REFUTABILITY AND -VALIDATABILITY]:
In an experimental specification context, a non-void and non-contradictory specification is strongly D-refutable iff it is strongly D-validatable.

Proof: Let *spec* be strongly D-refutable and *exp* as in the corresponding definition. If a system *sys* conforms to *spec,* then *exp* can yield with *sys* only outcomes in *valid(spec,exp)*. Thus each outcome is outside of *invalid(spec,exp)* and is, by strong refutability, only possible for conforming systems, hence validating.

The other direction, from strong D-validatability to strong D-refutability runs analogously. ■

THEOREM 5 [STRONG IMPLIES WEAK]:
For a non-void and non-contradictory specification, strong D-validatability implies weak D-validatability, and strong D-refutability implies weak D-refutability.

The proof is trivial. ■

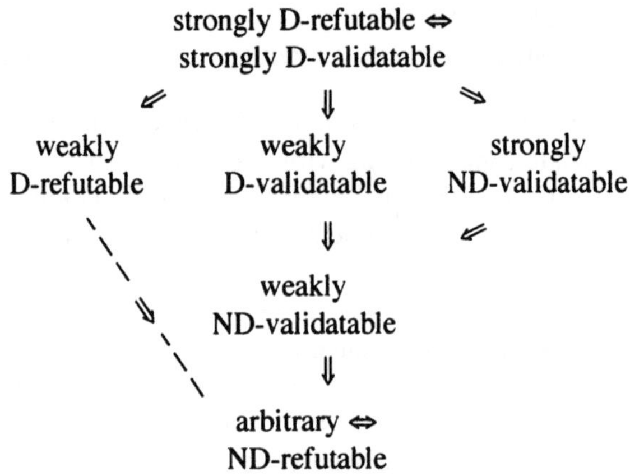

Figure 4: Implications between testability properties

THEOREM 6 [D IMPLIES ND]:
For a non-void and non-contradictory specification, strong D-validatability implies strong ND-validatability, and weak D-validatability implies weak ND-validatability.

Proof: For the "strong" property, take *Evidence* := {*exp*} × *validating*(*spec*,*exp*); the weak part is trivial. ■

Summarizing the dependencies found so far for non-degenerate specifications, we obtain the implications depicted in Figure 4.

5 TIME, BOUNDED AND UNBOUNDED

Up to this point, we have not dealt with questions of duration or complexity of observations, experiments or outcomes. It is certainly of practical interest for all parties involved, be they customers or contractors, implementors, users or testers, to assure that each observation or outcome is obtained in a finite time interval. This is particularly the case in commercial testing, where only a previously fixed period of time is available for the performance of test cases (experiments) and test suites (sets of experiments). Breaking off a test amounts to the performance of another, shorter, test. At any rate, an observation or outcome only achieved after an infinite time span is not achieved at all, or at least not in this world.

An interesting distinction is whether for every experiment there is a pre-defined time limit within which it leads to an outcome, or whether there is no such limit. The latter is possible, even though each experiment performed on any system leads to an outcome in finite time: imagine that there are infinitely many systems sys_1, sys_2, ... and that it takes n seconds for the outcome to appear if the experiment *exp* is performed on system sys_n. However, temporally open-ended testing, if it is undertaken at all, will usually be performed on the basis of payment per time unit.

Nobody would risk to work for payment only after delivery for an effort that may possibly not be completed during one's lifetime.

Due to the multitude of new notions arising in this context, this topic is treated in a separate paper [4].

6 CONCLUSION AND OUTLOOK

General remarks about some poorly defined phenomenon called testability are prone to ambiguity, if not even meaningless. The purpose of this paper was to define clearly various intuitive qualitative aspects of testability and to distinguish carefully among them.

Mathematically, our definitions could be formulated more elegantly in terms of co- and contravariant Galois connections, cf. [2,3]. But we did not want to presume foreknowledge in this area.

The testability properties identified in this paper do not relate very obviously with the legacy concepts outlined in Section 2, except with those in [7]. If we consider as our population of systems one given Mealy automaton in various start states, input sequences as experiments and output sequences as outcomes, and use the claims of a particular start state as specifications, then the initial-state determinability of [16] can be identified with the strong D-validatability of all specifications. It should be interesting to find out the "specification contexts" behind other testability definitions, i.e. which systems, observations and specifications were meant. Controllability in [16], and similarly observability in [19], seem to be mainly properties of all systems, or of *Systs*, rather than of single specifications: with their aid more specifications become testable, i.e. refutable or validatable in some of the senses introduced in this paper. Thus, testability has definitely even more flavours than those introduced in the present paper, no matter how obvious (and unfortunately numerous) they are.

It should be interesting to find clear testability notions not only for specifications, but also for system properties in *Props*, required in or derived from specifications. This could lead to a clear definition of the slightly vague concept of test purpose. Similarly, we would like to spend future efforts on relating FMCT notions [15]

As mentioned in the text, the oracle problem, with its aspects of computability, decidability and complexity, imposes limits to effective refutation and validation and therefore deserves closer investigation.

Future efforts should be spent on quantifying testability notions defined in this paper. Experiments and observations can be weighted by cost functions, such that minimum or maximum costs for refutation or validation may be computed. Moreover, if probability distributions are available for non-deterministic alternatives, then expected return-for-investment relationships might be computed for various testing strategies. These distributions may concern the probability of various systems to be submitted to testing against a given *spec*, or the probabilities of certain observations being made or outcomes turning up in an experiment.

The authors are indebted to Boris Beizer and Olaf Henniger for fruitful discussions.

REFERENCES

[1] B. Baumgarten: Die algebraische Spezifikation von Prozessen. Course notes, TH Darmstadt, WS 1992/93

[2] B. Baumgarten: Mathematical Aspects of Specification Theory. *Arbeitspapiere der GMD* Nr. 1066, 1997

[3] B. Baumgarten: The Observational Significance of System Requirements. Submitted for publication, 1998

[4] B. Baumgarten, O. Henniger: Testability with Unbounded Testing Strategies. Submitted for publication, 1998

[5] B. Beizer: *Software Testing Techniques*. 2nd ed., V. Nostrand Reinhold, 1990

[6] E. Brinksma et al: A Formal Approach to Conformance Testing. In: *Protocol Test Systems*, J. de Meer, L. Mackert, W. Effelsberg (eds.), Elsevier/North-Holland, pp 349-363, 1990

[7] R. Carnap: Testability and Meaning. *Philosophy of Science*, 3/4 , pp 420-471, 4/1, pp 1-40, 1936/37

[8] R. Dssouli et al.: Towards testable communication software. In: *Protocol Test Systems VIII*, A. Cavalli, S. Budkowski (eds.), Chapman&Hall, pp 237-251, 1996

[9] J.L. Fiadeiro, J.F.Costa: Mirror, Mirror in my Hand: a duality between specifications and models of process behaviour. *Mathematical Structures in Computer Science* 6/4, pp. 353-373, 1996

[10] R.S. Friedman: Testability of Software Components. *IEEE Trans. on Software Engineering*, 17/6, pp 553-564, 1991

[11] B. Ganter, R. Wille: *Formal Concept Analysis*. Springer, 1998

[12] M. Hennessy: *Algebraic theory of processes*. MIT-Press, 1988

[13] IEEE Standard Glossary of Software Engineering Terminology, 1990

[14] ISO/IEC IS 9646-1: *Conformance Testing Methodology and Framework*, Part 1: General Concepts, 2nd ed., 1994

[15] ISO/IEC DIS 13245/ ITU-T Z.500: *Formal Methods in Conformance Testing*, 1997

[16] R.E Kalman, P.L. Falb, M.A. Arbib: *Topics in Mathematical System Theory*. McGraw-Hill, 1969

[17] R.E Kalman: On the General Theory of Control Systems. *Proc. 1st Intl. Congr. on Automatic Control*, Butterworth, pp 481-492, 1961

[18] A. Petrenko, R. Dssouli, H. König: On Evaluation of Testability of Protocol Structures. In: *Protocol Test Systems VI*, O. Rafiq (ed.), Chapman&Hall, pp 111-123, 1994

[19] P.H. Starke: *Abstrakte Automaten*. Deutscher Verlag der Wissenschaften, 1969

[20] J.M. Voas, K.W. Miller: Software Testability: The New Verification. *IEEE Software*, May 1995

[21] S.T. Vuong, A.A.F. Loureiro, S.T. Chanson: A Framework for the Design for Testability of Communication Protocols. In: *Protocol Test Systems VI*, O. Rafiq (ed.), Chapman&Hall, pp 89-108, 1994

MIX
Papier aus verantwortungsvollen Quellen
Paper from responsible sources
FSC® C105338

If you have any concerns about our products,
you can contact us on
ProductSafety@springernature.com

In case Publisher is established outside the EU,
the EU authorized representative is:
Springer Nature Customer Service Center GmbH
Europaplatz 3, 69115 Heidelberg, Germany

Printed by Libri Plureos GmbH
in Hamburg, Germany